Maximum-Entropy and Bayesian Spectral Analysis and Estimation Problems

Fundamental Theories of Physics

A New International Book Series on the Fundamental Theories of Physics: Their Clarification, Development and Application

Maximum-Entropy and Bayesian Spectral Analysis and Estimation Problems

Proceedings of the Third Workshop on Maximum Entropy and Bayesian Methods in Applied Statistics, Wyoming, U.S.A., August 1-4, 1983

edited by

C. Ray Smith

U.S. Army Missile Command, Redstone Arsenal, Alabama, U.S.A.

and

Gary J. Erickson

Department of Electrical Engineering, Seattle University, Seattle, Washington, U.S.A.

D. Reidel Publishing Company

A MEMBER OF THE KLUWER ACADEMIC PUBLISHERS GROUP

Dordrecht / Boston / Lancaster / Tokyo

Library of Congress Cataloging in Publication Data

CIP

Maximum Entropy Workshop (3rd: 1983: Laramie, Wyo.)
Maximum-entropy and Bayesian spectral analysis and estimation problems.

(Fundamental theories of physics)
Includes index.
1. Entropy (Information theory)—Congresses. 2. Bayesian statistical decision theory—Congresses. I. Smith, C. Ray, 1933– . II. Erickson, Gary J. III. Title. IV, Series.
Q370.M385 1983 001.53′9 87–23228
ISBN 90–277–2579–9

Published by D. Reidel Publishing Company,
P.O. Box 17, 3300 AA Dordrecht, Holland.

Sold and distributed in the U.S.A. and Canada
by Kluwer Academic Publishers,
101 Philip Drive, Assinippi Park, Norwell, MA 02061, U.S.A.

In all other countries, sold and distributed
by Kluwer Academic Publishers Group,
P.O. Box 322, 3300 AH Dordrecht, Holland.

Printed in The Netherlands

To the memories of our fathers,
Robert Austin Smith and Phillip Christian Erickson

CONTENTS

PREFACE

This volume has its origin in the third "Workshop on Maximum-Entropy and Bayesian Methods in Applied Statistics," held at the University of Wyoming, August 1 to 4, 1983. It was anticipated that the proceedings of this workshop could not be prepared in a timely fashion, so most of the papers were not collected until a year or so ago. Because most of the papers are in the nature of advancing theory or solving specific problems, as opposed to status reports, it is believed that the contents of this volume will be of lasting interest to the Bayesian community.

The workshop was organized to bring together researchers from different fields to examine critically maximum-entropy and Bayesian methods in science, engineering, medicine, economics, and other disciplines. Some of the papers were chosen specifically to kindle interest in new areas that may offer new tools or insight to the reader or to stimulate work on pressing problems that appear to be ideally suited to the maximum-entropy or Bayesian method.

Certain facets of publishing a book are inherently unrewarding and frustrating. Or so it seems until the task is completed, and one has the pleasure of acknowledging publicly those who have helped along the way. Adequate thanks to Martha Stockton are impossible. The camera-ready copy prepared by Martha has benefited substantially by her editorial, proofreading, and drafting assistance. Dr. David Larner and Professor Alwyn van der Merwe, both affiliated with Reidel, provided encouragement and friendship at critical times. We are happy that Reidel has agreed to publish future proceedings of these workshops. Others who have made our work easier or more rewarding include Evelyn Haskell, Marce Mitchum, and our friends of the SDC Passive Sensors Division. Dr. Rabinder Madan of the Office of Naval Research has provided continual encouragement and assisted us in obtaining much-needed funding.

August 1987

C. Ray Smith
Gary J. Erickson

BAYESIAN SPECTRUM AND CHIRP ANALYSIS

E. T. Jaynes

Wayman Crow Professor of Physics
Washington University, St. Louis, MO 63130

We seek optimal methods of estimating power spectrum and chirp (frequency change) rate for the case that one has incomplete noisy data on values y(t) of a time series. The Schuster periodogram turns out to be a "sufficient statistic" for the spectrum, a generalization playing the same role for chirped signals. However, the optimal processing is not a linear filtering operation like the Blackman-Tukey smoothing of the periodogram, but rather a nonlinear operation. While suppressing noise/side lobe artifacts, it achieves the same kind of improved resolution that the Burg method did for noiseless data.

C. R. Smith and G. J. Erickson (eds.),
Maximum-Entropy and Bayesian Spectral Analysis and Estimation Problems, 1–37.

1. Introduction

The maximum entropy solution found by Burg [1967, 1975] has been shown to give the optimal spectrum estimate—by a rather basic, inescapable criterion of optimality—in one well defined problem [Jaynes, 1982]. In that problem we estimate the spectrum of a time series $\{y_1 \ldots y_N\}$, from incomplete data consisting of a few autocovariances $\{R_0 \ldots R_m\}$, $m < N$, measured from the entire time series, and there is no noise.

This is the first example in spectrum analysis of an exact solution, which follows directly from first principles without ad hoc intuitive assumptions and devices. In particular, we found that there was no need to assume that the time series was a realization of a "stationary Gaussian process." The maximum entropy principle automatically created the Gaussian form for us, out of the data. This indicated something that could not have been learned by assuming a distribution, namely, that the Gaussian distribution is the one that can be realized by Nature in more ways than can any other that agrees with the given autocovariance data. This classic solution will go down in history as the "hydrogen atom" of spectrum analysis theory.

But a much more common problem, also considered by Burg, is the one where our data consist, not of autocovariances, but of the actual values of $\{y_1 \ldots y_N\}$, a subset of the full time series, contaminated with noise. Experience has shown Burg's method to be very successful here also, if we first estimate m autocovariances from the data and then use them in the maximum entropy calculation. The choice of m represents our judgment about the noise magnitude, values too large introducing noise artifacts, values too small losing resolution. For any m, the estimate we get would be the optimal one if (a) the estimated autocovariances were known to be the exact values and (b) we had no other data beyond those m autocovariances.

Although the success of the method just described indicates that it is probably not far from optimal when used with good judgment about m, we have as yet no analytical theory proving this or indicating any preferred different procedure. One would think that a true optimal solution should (1) use all the information the data can give; i.e., estimate not just $m < N$ autocovariances from the data, but find our "best" estimate of all N of them and their probable errors; (2) then make allowance for the uncertainty of these estimates by progressively de-emphasizing the unreliable ones. There should not be any sharp break as in the procedure used now, which amounts to giving full credence to all autocovariance estimates up to lag m, zero credence to all beyond m.

In Jaynes [1982] we surveyed these matters very generally and concluded that much more analytical work needs to be done before we can know how close the present partly ad hoc methods are to optimal in problems with noisy and/or incomplete data. The following is a sequel, reporting the first stage of an attempt to understand the theoretical situation better, by a direct Bayesian analysis of the noisy data problem. In effect, we are trying to advance from the "hydrogen atom" to the "helium atom" of spectrum analysis theory.

One might think that this had been done already, in the many papers that study autoregressive (AR) models for this problem. However, as we have noted before [Jaynes, 1982], introducing an AR model is not a step toward solving a spectrum analysis problem, only a detour through an alternative way of formulating the problem. An AR connection can always be made if one wishes to do so, for any power spectrum determines a covariance function, which in turn determines a Wiener prediction filter, whose coefficients can always be interpreted as the coefficients of an AR model. But while this is always possible, it may not be appropriate (just as representing the function $f(x) = \exp(-x^2)$ by an infinite series of Bessel functions is always possible but not always appropriate).

Indeed, learning that spectrum analysis problems can be formulated in AR terms amounts to little more than discovering the Mittag-Leffler theorem of complex variable theory (under rather general conditions an analytic function is determined by its poles and residues).

In this field there has been some contention over the relative merits of AR and other models such as the MA (moving average) one. Mathematicians never had theological disputes over the relative merits of the Mittag-Leffler expansion and the Taylor series expansion. We expect that the AR representation will be appropriate (i.e., conveniently parsimonious) when all the poles happen to be close to the unit circle; it may be very inappropriate otherwise.

Better understanding should come from an approach that emphasizes logical economy by going directly to the question of interest. Instead of invoking an AR model at the beginning (which might bring in a lot of inappropriate and unnecessary detail, and also limits the scope of what can be done thereafter), let us start with a simpler, more flexible model that contains only the facts of data and noise, the specific quantities we want to estimate, and no other formal apparatus. If AR relations—or any other kind—are appropriate, then they ought to appear automatically, as a consequence of our analysis, rather than as initial assumptions.

This is what did happen in Burg's problem; maximum entropy based on autocovariance data led automatically to a spectrum estimator that could be expressed most concisely (and beautifully) in AR form, the Lagrange multipliers being convolutions of the AR coefficients: $\lambda_n = \Sigma_k a_k a_{n-k}$. The first reaction of some was to dismiss the whole maximum entropy principle as "nothing but AR," thereby missing the point of Burg's result. What was important was not the particular analytical form of the solution, but rather the logic and generality of his method of finding it.

The reasoning will apply equally well, generating solutions of different analytical form, in other problems far beyond what any AR model could cope with. Indeed, Burg's method of extrapolating the autocovariance beyond the data was identical in rationale, formal relations, and technique, with the means by which modern statistical mechanics predicts the course of an irreversible process from incomplete macroscopic data.

This demonstration of the power and logical unity of a way of thinking, across the gulf of what appeared to be entirely different fields, was of vastly greater, and more permanent, scientific value than merely finding the solution to one particular technical problem.

Quickly, the point was made again just as strongly by applying the same reasoning to problems of image reconstruction, of which the work of Gull and Daniell [1978] is an oustandingly concise, readable example.

I think that, 200 years from now, scholars will still be reading these works, no longer for technical enlightenment—for by then this method of reasoning will be part of the familiar cultural background of everybody--but as classics of the History of Science, which opened up a new era in how scientists think.

The actual reasoning had, in fact, been given by Boltzmann and Gibbs long before; but it required explicit, successful applications outside the field of thermodynamics before either physicists or statisticians could perceive its power or its generality.

In the study reported here we have tried to profit by these lessons in logical economy; at the beginning it was decided not to put in that fancy entropy stuff until we had done an absolutely conventional, plain vanilla Bayesian analysis--just to see what was in it, unobscured by all the details that appear in AR analyses. It turned out that so much surprising new stuff was in it that we are still exploring the plain vanilla Bayesian solution and have not yet reached the entropic phase of the theory!

The new stuff reported here includes what we think is the first derivation of the Schuster periodogram directly from the principles of probability theory, and an extension of spectrum analysis to include chirp analysis (rate of change of frequency). Before we had progressed very far it became evident that estimation of chirp is theoretically no more difficult than "stationary" spectrum estimation. Of course, this takes us beyond the domain of AR models, and could never have been found within the confines of an AR analysis.

Our calculations and results are straightforward and elementary; in a communication between experienced Bayesians it could all be reported in five pages and the readers would understand perfectly well what we had done, why we had done it, and what the results mean for data processing. Such readers will doubtless find our style maddeningly verbose.

However, hoping that this work might also serve a tutorial function, we have scattered throughout the text and appendices many pages of detailed explanation of the reasons for what we do and the meaning of each new equation as it appears.

Also, since consideration of chirp has not figured very much in past spectrum analysis, and Bayesian analysis has not been very prominent either, the next three sections survey briefly the history and nature of the chirp problem and review the Bayesian reasoning format. Our new calculation begins in Section 5.

2. Chirp Analysis

The detection and analysis of chirped signals in noise may be viewed as an extension of spectrum analysis to include a new parameter, the rate of change of frequency. We ask whether there exist principles for optimal data processing in such problems.

Chirped signals occur in many contexts: in quantum optics [Jaynes, 1973; Nikolaus and Grischkowsky, 1983]; the ionospheric "whistlers" of Helliwell [1965]; human speech; the sounds of birds, bats, insects, and slide trombonists; radio altimeters and frequency-modulated radar, etc. Any transient signal, propagating through a dispersive medium, emerges as a chirped signal, as is observed in optics, ultrasonics, and oceanography—and presumably also in seismology, although the writer has not seen explicit mention of it. Thus in various fields the detection and/or analysis of chirped signals in noise is a potentially useful adjunct to spectrum analysis.

Quite aside from applications, the problem is a worthy intellectual challenge. Bats navigate skillfully in the dark, avoiding obstacles by a kind of acoustical chirped radar [Griffin, 1958]. It appears that they can detect echoes routinely in conditions of signal/noise ratio where our best correlation filters would be helpless. How do they do it?

At first glance it seems that a chirped signal would be harder to detect than a monochromatic one. Why, then, did bats evolve the chirp technique? Is there some as yet unrecognized property of a chirped signal that makes it actually easier to detect than a monochromatic one?

Further evidence suggesting this is provided by the "whistle language" developed by the Canary Islanders. We understand that by a system of chirped whistles they are able to communicate fairly detailed information between a mountaintop and the village below, in conditions of range (a few miles) and wind where the human voice would be useless.

In the case of the bats, one can conjecture three possible reasons for using chirp: (a) Since natural noise is usually generated by some "stationary" process, a weak chirped signal may resemble natural noise less than does a weak monochromatic signal. (b) Prior information about the chirp rate possessed by the bat may be essential; it helps to know what you are looking for. (c) Our whole conceptual outlook, based on years of nonchirp thinking, may be wrong; bats may simply be asking a smarter question than we are.

Of course, for both the bats and the Canary Islanders, it may be that chirped signals are not actually easier to detect, only easier to recognize and interpret in strong noise.

After noting the existing "spectral snapshot" method of chirp analysis, we return to our general Bayesian solution for a model that represents one possible real situation, then generalize it in various ways on the lookout for evidence for or against these conjectures.

In this problem we are already far beyond what Box and Tukey have called the "exploratory phase" of data analysis. We already know that the bats, airplanes, and Canary Islanders are there, that they are emitting purposeful signals that it would be ludicrous to call "random," and that those signals are being corrupted by additive noise that we are unable to control or predict. Yet we do have some cogent prior information about both the signals and the noise, and so our job is not to ask, "What seems to go on here?" but rather to set up a model that expresses that prior information.

3. Spectral Snapshots

A method of chirp anaysis used at present, because it can be implemented with existing hardware, is to do a conventional Blackman-Tukey spectrum analysis of a run of data over an interval ($t_1 \pm T$), then over a later interval ($t_2 \pm T$), and so on. Any peak that appears to move steadily in this sequence of spectral snapshots is naturally interpreted as a chirped signal (the evanescent character of the phenomenon making the adjective "spectral" seem more appropriate than "spectrum").

The method does indeed work in some cases, and impressively in the feat of oceanographers [Barber and Ursell, 1948; Munk and Snodgrass, 1957] to correlate chirped ocean waves, with periods in the 10- to 20-sec range, with storms thousands of miles away. Yet it is evident that spectral snapshots do not extract all the relevant information from the data; at the very least, evidence contained in correlations between data segments is lost.

The more serious and fundamental shortcoming of this method is that if one tries to analyze chirped data by algorithms appropriate to detect monochromatic signals, the chirped signal of interest will be weakened, possibly disastrously, through phase cancellation (what physicists call Fresnel diffraction).

For this reason, if we adhere to conventional spectrum analysis algorithms, the cutting of the data into segments analyzed separately is not a correctible approximation. As shown in Appendix A, to detect a signal of chirp rate α—that is, a sinusoid $\cos(\omega t + \alpha t^2)$—with good sensitivity by that method, one must keep the data segments so short that $\alpha T^2 \lesssim 1$. If T is much longer than this, a large chirped signal—far above the noise level—can still be lost in the noise through phase cancellation. Appendix A also tries to correct some currently circulating misconceptions about the history of this method.

Our conclusion is that further progress beyond the spectral snapshot method is necessary and possible—and it must consist of finding new algorithms that (a) protect against phase cancellation, (b) extract more information from the data, and (c) make more use of prior information. But nobody's intuition has yet revealed the specific algorithm for this data analysis, so we turn for guidance to probability theory.

4. The Basic Reasoning Format

The principle we need is just the product rule of probability theory, $p(AB|C) = p(A|C)p(B|AC)$, which we note is symmetric in the propositions A and B. Therefore let I = prior information, H = any hypothesis to be tested, and D = data. Then $p(HD|I) = p(D|I)p(H|DI) = p(H|I)p(D|HI)$ or, if $p(D|I) > 0$ (that is, if the data set is a possible one),

$$p(H|DI) = p(H|I) \frac{p(D|HI)}{p(D|I)}, \tag{1}$$

which is Bayes' theorem, showing how the prior probability $p(H|I)$ of H is updated to the posterior probability $p(H|DI)$ as a result of acquiring the new information D. Bayesian analysis consists of the repeated application of this rule.

Progress in scientific inference was held up for decades by a belief that the equations of probability theory were only rules for calculating frequencies, not for conducting inference. However, we now have many analyses (B. de Finetti, H. Jeffreys, R. T. Cox, A. Wald, L. J. Savage, D. V. Lindley, and others) showing that these equations are also the uniquely "right" rules for conducting inference.

That is, it is a theorem that anyone who represents degrees of plausibility by real numbers—and then reasons in a way not reducible to these equations—is necessarily violating some very elementary qualitative desiderata of rationality (transitivity, strong domination, consistency, coherence, etc.). We are concerned simply with the logic of consistent plausible reasoning; there is no necessary connection with frequencies or random experiments.

Put differently, sufficiently deep and careful intuitive thinking will, after all inconsistencies have been detected and removed, necessarily converge eventually to the Bayesian conclusions from the same information. Recognizing this only enables us to reach those conclusions more quickly. How much more quickly we shall see presently.

New demonstrations of the power of Bayesian inference in real problems—yielding in a few lines important results that decades of "frequentist" analysis or intuitive thinking had not found—have been appearing steadily for about 20 years, the present work providing another example.

However, before we can apply Eq. (1) quantitatively, our problem must have enough structure so that we can determine the term $p(D|HI)$. In its dependence on D for fixed H, this is the "sampling distribution"; in its dependence on H for fixed D, it is the "likelihood function." In the exploratory phase of a problem such structure may not be at hand.

Fortunately, our present problem is free of this difficulty. We shall apply Eq. (1) in which H stands typically for the statement that a multidimensional parameter lies in a certain specified region of the parameter space. Deploring, but nevertheless following, the present common custom, we use the same symbol $p(x|y)$ for a probability or a probability density; the distinction must be read from the context.

5. A Simple Bayesian Model

In this section which is possibly the first direct Bayesian analysis of the problem, attempts at conceptual innovation are out of order, and we wish only to learn the consequences of an absolutely standard kind of model. We follow slavishly the time-worn procedure of "assuming the data contaminated with additive white Gaussian noise," which is time-worn just because it has so many merits. As has long been recognized, not only is it the most realistic choice one can make in most problems, but the solutions will be analytically simple and not far from optimal in others.

But there is an even more cogent reason for choosing this probability assignment. In most real problems, the only prior information we have about the noise is its mean square value; often not even that. Then because it has maximum entropy for a given mean square noise level, the independent "white" Gaussian distribution will be the safest, most conservative one we can use; that is, it protects us most strongly from drawing erroneous conclusions. But since there are still many people who simply do not believe this, let us amplify the point.

The reason for assigning a prior distribution to the noise is to define the range of possible variations of the noise vector $e = (e_1 \ldots e_n)$ that we shall make allowance for in our inference. As is well known in the literature of information theory, the entropy of a distribution is an asymptotic measure of the size of the basic "support set" W of that distribution, in our case the n-dimensional "volume" occupied by the reasonably probable noise vectors. The maximum entropy principle tells us—as does elementary common sense—that the distribution that most honestly represents what we know is the one with the largest support set W_{max} permitted by our information.

Unless we have very specific prior information in addition to the mean square value, so that we know the particular way in which the noise departs from white Gaussian, it would be dangerous to use any other distribution in our inference. According to the maximum entropy principle, to do so would necessarily be making an assumption, which restricts our considerations to some arbitrary subset $W \subset W_{max}$, in a way not justified by our prior information about the noise.

The price we would pay for this indiscretion is that if the true noise vector happened to lie in the complementary set $W' = W_{max} - W$, then we would be misled into interpreting as a real effect what is only an artifact of the noise. And the chance of this happening is not small, as shown by the entropy concentration theorem [Jaynes, 1982]. To assign a distribution with entropy only slightly smaller than the maximum may contract the volume of the support set by an enormous factor, often more than 10^{10}. Then virtually every possible noise vector would lie in W', and we would be seeing things that are not there in almost every data set. Gratuitous assumptions in assigning a noise distribution can be very costly.

Indeed, unless we can identify and correctly understand some specific defect in this simplest independent Gaussian model, we are hardly in a position to invent a better one.

However, these arguments apply only to the noise, for the noise is completely unknown except for its mean square value. The signal of interest is of course something about which we know a great deal in advance (just the reason that it is of interest). We should take into account all the prior information we have about its structure (functional form), which affords us our best means of finding it in the noise.

To represent the signal as a "sample from a Gaussian random process" would be, in effect, to change the problem into that of estimating the parameters in an imaginary Gaussian distribution of all possible signals. That is not our aim here; we want to estimate a property of the real world, the spectrum of the specific real signal that generated our data.

Our only seemingly drastic simplification is that for the time being we suppose it known in advance that the signal contains only a single term, of the form

$$f(t) = A\cos(\omega t + \alpha t^2 + \theta), \tag{2}$$

and so the problem reduces to estimating the four parameters (A,ω,α,θ). However, this is a fairly realistic assumption for the oceanographic chirp problem discussed in Appendix A, where it is unlikely, although not impossible, that waves from two different storms are arriving simultaneously.

In the end it will develop that for purposes of estimating the power spectral density this assumption of a single signal is hardly an assumption at all; the resulting solution remains valid, with only a slight reinterpretation and no change in the actual algorithm, however many signals may be present. It is a restrictive assumption only when we ask more detailed questions than "What is your best estimate of the spectral density?"

Other assumptions (such as constant amplitude and chirp rate) turn out also to be removable. In fact, once we understand the solution to this simplest problem, it will be evident that it can be generalized to detection of any signal of known functional-parametric form, sampled at arbitrary times, in nonwhite noise.

But for the present our true signal, Eq. (2), is contaminated with the aforementioned white Gaussian noise e(t), so the observable data are values of the function

$$y(t) = f(t) + e(t) . \tag{3}$$

In practice we shall have these data only at discrete times, which we suppose for the moment to be equally spaced at integer values of t, and over a finite interval 2T. Thus our data consist of $N = (2T + 1)$ values

$$D = \{\, y(t), \quad -T \leq t \leq T \,\}, \tag{4}$$

and we assign the aforementioned independent Gaussian joint probability distribution for the values of the noise e(t) at the corresponding times, taking each $e(t) \sim N(0,\sigma)$ where the variance σ^2 is supposed known.

We have gone to some length to explain our basis for choosing this noise distribution because it is a matter of much confusion, different schools of thought holding diametrically opposite views as to whether this is or is not a restrictive assumption. In fact, this confusion is so great that the rationale of our choice still requires further discussion, continued in Appendix B.

Whatever school of thought one favors, our equations will be the same; only our judgments of their range of validity will differ. Given any true signal f(t), the probability (density) that we shall obtain the data set D = {y(t)} is just the probability that the noise values {e(t)} will make up the difference:

$$p(D|A,\omega,\alpha,\theta,\sigma) = \prod_{t=-T}^{T} \frac{1}{\sqrt{2\pi\sigma^2}} \exp\{-\frac{1}{2\sigma^2}[y(t) - f(t)]^2\}, \tag{5}$$

which is our sampling distribution. Conversely, given σ and the data D, the joint likelihood of the unknown parameters is

$$L(A,\omega,\alpha,\theta) \propto \exp\{-\frac{1}{2\sigma^2}\sum_{t=-T}^{T}[y(t) - A\cos(\omega t + \alpha t^2 + \theta)]^2\}. \tag{6}$$

In analysis of discrete time series, the mathematics has tended to get cluttered with minute details about "end effects" associated with the exact limits of summation. But this is only a notational problem; we can remove the clutter with no loss of precision if we adopt the convention (call it "infinite padding with zeros" if you like):

$$y_t = y(t) = 0, \quad |t| \geq T. \tag{7}$$

Then all our sums of functions $K(y_t)$ over time indices can take the form ΣK_t, understood to run over $(-\infty < t < \infty)$. In this notation, which we use henceforth, all those little details are present automatically, but kept out of sight.

Usually, the absolute phase θ is of no interest to us and we have no prior information about it; that is, it is a "nuisance parameter" that we want to eliminate. We may integrate it out with respect to a uniform prior probability density, getting a marginal quasi-likelihood

$$L(A,\omega,\alpha) = \frac{1}{2\pi}\int_0^{2\pi} L(A,\omega,\alpha,\theta)\, d\theta \tag{8}$$

that represents the contribution from the data to the joint marginal posterior distribution of (A,ω,α). This is the course we shall pursue in the present work.

But an important exception occurs if our ultimate objective is not to estimate the parameters (A,ω,α) of the "regular" signal f(t) but rather to estimate the particular "irregular" noise sequence $\{e(t)\}$ that occurred during our observation period. This is the problem of seasonal adjustment, where an estimate of θ is also needed, and our data processing algorithm will be quite different. The Bayesian theory of seasonal adjustment, to be given elsewhere, yields a new demonstration of the power of prior information to improve our estimates; we invite intuitionists to discover it without Bayesian methods.

6. The Phaseless Likelihood Function

We shall consider the exact relations later when we generalize to many signals, but for the moment we make an approximation that we believe to be generally accurate and harmless (although with obvious exceptions like $\omega = \alpha = 0$, or $\omega = \pi$, $\alpha = 0$):

$$\sum_t \cos^2(\omega t + \alpha t^2 + \theta) \simeq \frac{2T+1}{2} = \frac{N}{2} \; . \tag{9}$$

Values of α differing by 2π are indistinguishable in this discrete sampled data; that is, chirp aliasing, like frequency aliasing, confines us to the domain $(-\pi < \alpha \leq \pi)$.

The joint likelihood of the four signal parameters is then

$$L(A,\omega,\theta) = \exp\left\{ \frac{A}{\sigma^2} \sum_t y_t \cos(\omega t + \alpha t^2 + \theta) - \frac{NA^2}{4\sigma^2} \right\}, \tag{10}$$

in which, since only the dependence on (A,ω,α,θ) matters, we may discard any factor not containing them.

The integration in Eq. (8) over θ, carried out in Appendix C, yields the phaseless likelihood (or quasi-likelihood) function

$$L(A,\omega,\alpha) = e^{-NA^2/4\sigma^2}\, I_0\left(\frac{A\sqrt{NC(\omega,\alpha)}}{\sigma^2}\right), \tag{11}$$

where $I_0(x) = -iJ_0(ix)$ is a Bessel function and

$$C(\omega,\alpha) \equiv N^{-1} \sum_{ts} y_t y_s \cos[\omega(t-s) + \alpha(t^2 - s^2)] \; . \tag{12}$$

The form of Eq. (11) already provides some (at least to the writer) unexpected insight. Given any sampling distribution, a likelihood or quasi-likelihood function, in its dependence on the parameters, contains all the information in the data that is relevant for any inference about those parameters—whether it be joint or individual point estimation, interval estimation, testing any hypotheses concerning them, etc. But the only data dependence in $L(A,\omega,\alpha)$ comes from the function $C(\omega,\alpha)$. Therefore, C plays the role of a "sufficient statistic"; this function summarizes all the information in the data that is relevant for inference about (A,ω,α).

Because of its fundamental importance, $C(\omega,\alpha)$ should be given a name. We shall call it the chirpogram of the data, for a reason that will appear in Eq. (13) below. It seems, then, that whatever specific question we seek to answer about a chirped signal, the first step of data analysis will be to determine the chirpogram of the data.

Of course (to answer a recent criticism), in setting up a model, the Bayesian—like any other theoretician—is only formulating a working hypothesis, to find out what its consequences would be. He has not thereby taken a vow of theological commitment to believe it forever in the face of all new evidence. Having got this far in our calculation, nothing in Bayesian principles forbids us to scan that chirpogram by eye, on the lookout for any unusual features (such as a peak stretched out diagonally, which suggests a change in chirp rate during the data run) that we had not anticipated when setting up our model.

Indeed, we consider the need for such elementary precautions so obvious and trivial that it would never occur to us that anyone could fail to see it. If we do not stress this constantly in our writings, it is because we have more substantive things to say.

7. Discussion: Meaning of the Chirpogram

The chirpogram appears less strange if we note that when $\alpha = 0$, it reduces to

$$
\begin{aligned}
C(\omega,0) &= N^{-1} \sum_{ts} y_t y_s \cos\omega(t - s) \\
&= N^{-1} \Big| \sum_t y_t \, e^{i\omega t} \Big|^2 \\
&= \sum_t R(t) \cos\omega t
\end{aligned}
\tag{13}
$$

where $R(t)$ is the data autocovariance:

$$
R(t) = N^{-1} \sum_s y_s y_{s+t} \quad . \tag{14}
$$

Thus $C(\omega,0)$ is just the periodogram of Schuster [1897].

For nearly a century, therefore, the calculation of $C(\omega,0)$ has seemed, intuitively, the thing to do in analyzing a stationary power spectrum. However, the results were never satisfactory. At first one tried to interpret $C(\omega,0)$ as an estimate of the power spectrum of the sampled signal. But the periodograms of real data appeared to the eye as too wiggly to believe, and in some problems the details of those wiggles varied erratically from one data set to another.

Intuition then suggested that some kind of smoothing of the wiggles is called for. Blackman and Tukey [1958], hereinafter denoted B-T, recognized the wiggles as in part spurious side-lobes, in part beating between "outliers" in the data, and showed that in some cases one can make an estimated power spectrum with a more pleasing appearance, which one therefore feels has more truth in it, by introducing a lag window function $W(t)$, which cuts off the contributions of large t in Eq. (13), giving the B-T spectrum estimate

$$\hat{P}(\omega)_{BT} = \sum_{t=-m}^{m} W(t)\, R(t) \cos\omega t \,, \tag{15}$$

in which we use only autocovariances determined from the data up to some lag m, which may be a small fraction of the record length N.

That this estimate disagrees with the data [the measured autocovariance $R(t)$] at every lag t for which $W(t) \neq 1$, does not seem to have troubled anyone until Burg pointed it out 17 years later. He termed it a "willingness to falsify" the data, and advocated instead the maximum entropy estimate, which was forced by the constraints to agree with the data, the wiggles being removed by a totally different method, the "smoothest" extrapolation of $R(t)$ beyond the data. Others, including this writer, quickly echoed Burg's argument with enthusiasm; but as we shall see presently, this was not the end of the story.

In any event, a lag window $W(t)$ smoothly tapered to zero at $t = m$ does reduce the unwanted wiggles—at a price. It leads to a reasonable estimate in the case of a broad, featureless spectrum; but of course, in that case the contributions from large t were small anyway, and the window had little effect. But lag window smoothing (equivalent to a linear filtering operation that convolves the periodogram with the Fourier transform of the lag window function, thus smearing out the wiggles sideways) necessarily loses resolution and makes it impossible to represent sharp spectrum lines correctly.

One wonders, then, why B-T stopped at this point, for other procedures were available. To put it with a cynicism that we shall correct later: once one has been willing to falsify the data in one way, then his virtue is lost, and he should have no objection to falsifying them also in other ways. Why, then, must we use the same window function at all frequencies? Why must we process $R(t)$ linearly? Why must we set all $R(t)$ beyond m equal to zero, when we know perfectly well that this is almost certainly wrong? There were dozens of other ad hoc procedures, which would have corrected the failure of the B-T method to deal with sharp lines without venturing into the forbidden realm of Bayesianity.

But history proceeded otherwise; and now finally, the first step of a Bayesian analysis has told us what a century of intuitive ad hockery did not. The periodogram was introduced previously only as an intuitive spectrum estimate; but now that it has been derived from the principles of probability theory we see it in a very different light. Schuster's periodogram is indeed fundamental to spectrum analysis; but not because it is itself a satisfactory spectrum estimator, nor because any linear smoothing can convert it into one in our problem.

The importance of the periodogram lies rather in its information content; in the presence of white Gaussian noise, it conveys all the information the data have to offer about the spectrum of f(t). As noted, the chirpogram has the same property in our more general problem.

It will follow from Eq. (11) [see Eq. (28) below], that the proper algorithm to convert $C(\omega,0)$ into a power spectrum estimate is a complicated nonlinear operation much like exponentiation followed by renormalization, a crude approximation being

$$\hat{P}(\omega) \approx A \exp[C(\omega,0)/\sigma^2] \quad . \tag{16}$$

This will suppress those spurious wiggles at the bottom of the periodogram as well as did the B-T linear smoothing; but it will do it by attenuation rather than smearing, and will therefore not lose any resolution. The Bayesian nonlinear processing of $C(\omega,0)$ will also yield, when the data give evidence for them, arbitrarily sharp spectral line peaks from the top of the periodogram that linear smoothing cannot give.

It is clear from Eq. (11) why a nonlinear processing of C is needed. The likelihood involves not just C, but C in comparison with the noise level σ^2. The B-T procedure, Eq. (15), smears out all parts of C equally, without considering where they stand relative to any noise. The Bayesian nonlinear processing takes the noise level into account; wiggles below the noise level are almost certainly artifacts of the noise and are suppressed, while peaks that rise above the noise level are believed and emphasized.

It may seem at this point surprising that intuition did not see the need for this long ago. Note, however, that Blackman and Tukey had in mind a very different problem than ours. For them the whole data y(t) were a sample from a "stochastic process" with a multivariate Gaussian distribution. From the standpoint of our present problem, we might interpret the B-T work as a preliminary study of the noise spectrum before the purposeful signal was added. So for them the notion of "C in comparison with the noise level" did not exist.

To emphasize this, note that B-T considered the periodogram to have a sampling distribution that was chi-squared with two degrees of freedom, independently of the sample size. That would not be the case in the problem we are studying unless the signal f(t) were absent.

From this observation there follows a point that has not been sufficiently stressed—or even noticed—in the literature: the B-T efforts were not directed at all toward the present problem of estimating the power spectrum of a signal f(t) from data that are contaminated with noise.

Confusion over "What is the problem?" has been rampant here. We conceded, after a theoretical study [Jaynes, 1981], that pure maximum entropy is not optimal for estimating the spectrum of a signal in the presence of noise; but we failed to see the point just noted. Immediately, Tukey and Brillinger [1982] proceeded to stress the extreme importance of noise in real problems and the necessity of taking it into account. But they failed to note the robustness of maximum entropy with respect to noise (that is, its practical success in problems where noise is present), or that a given procedure may solve more than one problem, and in fact maximum entropy is also the optimal solution to a Gaussian problem.

Although we agree with the need to take noise into account (as the present work demonstrates), we can hardly see that as an argument in favor of B-T methods in preference to maximum entropy in any problem. For we must distinguish between the B-T problem (spectrum of Gaussian noise) and the B-T procedure, Eq. (15), which has not been derived from, or shown to have any logical connection at all to, that problem.

Indeed, Burg's original derivation of the maximum entropy algorithm started from just the B-T assumption that the data are Gaussian noise; and we showed [Jaynes, 1982] that pure maximum entropy from autocovariance data leads automatically to a Gaussian predictive distribution for future data. Thus it appears that the "best" solution to the B-T problem is not the B-T tapering procedure, Eq. (15), but the Burg procedure.

But strangely enough, this enables us to take a more kindly view toward B-T methods. The procedure in Eq. (15) cannot be claimed as the "best" one for any spectrum analysis problem; yet it has a place in our toolbox. As a procedure it is applicable to any data, and is dependent on no hypotheses of Gaussianity. Whatever the phenomenon, if nothing is known in advance about its spectrum, the tapering Eq. (15) is a quick and easy way to wash out the wiggles enough to get a preliminary view of the broad features of the spectrum, helpful in deciding whether a more sophisticated data analysis is called for. The most skilled precision machinist still has frequent use for a jackknife.

Any tapering clearly does falsify the data, throwing away usable information. But data contaminated with noise are themselves in part "false." The valid criticism from the standpoint of our present problem is that when the noise goes away, the falsification in Eq. (15) remains; that is the case where Burg's pure maximum entropy solution was clearly the optimal one. But in the different problem envisaged by B-T, which led to Eq. (15), the noise cannot go away because the noise and the data are identical.

Thus from inspection of Eq. (11), we can see already some clarification of these muddy waters. The Bayesian method is going to give us, for spectrum estimation of a signal in the presence of noise, the same kind of improvement in resolution and removal of spurious features (relative to the periodogram) that the maximum-entropy formalism did in the absence of noise; and it will do this as well for chirped or nonchirped signals. There is no meaningful comparison with B-T methods at all, for they do not address the same problem.

Realizing these things changed the direction of the present work. We started with the intention only of getting a quick, preliminary glimpse at what Bayesian theory has to say about monochromatic spectrum analysis in the presence of noise, before proceeding to entropy considerations. But as soon as the phaseless likelihood function Eq. (11) appeared, it was realized that (a) the status of B-T methods in this problem is very different from what we and others had supposed, (b) monochromatic spectrum estimation from noisy data must be radically revised by these results, and (c) given that revision, the extension to chirp is almost trivial.

Therefore, we now reconsider at some length the "old" problem of conventional pre-entropy spectrum estimation of a signal in noise, from this new viewpoint.

8. Power Spectrum Estimates

The terms "power spectrum" and "power spectrum estimator" can be defined in various ways. Also, we need to distinguish between the quite different goals of estimating a power spectral density, estimating the power in a spectrum line, and estimating the frequencies present.

In calling $\hat{P}(\omega)$ an estimate of the power spectral density, we mean that

$$\int_a^b \hat{P}(\omega)\, d\omega \tag{17}$$

is the expectation, over the joint posterior distribution of all the unknown parameters, of the energy carried by the signal f(t), not the noise, in the frequency band $(a < \omega < b)$, in the observation time $N = 2T + 1$. The true total energy is $NA^2/2$, and given data D we can write its expectation as

$$\frac{N}{2}\, E(A^2|D,I) = \frac{N}{2} \int_{-\pi}^{\pi} d\omega \int_0^{\infty} dA\, A^2 \int_{-\pi}^{\pi} d\alpha\; p(A,\omega,\alpha,|D,I)\,. \tag{18}$$

For formal reasons it is convenient to define our spectrum as extending over both positive and negative frequencies; thus Eq. (18) should be equated to the integral (17) over $(-\pi,\pi)$. Therefore our power spectrum estimate is

$$\hat{P}(\omega) = \frac{N}{2} \int_0^{\infty} dA\, A^2 \int_{-\pi}^{\pi} d\alpha\; p(A,\omega,\alpha,|D,I)\,, \quad (-\pi < \omega < \pi)\,. \tag{19}$$

To define a power spectrum only over positive frequencies (as would be done experimentally), one should take instead $\hat{P}_+(\omega) = 2\hat{P}(\omega)$, $0 \leq \omega < \pi$.

In Eqs. (18) and (19), $p(A,\omega,\alpha|D,I)$ is the joint posterior distribution

$$p(A,\omega,\alpha|D,I) = B\, p(A,\omega,\alpha|I)\, L(A,\omega,\alpha)\,, \tag{20}$$

in which B is a normalization constant, I stands for prior information, and $p(A,\omega,\alpha|I)$ is the joint prior probability density for the three parameters.

If we have any prior information about our parameters, this is the place to put it into our equations, in the prior probability factors; if we do not, then a noninformative, flat prior will express that fact and leave the decision to the evidence of the data in the likelihood function (thus realizing R. A. Fisher's goal of "letting the data speak for themselves").

Of course, it is not required that we actually have, or believe, a piece of prior information I or data D before putting it into these equations; we may wish to find out what would be the consequences of having or not having a certain kind of prior information or data, to decide whether it is worth the effort to get it.

Therefore, by considering various kinds of prior information and data in Eq. (19) we can analyze a multitude of different special situations, real or hypothetical, of which we can indicate only a few in the present study.

First, let us note the form of traditional spectrum analysis, which did not contemplate the existence of chirp, that is contained in this "formalism." In many situations we know in advance that our signals are not chirped; that is, the prior probability in Eq. (20) is concentrated at $\alpha = 0$. Our equations will then be the same as if we had never introduced chirp at all; that is, failure to note its possible existence is equivalent to asserting (unwittingly) the prior information: "$\alpha = 0$: there is no chirp."

(We add parenthetically that much of the confusion in statistics is caused by this phenomenon. In other areas of applied mathematics, failure to notice all the possibilities means only that not all possibilities will be studied. In probability theory, failure to notice some possibilities is mathematically equivalent to making unconscious, and often very strong, assumptions about them.)

Setting $\alpha = 0$ in Eq. (11) gives the joint likelihood $L(\alpha,\omega)$. Not using prior infomation (that is, using flat prior densities for A,ω), our spectrum estimator is then

$$\hat{P}(\omega) = \frac{\dfrac{N}{2}\displaystyle\int_0^\infty dA\; A^2 L(A,\omega)}{\displaystyle\int_{-\pi}^{\pi} d\omega \int_0^\infty dA\; L(A,\omega)}\,. \tag{21}$$

Note that this can be factored:

$$\hat{P}(\omega) = \left[\frac{\frac{N}{2}\int_0^\infty dA\ A^2 L(A,\omega)}{\int_0^\infty dA\ L(A,\omega)}\right] \cdot \left[\frac{\int_0^\infty dA\ L(A,\omega)}{\int_{-\pi}^{\pi} d\omega \int_0^\infty dA\ L(A,\omega)}\right] \tag{22}$$

or

$$\hat{P}(\omega) = (N/2)\ E(A^2|\omega DI)\ p(\omega|DI)\ . \tag{23}$$

In words: $\hat{P}(\omega)$ = (conditional expectation of energy given ω) × (posterior density of ω given the data). Both factors may be of special interest, so we evaluate them separately. In Appendix C we derive

$$\frac{N}{2}\ E(A^2|\omega DI) = \sigma^2\left[1 + 2q + 2q\ \frac{I_1(q)}{I_0(q)}\right], \tag{24}$$

where

$$q(\omega) \equiv C(\omega,0)/2\sigma^2\ . \tag{25}$$

This has two limiting forms:

$$\frac{N}{2}\ E(A^2|\omega DI) \rightarrow \begin{cases} 2C(\omega,0), & C(\omega,0) > 2\sigma^2 \\ \sigma^2 + C(\omega,0), & C(\omega,0) \ll \sigma^2 \end{cases} \tag{26}$$

The second case is unlikely to arise, because it would mean that the signal and noise have, by chance, nearly canceled each other out. If this happens, Bayes' theorem tells us to hedge our bets a little. Something strange is afoot.

From Appendix C, the second factor in Eq. (22) is

$$p(\omega|DI) = \frac{\exp(q) I_0(q)}{\int_{-\pi}^{\pi} \exp(q)\ I_0(q)\ d\omega}\ . \tag{27}$$

This answers the question: "Conditional on the data, what is the probability that the signal frequency lies in the range $(\omega,\ \omega+d\omega)$ irrespective of its amplitude?" In many situations it is this question, and not the power spectrum density, that matters, for the signal amplitude may have been corrupted en route by all kinds of irrelevant circumstances and the frequency alone may be of interest.

Combining Eqs. (24) and (27), we have the explicit Bayesian power spectrum estimate

$$\hat{P}(\omega) = \sigma^2 \frac{[(1+2q)I_0(q) + 2qI_1(q)]\exp(q)}{\int_{-\pi}^{\pi} \exp(q)\, I_0(q)\, d\omega} . \tag{28}$$

A graph of the nonlinear processing function

$$f(q) = [(1+2q)I_0(q) + 2qI_1(q)]\exp(q) \tag{29}$$

shows it close to the asymptotic form $(8/\pi q)^{1/2}\exp(2q)$ over most of the range likely to appear; in most cases we would not make a bad approximation if we replaced Eq. (28) by

$$\hat{P}(\omega) = \sigma^2 \frac{4q\exp(2q)}{\int_{-\pi}^{\pi} \exp(2q)\, d\omega} . \tag{30}$$

Whenever the data give evidence for a signal well above the noise level (that is, $C(\omega,0)$ reaches a global maximum $C_{max} = C(\nu,0) \gg \sigma^2$), then $q_{max} = C_{max}/\sigma^2 \gg 1$ and most of the contribution to the integral in the denominator of Eq. (28) will come from the neighborhood of this greatest peak. Expanding

$$q(\omega) = q_{max} - q''(\omega - \nu)^2/2 + \ldots \tag{31}$$

and using saddle-point integration, we have then

$$\int_{-\pi}^{\pi} e^q\, I_0(q)\, d\omega \simeq 2(q_{max}q'')^{-1/2}\exp(2q_{max}) , \tag{32}$$

the factor of 2 coming from the equal peak at $(-\nu)$. Near the greatest peak, the positive frequency estimator reduces to

$$\hat{P}_+(\omega) \simeq 2C_{max} \frac{1}{\sqrt{2\pi(\delta\omega)^2}} \exp\left[-\frac{(\omega-\nu)^2}{2(\delta\omega)^2}\right] \tag{33}$$

where $\delta\omega = (q'')^{-1/2}$ is the accuracy with which the frequency can be estimated. Comparing with the factorization Eq. (23), we see that the Gaussian is the posterior distribution of ω, while the first factor, the estimated total energy in the line, is

$$\int \hat{P}_+(\omega)\, d\omega = 2C_{max} \tag{34}$$

in agreement with Eq. (26).

As a further check, suppose we have a pure sinusoid with very little noise:

$$y_t = A \cos\nu t + e_t\,, \qquad A \gg \sigma\ . \tag{35}$$

Then the autocovariance (14) is approximately

$$R(t) \simeq (A^2/2) \cos\nu t \tag{36}$$

and so from Eq. (13)

$$C_{max} = C(\nu,0) = \frac{A^2}{2} \sum_t \cos^2\nu t = \frac{NA^2}{4} \tag{37}$$

so $2C_{max} = NA^2/2$ is indeed the correct total energy carried by the signal. Likewise,

$$\sigma^2 q'' = \sum_t t^2 R(t) \cos\nu t \simeq \frac{N^3A^2}{48} \tag{38}$$

which gives the width

$$\delta\omega \simeq \frac{\sigma}{N}\sqrt{\frac{12}{C_{max}}}\ . \tag{39}$$

These approximate relations illustrate what we stressed above; in this problem the periodogram $C(\omega,0)$ is not even qualitatively an estimate of the power spectral density. Rather, when C reaches a peak above the noise level, thus indicating the probable presence of a spectrum line, its peak value $2C_{max}$ is an estimate of the total energy in the line.

9. Extension to Chirp

Let $\hat{P}(\omega,\alpha)d\omega d\alpha$ be the expectation of energy carried by the signal f(t) in an element $d\omega d\alpha$ of the frequency-chirp plane. From Eq. (14), the frequency-chirp density estimator that does not use any prior information about (A,ω,α) is simply

$$\hat{P}(\omega,\alpha) = \frac{f(q)}{\int d\omega \int d\alpha \exp(q) I_0(q)} \tag{40}$$

where f(q) is the same nonlinear processing function (29), except that now in place of Eq. (25) we have

$$q \equiv q(\omega,\alpha) = \frac{C(\omega,\alpha)}{2\sigma^2} . \tag{41}$$

As in Eq. (28), any peaks of $C(\omega,\alpha)$ that rise above the noise level will be strongly emphasized, indicating high probability of a signal.

If we have indeed no prior information about the frequencies and chirp rates to be expected, but need to be ready for all contingencies, then there seems to be no way of avoiding the computation in determining $C(\omega,\sigma)$ over the entire plane $(-\pi < \omega,\alpha < \pi)$. Note that, while $\hat{P}(\omega,0)$ equals $\hat{P}(-\omega,0)$ by symmetry, when chirp is present we have inversion symmetry, $\hat{P}(\omega,\alpha) = \hat{P}(-\omega,-\alpha)$, so half of the $(\omega - \alpha)$ plane needs to be searched if no prior information about the signal location is at hand.

But noting this shows how much reduction in computation can be had with suitable prior information. That bat, knowing in advance that it has emitted a signal of parameters ω_0,α_0, and knowing also what frequency interval (determined by its flight speed v and the range of important targets) is of interest, does not need to scan the whole plane. It need scan only a portion of the line $\alpha \simeq \alpha_0$ extending from about $\omega_0(1 - v/c)$ to $\omega_0(1 + 4v/c)$, where c = velocity of sound, to cover the important contingencies (a target that is approaching is a potential collision; one dead ahead approaching at the flight speed is a stationary object, a potential landing site; one moving away rapidly is uninteresting; one moving away slowly may be a moth, a potential meal).

10. Many Signals

In the above we made what seems a strong assumption, that only one signal $f(t) = A \cos(\omega t + \alpha t^2 + \theta)$ is present, and all our results were inferences as to where in the parameter space of (A,ω,α) that one signal might be. This is realistic in some problems (that is, oceanographic chirp, or only one bat is in our cage, etc.), but not in most. In what way has this assumption affected our final result?

Suppose for the moment that the chirp rate α equals 0. Then the power spectrum estimate $\hat{P}(\omega)\, d\omega$ in Eq. (28) represents, as we noted, the answer to:

> Question A: What is your "best" estimate (expectation) of the energy carried by the signal f(t) in the frequency band $d\omega$, in the interval of observation?

But had we asked a different question,

> Question B: What is your estimate of the product of the energy transports in the two nonoverlapping frequency bands $(a < \omega < b)$ and $(c < \omega < d)$?

the answer would be zero; a single fixed frequency ω cannot be in two different bands simultaneously. If our prior information I tells us that only one frequency can be present, then the joint posterior probability $p(E_{ab}, E_{cd}|D,I)$ of the events

$$
\begin{aligned}
E_{ab} &\equiv (a < \omega < b) \\
E_{cd} &\equiv (c < \omega < d)
\end{aligned}
\tag{42}
$$

is zero; in orthodox language, the "fluctuations" in nonoverlapping frequency bands must be perfectly negatively correlated in our posterior distribution.

Now if two frequencies can be present, it turns out that the answer to question (A) will be essentially the same. But the answer to question (B)—or any other question that involves the joint probabilities in two different frequency bands—will be different; for then it is possible for power to be in two different bands simultaneously, and it is not obvious without calculation whether the correlations in energy transport in different bands are positive or negative.

If now we suppose that three signals may be present, the answers to questions (A) and (B) will not be affected; it makes a difference only when we ask a still more complicated question, involving joint probabilities for three different frequency bands. For example,

> Question C: What is your estimate of the power carried in the frequency band $(a < \omega < b)$, given the powers carried in $(c < \omega < d)$ and $(e < \omega < f)$?

and so on! In conventional spectrum estimation one asks only question (A); and for that our one-signal solution (28) requires no change. However, many signals may be present. Our seemingly unrealistic assumption makes a difference only when we ask more complicated questions.

In this section we prove these statements, or at least give the general solution from which they can be proved. For the moment we leave out the chirp; it is now clear how we can add it easily to the final results. The signal to be analyzed is a superposition of n signals like Eq. (2):

$$f(t) = \sum_{j=1}^{n} A_j \cos(\omega_j t + \theta_j) \tag{43}$$

sampled at instants t_m, $1 \leq m \leq N$, which need not be uniformly spaced (our previous equally spaced version is recovered if we make the particular choice $t_m = -T + m - 1$). Use the notation

$$f_m = f(t_m) . \tag{44}$$

The joint likelihood of the parameters is now

$$L(A_j \omega_j \theta_j \sigma) = \sigma^{-N} \exp\left[-\frac{1}{2\sigma^2} \sum_{m=1}^{N} (y_m - f_m)^2\right]$$

$$= \sigma^{-N} \exp\left[-\frac{N}{2\sigma^2} (\overline{y^2} + Q)\right] \tag{45}$$

with the quadratic form

$$Q(A_j \omega_j \theta_j) = \overline{f^2} - 2\overline{yf} \tag{46}$$

in which the bars denote averages over the data:

$$\overline{y^2} \equiv N^{-1} \sum_{m=1}^{N} y_m^2 \tag{47}$$

$$\overline{yf} \equiv N^{-1} \sum_{m} y_m f_m \tag{48}$$

$$\overline{f^2} \equiv N^{-1} \sum_{m} f_m^2 . \tag{49}$$

To rewrite Q as an explicit function of the parameters, define the function

$$x(\omega) = N^{-1} \sum_{m=1}^{N} y_m \exp(i\omega t_m) , \tag{50}$$

which is the "complex square root of the periodogram"; its projections

$$d_j \equiv \mathrm{Re}[x(\omega_j)\, e^{i\theta_j}]$$

$$= N^{-1}\sqrt{NC(\omega_j,0)}\, \cos(\theta_j + \psi_j)\,, \quad 1 \le j \le n\,, \tag{51}$$

where we have used Appendix C to write it in terms of the periodogram; and the matrix

$$M_{jk} \equiv N^{-1} \sum_{m=1}^{N} \cos(\omega_j t_m + \theta_j) \cos(\omega_k t_m + \theta_k)\,, \quad 1 \le j,\; k \le n\,. \tag{52}$$

Then

$$\overline{yf} = \sum_{j=1}^{n} d_j A_j \tag{53}$$

$$\overline{f^2} = \sum_{jk=1}^{n} M_{jk} A_j A_k \tag{54}$$

and the quadratic form (46) is

$$Q = \sum_{jk} M_{jk} A_j A_k - 2 \sum_{j} d_j A_j\,. \tag{55}$$

The likelihood Eq. (45) will factor in the desired way if we complete the square in Q. Define the quantities $\hat{A}_k$ by

$$d_j = \sum_{k=1}^{n} M_{jk} \hat{A}_k\,, \quad 1 \le j \le n\,, \tag{56}$$

and note that, if the inverse matrix M^{-1} exists,

$$\sum_j d_j \hat{A}_j = \sum_{jk} M_{jk} \hat{A}_j \hat{A}_k = \sum_{kj} M^{-1}{}_{kj} d_k d_j\,. \tag{57}$$

Then

$$Q = \sum_{jk} M_{jk}(A_j - \hat{A}_j)(A_k - \hat{A}_k) - \sum_{jk} M_{jk}\hat{A}_j\hat{A}_k \tag{58}$$

and the joint likelihood function splits into three factors:

$$L = L_1 L_2 L_3 \tag{59}$$

with

$$L_1 = \sigma^{-N} \exp(-N\overline{y^2}/2\sigma^2) \tag{60}$$

$$L_2 = \exp\{-\frac{N}{2\sigma^2} \sum_{jk} M_{jk}(A_j - \hat{A}_j)(A_k - \hat{A}_k) \tag{61}$$

$$L_3 = \exp\{+\frac{N}{2\sigma^2} \sum_{jk} M_{jk}\hat{A}_j\hat{A}_k)\} . \tag{62}$$

If σ is known, the factor L_1 may be dropped, since it will be absorbed into the normalization constant of the joint posterior distribution of the other parameters. If σ is unknown, then it becomes a "nuisance parameter" to be integrated out of the problem with respect to whatever prior probability $p(\sigma|I)$ describes our prior information about it. The most commonly used case is that where we are initially completely ignorant about σ—or wish to proceed as if we were, to see what the consequences would be. As expounded in some detail elsewhere [Jaynes, 1980], the Jeffreys prior probability assignment $p(\sigma|I) = 1/\sigma$ is uniquely determined as the one that expresses "complete ignorance" of a scale parameter.

One of the fascinating things about Bayes' theorem is its efficiency in handling this situation. The noise level σ is highly relevant to our inferences; so if it is initially completely unknown, then it must be estimated from the data. But Bayes' theorem does this for us automatically in the process of integrating σ out of the problem. To see this, note that from Eq. (45) when we integrate out σ the quasi-likelihood of the remaining parameters becomes

$$\int_0^\infty \frac{L d\sigma}{\sigma} \propto [1 + (Q/\overline{y^2})]^{-N/2} , \tag{63}$$

but when N is reasonably large (that is, when we have enough data to permit a reasonably good estimate of σ), this is nearly the same as

$$\exp(-NQ/2\overline{y^2}) \quad . \tag{64}$$

In its dependence on $\{A_i \omega_i \theta_i\}$ this is just Eq. (45) with σ^2 replaced by $\overline{y^2}$. Thus, in effect, if we tell Bayes' theorem: "I'm sorry, I don't know what σ^2 is!" it replies to us, "That's all right, don't worry. We'll just replace σ^2 by the best estimate of σ^2 that we can make from the data, namely $\overline{y^2}$." After doing this, we shall have the same quadratic form Q as before, and its minimization will locate the same "best" estimates of the other parameters as before. The only difference is that for small N the peak of Eq. (63) will not be as sharp as that of the Gaussian (45), so we are not quite so sure of the accuracy of our estimates; but that is the only price we paid for our ignorance of σ.

Therefore, if N is reasonably large it hardly matters whether σ is known or unknown. Supposing for simplicity, as we did before, that σ is known, the joint posterior density of the other parameters $\{A_i \omega_i \theta_i\}$ factors:

$$p(\{A_i \omega_i \theta_i\} | DI) = p(A_i | \omega_i \theta_i DI)\, p(\omega_i \theta_i | DI) \tag{65}$$

in which we have the joint conditional probability of the amplitudes A_i given the frequencies and phases:

$$p(A_j | \omega_j \theta_j DI) \propto \exp\{- \frac{N}{2\sigma^2} \sum_{jk} M_{jk}(A_j - \hat{A}_j)(A_k - \hat{A}_k)\} \tag{66}$$

and the joint marginal posterior density of the frequencies and phases:

$$p(\{\omega_j \theta_j\} | DI) \propto \exp\{+ \frac{N}{2\sigma^2} \sum_{jk} M_{jk} \hat{A}_j \hat{A}_k\} \quad . \tag{67}$$

Equation (66) says that $\hat{A}_j$ is the estimate of the amplitude A_j that we should make, given the frequencies and phases $\{\omega_j \theta_j\}$, and Eq. (67) says that the most probable values of the frequencies and phases are those for which the estimated amplitudes are large.

The above relations are, within the context of our model, exact and quite general. The number n of possible signals and the sampling times t_m may be chosen arbitrarily. Partly for that reason, to explore all the results that are in Eqs. (65) to (67) would require far more space than we have here. We shall forego working out the interesting details of what happens to our conclusions when the sampling times are not equally spaced, what the answers to those more complicated questions like (B) and (C) are, and what happens to the matrix M in the limit when two frequencies coincide.

Somehow, as $|\omega_j - \omega_k| \to 0$, it must be that A_j and A_k become increasingly confounded (indistinguishable). In the limit there is only one amplitude where two used to be. Then will the rank of M drop from n to (n-1)? It is a recommended exercise to work this out in detail for the case n = 2 and see how, as the two signals merge continuously into one, there is a mixing of the eigenvectors of M much like the "level crossing" phenomenon of quantum theory. At every state the results make sense in a way that we do not think anybody's intuition can foresee, but which seems obvious after we contemplate what Bayes' theorem tells us.

Here we shall examine only the opposite limit, that of frequencies spaced far enough apart to be resolved, which of course requires that the number of signals is not greater than the number N of observations. If our sampling points are equally spaced:

$$t_m = -(T+1) + m, \qquad 1 \le m \le N, \quad 2T + 1 = N\,, \tag{68}$$

then M_{jk} reduces to

$$M_{jk} = N^{-1} \sum_{t=-T}^{T} \cos(\omega_j t + \theta_j) \cos(\omega_k t + \theta_k) \ . \tag{69}$$

The diagonal elements are

$$M_{jj} = \frac{1}{2} + \frac{\sin N\omega_j}{2N \sin\omega_j} \cos 2\theta_j\,, \qquad 1 \le j \le n\,, \tag{70}$$

in which the second term becomes appreciable only when $\omega_j \to 0$ or $\omega_j \to \pi$. If we confine our frequencies to be positive, in $(0,\pi)$, then the terms in Eq. (69) with $(\omega_j + \omega_k)$ will never become large, and the off-diagonal elements are

$$M_{jk} \simeq \frac{\sin Nu}{2N \sin u} \cos(\theta_j - \theta_k) + O(N^{-1}) \tag{71}$$

where $u \equiv (\omega_j - \omega_k)/2$. This becomes of order unity only when the two frequencies are too close to resolve and that merging phenomenon begins. Thus as long as the frequencies $\{\omega_j\}$ are so well separated that

$$N|\omega_j - \omega_k| \gg 1\,, \tag{72}$$

a good approximation will be

$$M_{jk} \simeq (1/2)\,\delta_{jk} \tag{73}$$

and the above relations simplify drastically. The amplitude estimates reduce to, from Eq. (51),

$$\hat{A}_j = 2d_j = 2N^{-1}\sqrt{NC(\omega_j,0)}\,\cos(\theta_j + \psi_j) \quad . \tag{74}$$

The joint posterior density of the frequencies and phases is

$$p(\{\omega_j\theta_j\}|DI) \propto \exp[\frac{N}{\sigma^2}\sum_j d_j^2]$$

$$= \exp[\sigma^{-2}\sum_j C(\omega_j,0)\cos^2(\theta_j + \psi_j)] \tag{75}$$

and the joint posterior density of all the parameters is

$$p(\{A_j\omega_j\theta_j\}|DI)$$

$$= \exp\{\sigma^{-2}\sum_{j=1}^{n}[A_j\sqrt{NC(\omega_j,0)}\,\cos(\theta_j + \psi_j) - NA_j^2/4]\} \ , \tag{76}$$

which is a product of independent distributions. If the phases θ_j were initially correlated in their joint prior distribution, there would be an extra factor $p(\theta_1 \ldots \theta_n|I)$ in Eq. (76) that removes this independence, and might make an appreciable difference in our conclusions.

This could arise, for example, if we knew that the entire signal is a wavelet originating in a single event some time in the past; and at that initial time all frequency components were in phase. Then integrating the phases out of Eq. (76) will transfer that phase correlation to a correlation in the A_j. As a result, the answers to questions such as (B) and (C) above would be changed in a possibly more important way. This is another interesting detail that we cannot go into here, but merely note that all this is inherent in Bayes' theorem.

But usually our prior probability distribution for the phases is independent and uniform on $(0,2\pi)$; that is, we have no prior information about either the values of, or connections between, the different phases. Then the best inference we shall be able to make about the amplitudes and frequencies is obtained by integrating all the θ_j out of Eqs. (75) and (76) independently. This will generate just the same $I_0(q)$ Bessel functions as before; and writing $q_j \equiv C(\omega_j,0)/2\sigma^2$, our final results are:

$$p(\{\omega_j\}|DI) \propto \prod_{j=1}^{n} \exp(q_j)\, I_0(q_j) \tag{77}$$

and

$$p(\{A_j\omega_j\}|DI) \propto \prod_{j=1}^{n} \exp(-NA_j^2/4\sigma^2)\, I_0[A_j \sqrt{NC(\omega_j,0)}/\sigma^2\,] \;. \tag{78}$$

But these are just products of independent distributions identical with our previous single-signal results, Eqs. (11) and (27). We leave it as an exercise for the reader to show from this that our previous power spectrum estimate (28) will follow. Thus as long as we ask only question (A) above, our single-signal assumption was not a restriction after all.

At this point it is clear also that if our n signals are chirped, we need only replace $C(\omega_j,0)$ by $C(\omega_j,\alpha_j)$ in these results, and we shall get the same answers as before to any question about frequency and chirp that involves individual signals, but not correlations between different signals.

11. Conclusion

Although the theory presented here is only the first step of the development that is visualized, we have thought it useful to give an extensive exposition of the Bayesian part of the theory.

No connection with AR models has yet appeared; but we expect this to happen when additional prior information is put in by entropy factors. In a full theory of spectrum estimation in the presence of noise, in the limit as the noise goes to zero the solution should reduce to something like the original Burg pure maximum entropy solution (it will not be exactly the same, because we are assuming a different kind of data).

For understanding and appreciating Bayesian inference, no theorems proving its secure theoretical foundations can be quite as effective as seeing it in operation on a real problem. Every new Bayesian solution like the present one gives us a new appreciation of the power and sophistication of Bayes' theorem as the true logic of science. It seeks out every factor in the model that has any relevance to the question being asked, tells us quantitatively how relevant it is—and relentlessly exposes how crude and primitive other methods were.

We could expand the example studied here to a large volume without exhausting all the interesting and useful detail contained in the general solution—almost none of which was anticipated by sampling theory or intuition.

12. Appendix A: Oceanographic Chirp

We illustrate the phase cancellation phenomenon, for the spectral snapshot method described in the text, as follows. That method essentially calculates the periodogram of a data set $\{y_t: -T \lesssim t \lesssim T\}$, or possibly a Blackman-Tukey smoothing of it (the difference is not crucial for the point to be made here, affecting only the quantitative details). The periodogram is

$$X(\omega) = N^{-1}\left|\sum_t y_t\, e^{i\omega t}\right|^2 . \tag{A1}$$

If y_t is a sinusoid of fixed frequency ν,

$$y_t = A\cos(\nu t + \theta)\,, \tag{A2}$$

then the periodogram reaches its peak value at or very near the true frequency,

$$X(\nu) = \frac{NA^2}{4} . \tag{A3}$$

But if the signal is chirped,

$$y_t = A\cos(\nu t + \alpha t^2 + \theta)\,, \tag{A4}$$

then the periodogram (A1) is reduced, broadened, and distorted. Its value at the center frequency is only about

$$X(\nu) = \frac{NA^2}{4}\left|N^{-1}\sum_t e^{i\alpha t^2}\right|^2, \tag{A5}$$

which is always less than (A3) if $\alpha \neq 0$. As a function of α, (A5) is essentially the Fresnel diffraction pattern of N equally spaced narrow slits. To estimate the phase cancellation effect, note that when $\alpha T < 1$ the sum in (A5) may be approximated by a Fresnel integral, from whose analytic properties we may infer that, when $\alpha T^2 \gtrsim 1$, $X(\nu)$ will be reduced below (A3) by a factor of about

$$\frac{\pi}{4\alpha T^2}\left[1 - \left(\frac{2}{\pi\alpha T^2}\right)^{1/2}\cos(\alpha T^2 + \pi/4) + O(T^{-2})\right] . \tag{A6}$$

It is clear from (A5) that the reduction is not severe if $\alpha T^2 \lesssim 1$, but (A6) shows that it can essentially wipe out the signal if $\alpha T^2 \gg 1$.

Also, in the presence of chirp the periodogram (A1) exhibits "line broadening" and "line splitting" phenomena; for some values of αT^2 there appear to be two or more lines of different amplitudes. For graphs demonstrating this, see Barber and Ursell [1948].

Discovery of chirped ocean waves, originating from storms thousands of miles away, is often attributed to Munk and Snodgrass [1957]. In Tukey et al. [1980] their feat was termed "one of the virtuoso episodes in the annals of power spectrum analysis," showing the value of alertness to small unexpected things in one's data. It has been suggested that a Bayesian wears some kind of blinders that make him incapable of seeing such things, and that discovery of the phenomenon might have been delayed by decades—or even centruies—if Munk and Snodgrass had used Bayesian, AR, or maximum entropy methods.

The writer's examination of the Munk-Snodgrass article has led him to a different picture of these events. The chirped signals they found were not small; in the frequency band of interest they were the most prominent feature present. The signals consisted of pressure variations measured off the California coast at a depth of about 100 meters, attributed to a storm in the Indian Ocean about 9000 miles away. The periods were of the order of 20 seconds, decreasing at a rate of about 10% per day for a few days.

They took measurements every 4 seconds, accumulating continuous records of length $N = 2T = 6000$ observations, or 6-2/3 hours. Thus, from their grand average measured chirp rate of about $\alpha = 1.6 \times 10^{-7}$ sec^{-2} we get $\alpha T^2 = 24$, so, with the sum in (A5) wrapped several times around the Cornu spiral and from (A6), phase cancellation must have reduced the apparent signal power at the center frequency to about 3% of its real value (this would have been about their best case, since nearer sources would give proportionally larger αT^2).

Such a phase cancellation would not be enough to prevent them from seeing the effect altogether, but it would greatly distort the line shape, as seems to have happened. Although they state that the greater length of their data records makes possible a much higher resolution (of the order of one part in 3000) than previously achieved, their actual lines are of complicated shape, and over 100 times wider than this.

As to date of discovery, this same phenomenon had been observed by Barber and Ursell in England, as early as 1945. A decade before Munk and Snodgrass, they had correlated chirped ocean waves arriving at the Cornwall coast with storms across the Atlantic, and as far away as the Falklands. In Barber and Ursell [1948] we find an analysis of the phase cancellation effect, which led them to limit their data records to 20 minutes, avoiding the difficulty. Indeed, the broad and complicated line shapes published by Munk and Snodgrass look very much like the spectra calculated by Barber and Ursell to illustrate the disastrous effects of phase cancellation when one uses too long a record.

The theory of this phenomenon, relating the chirp rate to the distance r to the source, was given in 1827 by Cauchy. In our notation, α equals g/4r, where g is the acceleration of gravity. For example, in the summer of 1945

Barber and Ursell observed a signal whose period fell from 17.4 sec to 12.9 sec in 36 hours. These data give $\alpha = 5 \times 10^{-7}\ \text{sec}^{-2}$, placing the source about 3000 miles away, which was verified by weather records. In May 1946 a signal appeared, whose period fell from 21.0 sec to 13.9 sec in 4 days, and came from a source 6000 miles away.

Up to 1947 Barber and Ursell had analyzed some 40 instances like this, without using anybody's recommended methods of spectrum analysis to measure those periods. Instead they needed only a home-made analog computer, putting a picture of their data on a rotating drum and noting how it excited a resonant galvanometer as the rotation speed varied!

Munk and Snodgrass surely knew about the phase cancellation effect, and were not claiming to have discovered the phenomenon, for they make reference to Barber and Ursell. It appears to us that they did not choose their record lengths with these chirped signals in mind, simply because they had intended to study other phenomena. But the signals were so strong that they saw them anyway—not as a result of using a data analysis method appropriate to find them, but in spite of an inappropriate method.

Today, it would be interesting to re-analyze their original data by the method suggested here. If there is a constant amplitude and chirp rate across the data record, the chirpogram should reach the full maximum (A3), without amplitude degradation or broadening, at the true center frequency and chirp rate, and shoud therefore provide a much more sensitive and accurate data analysis procedure.

13. Appendix B: Why Gaussian Noise?

In what Savage [1954] called the "objectivist" school of statistical thought (nowadays more often called "orthodox" or "sampling theory"), assigning a noise distribution is interpreted as asserting or hypothesizing a statement of fact, that is, a physically real property of the noise. It is, furthermore, a widely believed "folk-theorem" that if the actual frequency distribution of the noise differs from the probability distribution that we assigned, then all sorts of terrible things will happen to us; we shall be misled into drawing all sorts of erroneous conclusions. We wish to comment on both of these beliefs.

We are aware of no real problem in which we have the detailed information that could justify such a strong interpretation of our noise distribution at the beginning of a problem; nor do we ever acquire information that could verify such an interpretation at the end of the problem.

As noted in the text, an assigned noise distribution is a joint distribution of all the errors, that is, a probability $p(e|I)$ assigned to the total noise vector $e = (e_1 \ldots e_n)$ in an n-dimensional space. Obviously, this cannot be a statement of verifiable fact, for the experiment generates only one noise vector. Our prior distribution $p(e|I)$ defines rather the range of different possible noise vectors that we wish to allow for, only one of which will actually be realized.

In the problems considered here, the information that would be useful in improving our spectrum estimates consists of correlations between the e_i (nonwhite noise). Such correlations cannot be described at all in terms of the frequencies of individual noise values. A correlation extending over a lag m is related to frequencies only of noise sequences of length greater than or equal to m. Even for $m = 3$, the number of possible sequences of length m is usually far greater than the length of our data record; so it is quite meaningless to speak of the "frequencies" with which the different sequences of length $m = 3$ appear in our data, and therefore equally meaningless to ask whether our noise probability distribution correctly describes those frequencies.

As these considerations indicate, the function of $p(e|I)$ cannot be to describe the noise, but rather to describe our state of knowledge about the noise. It is related to facts to this extent: we want to be fairly sure that we choose a set of possible vectors big enough to include the true one. This is a matter of being honest about just how much prior information we actually have, that is, of avoiding unwarranted assumptions.

If in our ignorance we assign a noise distribution that is "wider" (for example, that supposes a greater mean-square error) than the actual noise vector, then we have only been more conservative—making allowance for a greater range of possibilities—than we needed to be. But the result is not that we shall see erroneous "effects" that are not there; rather, we shall have less discriminating power to detect small effects than we might have enjoyed had we more accurate prior knowledge about the noise.

If we assign a noise distribution so "narrow" that the true noise vector lies far outside the set thought possible, then we have been dishonest and made unwarranted assumptions, for valid prior information could not have justified such a narrow distribution. Then, as noted in the text, we shall indeed pay the penalty of seeing things that are not there. The goal of stating, by our prior distribution, what we honestly do know—and nothing more—is the means by which we protect ourselves against this danger.

Tukey et al. [1980] comment: "Trying to think of data analysis in terms of hypotheses is dangerous and misleading. Its most natural consequences are (a) hesitation to use tools that would be useful because 'we do not know that their hypotheses hold' or (b) unwarranted belief that the real world is as simple and neat as these hypotheses would suggest. Either consequence can be very costly. . . . A procedure does not have hypotheses—rather there are circumstances where it does better and others where it does worse."

We are in complete agreement with this observation, and indeed would put it more strongly. Although some hypotheses about the nature of the phenomenon may suggest a procedure—or even uniquely determine a procedure—the procedure itself has no hypotheses, and the same procedure may be suggested by many very different hypotheses. For example, as we noted in the text, the Blackman-Tukey window-smoothing procedure was associated by them with the hypothesis that the data were a realization of a "stationary Gaussian random process." But of course nothing prevents one from applying the procedure itself to any set of data whatsoever, whether or not "their hypotheses hold." And indeed, there are "circumstances where it does better and others where it does worse."

But we believe also that probability theory incorporating Bayesian maximum entropy principles is the proper tool—and a very powerful one—for (a) determining those circumstances for a given procedure, and (b) determining the optimal procedure, given what we know about the circumstances. This belief is supported by decades of theoretical and practical demonstrations of that power.

Clearly, while striving to avoid gratuitous assumption of information that we do not have, we ought at the same time to use all the relevant information that we actually do have; and so Tukey has also wisely advised us to think very hard about the real phenomenon being observed, so that we can recognize those special circumstances that matter and take them into account. As a general statement of policy, we could ask for nothing better; so our question is: How can we implement that policy in practice?

The original motivation for the principle of maximum entropy [Jaynes, 1957] was precisely the avoidance of gratuitous hypotheses, while taking account of what is known. It appears to us that in many real problems the procedure of the maximum entropy principle meets both of these requirements, and thus represents the explicit realization of Tukey's goal.

If we are so fortunate as to have additional information about the noise beyond the mean-square value supposed in the text, we can exploit this to make the signal more visible, because it reduces the measure W, or "volume," of the support set of possible noise variations that we have to allow for. For example, if we learn of some respect in which the noise is not white, then it becomes in part predictable and some signals that were previously indistinguishable from the noise can now be separated.

The effectiveness of new information in thus increasing signal visibility is determined by the reduction it achieves in the entropy of the joint distribution of noise values—essentially, the logarithm of the ratio W'/W by which that measure is reduced. The maximum entropy formalism is the mathematical tool that enables us to locate the new contracted support set on which the likely noise vectors lie.

It appears to us that the evidence for the superior power of Bayesian maximum entropy methods over both intuition and "orthodox" methods is now so overwhelming that nobody who is concerned with data analysis—in any field—can afford to ignore it. In our opinion these methods, far from conflicting with the goals and principles expounded by Tukey, represent their explicit quantitative realization, which intuition could only approximate in a crude way.

14. Appendix C: Details of Calculations

Derivation of the Chirpogram. Expanding the cosine in Eq. (10), we have

$$\sum_t y_t \cos(\omega t + \alpha t^2 + \theta) = P\cos\theta - Q\sin\theta$$

$$= (P^2 + Q^2)^{1/2} \cos(\theta + \tan^{-1}Q/P) \quad \text{(C1)}$$

where

$$P \equiv \sum_t y_t \cos(\omega t + \alpha t^2) \quad \text{(C2)}$$

$$Q \equiv \sum_t y_t \sin(\omega t + \alpha t^2) \; . \quad \text{(C3)}$$

But $(P^2 + Q^2)$ can be written as the double sum

$$\sum_{ts} y_t y_s[\cos(\omega t + \alpha t^2)\cos(\omega s + \alpha s^2) + \sin(\omega t + \alpha t^2)\sin(\omega s + \alpha s^2)]$$

$$= \sum_{ts} y_t y_s[\cos\omega(t - s) + \alpha(t^2 - s^2)] \; . \quad \text{(C4)}$$

Therefore, defining $C(\omega,\alpha)$ as

$$C(\omega,\alpha) \equiv N^{-1}(P^2 + Q^2) \quad \text{(C5)}$$

and substituting Eqs. (C1) and (C4) into Eq. (10), the integral (8) over θ is the standard integral representation of the Bessel function:

$$I_0(x) = (2\pi)^{-1}\int_0^{2\pi} e^{x\cos\theta}\, d\theta \quad \text{(C6)}$$

which yields the result (11) of the text.

Power Spectrum Derivations. From the integral formula

$$Z(a,b) = \int_0^\infty e^{-ax^2} I_0(bx)\, dx = (\pi/4a)^{1/2} \exp(b^2/8a)\, I_0(b^2/8a) \qquad \text{(C7)}$$

we obtain

$$\int_0^\infty x^2 e^{-ax^2} I_0(bx)\, dx = -\frac{\partial Z}{\partial a}$$

$$= \frac{1}{4}\sqrt{\frac{\pi}{a^3}}\,[(1+2q)\, I_0(q) + 2q\, I_1(q)]\, e^q \qquad \text{(C8)}$$

where $q = b^2/8a$. In the notation of Eq. (13) we have $x = A$, $a = N/4\sigma^2$, $b^2 = NC(\omega,0)/\sigma^4$; therefore $q = C(\omega,0)/2\sigma^2$. Thus Eqs. (C5) and (C6) become

$$\int_0^\infty dA\, L(A,\omega) = \sqrt{\frac{\pi\sigma^2}{N}}\, e^q\, I_0(q) \qquad \text{(C9)}$$

$$\int_0^\infty dA\, A^2 L(A,\omega) = 2\sigma^3\sqrt{\frac{\pi}{N^3}}\,[(1+2q) I_0(q) + 2q I_1(q)] e^q, \qquad \text{(C10)}$$

from which Eqs. (24) and (27) of the text follow.

15. References

Barber, N. F., and F. Ursell (1948), "The generation and propagation of ocean waves and swell," Phil. Trans. Roy. Soc. London A240, pp. 527-260.

Blackman, R. B., and J. W. Tukey (1958), The Measurement of Power Spectra, Dover Publications, New York.

Burg, J. P. (1967), "Maximum entropy spectral analysis," in Proc. 37th Meet. Soc. Exploration Geophysicists. Reprinted (1978) in Modern Spectrum Analysis, D. Childers, ed., IEEE Press, New York.

Burg, J. P. (1975), "Maximum Entropy Spectral Analysis," Ph.D. Thesis, Stanford University.

Griffin, D. R. (1958), Listening in the Dark, Yale University Press, New Haven; see also R. H. Slaughter and D. W. Walton, eds. (1970), About Bats, SMU Press, Dallas, Texas.

Nikolaus, B., and D. Grischkowsky (1983), "90 fsec tunable optical pulses obtained by two-stage pulse compression," Appl. Phys. Lett. 43, pp. 228-230.

Gull, S. F., and G. J. Daniell (1978), "Image reconstruction from incomplete and noisy data," Nature 272, pp. 686-690.

Helliwell, R. A. (1965), Whistlers and Related Ionospheric Phenomena, Stanford University Press, Palo Alto, California.

Jaynes, E. T. (1957), "Information theory and statistical mechanics," Phys. Rev. 106, pp. 620-630.

Jaynes, E. T. (1973), "Survey of the present status of neoclassical radiation theory," in Proceedings of the 1972 Rochester Conference on Optical Coherence, L. Mandel and E. Wolf, eds., Pergamon Press, New York.

Jaynes, E. T. (1980), "Marginalization and prior probabilities," reprinted in E. T. Jaynes (1982), Papers on Probability, Statistics, and Statistical Physics, a reprint collection, D. Reidel, Dordrecht-Holland.

Jaynes, E. T. (1981), "What is the problem?" Proceedings of the Second ASSP Workshop on Spectrum Analysis, S. Haykin, ed., McMaster University.

Jaynes, E. T. (1982), On the rationale of maximum entropy methods," Proc. IEEE 70, pp. 939-952.

Munk, W. H., and F. E. Snodgrass (1957), "Measurement of southern swell at Guadalupe Island," Deep-Sea Research 4, pp. 272-286.

Savage, L. J. (1954), The Foundations of Statistics, Wiley & Sons, New York.

Schuster, A. (1897), "On lunar and solar periodicities of earthquakes," Proc. Roy. Soc. 61, pp. 455-465.

Tukey, J. W., P. Bloomfield, D. Brillinger, and W. S. Cleveland (1980), The Practice of Spectrum Analysis, notes on a course given in Princeton, N.J., in December 1980.

Tukey, J. W., and D. Brillinger (1982), unpublished.

ON ENTROPY RATE

Athanasios Papoulis

Polytechnic Institute of New York, Route 110, Farmingdale, NY 11735

The concept of maximum entropy is reexamined in the context of spectral estimation. Applications include problems involving nonconsecutive correlation constraints and constraints in terms of system responses.

C. R. Smith and G. J. Erickson (eds.),
Maximum-Entropy and Bayesian Spectral Analysis and Estimation Problems, 39–49.

1. Introduction

We reexamine the concept of entropy and show that it can be used to simplify the solution of a variety of problems related to spectral estimation. In this section, we review, briefly, the fundamentals.

Suppose that $\mathbf{x}[n]$ is a real, discrete-time, stationary process with Nth order density the function $f(\mathbf{x}_1, \ldots, \mathbf{x}_N)$. This function is the joint density of N consecutive samples

$$\mathbf{x}[n], \mathbf{x}[n-1], \ldots, \mathbf{x}[n-N+1] \tag{1}$$

of $\mathbf{x}[n]$. The joint entropy

$$H(\mathbf{x}_1, \ldots, \mathbf{x}_N) = E\{-\ln f(\mathbf{x}_1, \ldots, \mathbf{x}_N)\} \tag{2}$$

of these samples is the Nth order entropy of the process $\mathbf{x}[n]$, and the ratio of $H(\mathbf{x}_1, \ldots, \mathbf{x}_N)/N$ is the average uncertainty per sample in a block of N samples. The limit of this ratio is the entropy rate $\overline{H}(\mathbf{x})$ of the process $\mathbf{x}[n]$. Thus,

$$\overline{H}(\mathbf{x}) = \lim_{N\to\infty} \frac{1}{N} H(\mathbf{x}_1, \ldots, \mathbf{x}_N) . \tag{3}$$

The conditional entropy

$$H(\mathbf{x}[n] | \mathbf{x}[n-1], \ldots, \mathbf{x}[n-N]) \tag{4}$$

is the uncertainty about the presence of $\mathbf{x}[n]$ under the assumption that its N most recent past values have been observed. The above conditional entropy is a nonincreasing sequence of N, and its limit

$$H_c(\mathbf{x}) = \lim_{N\to\infty} H(\mathbf{x}[n] | \mathbf{x}[n-1], \ldots, \mathbf{x}[n-N]) \tag{5}$$

is the <u>conditional entropy</u> of the process $\mathbf{x}[n]$.

We maintain that

$$H_c(\mathbf{x}) = \overline{H}(\mathbf{x}) . \tag{6}$$

Indeed, since we know that the limit of the sequence in Eq. (5) can be written as the limit of its Cesaro sum

$$H_c(\mathbf{x}) = \lim_{N\to\infty} \frac{1}{N} \sum_{m=1}^{N} H(\mathbf{x}[n] | \mathbf{x}[n-1], \ldots, \mathbf{x}[n-m]) \tag{7}$$

and since (chain rule)

$$H(\mathbf{x}_1, \ldots, \mathbf{x}_N) = \sum_{m=1}^{N} H(\mathbf{x}[n] | \mathbf{x}[n-1], \ldots, \mathbf{x}[n-m]) , \tag{8}$$

Eq. (6) follows if we insert Eq. (8) into Eq. (7).

Normal processes. Suppose, now, that $\mathbf{x}[n]$ is a normal process. In this case [Papoulis, 1984],

$$H(\mathbf{x}[n] | \mathbf{x}[n-1], \ldots, \mathbf{x}[n-N]) = \ln \sqrt{2\pi e \frac{\Delta_{N+1}}{\Delta_N}} \tag{9}$$

where Δ_N is the correlation determinant of $\mathbf{x}[n]$. As is known, the ratio Δ_{N+1}/Δ_N equals the mean square error P_N of the estimate of $\mathbf{x}[n]$ in terms of its N most recent past values. Hence,

$$\lim_{N\to\infty} \frac{\Delta_{N+1}}{\Delta_N} = \lim_{N\to\infty} P_N = P . \tag{10}$$

The constant P is the mean square value of the estimate of $\mathbf{x}[n]$ in terms of its entire past. Hence (Kolmogoroff-Szëgo formula) [Papoulis, 1984]:

$$P = \exp \left[\frac{1}{2\pi} \int_{-\pi}^{\pi} \ln S(e^{j\omega}) \, d\omega \right] \tag{11}$$

where $S(e^{j\omega})$ is the power spectrum of $\mathbf{x}[n]$.

From the above it follows that the entropy rate $\overline{H}(\mathbf{x})$ of a normal process is given by

$$\overline{H}(\mathbf{x}) = \frac{1}{4\pi} \int_{-\pi}^{\pi} \ln S(e^{j\omega}) \, d\omega . \tag{12}$$

(We have omitted the additive constant $\ln \sqrt{2\pi e}$.)

2. Maximum Entropy with Arbitrary Constraints

We shall consider the problem of estimating the power spectrum

$$S(e^{j\omega}) = \sum_{m=-\infty}^{\infty} R[m] \, e^{-jm\omega} \tag{13}$$

of a process $\mathbf{x}[n]$ under the assumption that its autocorrelation

$$R[m] = E\{\mathbf{x}[n+m]\,\mathbf{x}[n]\} \tag{14}$$

is specified for every m in a set D of integers

$$R[m] = \alpha_m, \qquad m \in D, \tag{15}$$

not necessarily consecutive, starting with $m = 0$ and ending with $m = M$. This problem has been investigated extensively. However, solutions are obtained mainly for consecutive data. In this section we consider the general case.

Our objective is to determine the maximum entropy (ME) estimate of $S(e^{j\omega})$ under the constraints (15). This involves the maximization of the Nth order entropy $H(\mathbf{x}_1, \ldots, \mathbf{x}_N)$ of the underlying process $\mathbf{x}[n]$ for any N and, hence, also its entropy rate $\overline{H}(\mathbf{x})$. Since the given constraints are in the form of second order moments, $H(\mathbf{x}_1, \ldots, \mathbf{x}_N)$ is a maximum only if the joint density of $\mathbf{x}[n]$ of any order is normal. From this it follows that the process $\mathbf{x}[n]$ is normal; hence, its entropy rate equals

$$\overline{H}(\mathbf{x}) = \frac{1}{4\pi}\int_{-\pi}^{\pi} \ln S(e^{j\omega})\, d\omega . \tag{16}$$

Thus, to determine the ME estimate of $S(e^{j\omega})$, we must choose the unspecified values of $R[m]$ so as to maximize the above integral. This yields

$$\frac{\partial \overline{H}}{\partial R[m]} = 0 = \frac{1}{4\pi}\int_{-\pi}^{\pi} \frac{1}{S(e^{j\omega})}\, e^{-jm\omega}\, d\omega , \qquad m \in \overline{D}, \tag{17}$$

from which it follows that

$$\frac{1}{S(e^{j\omega})} = \sum_{|m| \in D} c_m\, e^{-jm\omega} . \tag{18}$$

To complete the determination of $S(e^{j\omega})$, it suffices, therefore, to express the above coefficients c_m in terms of the given data α_m. We shall do so using a steepest ascent method that utilizes the numerical simplicity of Levinson's algorithm. As a preparation, we review, briefly, the well known solution for consecutive data.

Consecutive data. Suppose, first, that the set D consists of the M + 1 integers 0,1, ..., M. Since $S(e^{j\omega}) \geq 0$, it follows from the Fejér-Riesz theorem that the ME spectrum, Eq. (18), can be written as a square [Papoulis, 1977]

$$S(e^{j\omega}) = \frac{1}{\sum_{m=-M}^{M} c_m e^{-jm\omega}} = \frac{P}{|1 - \sum_{m=0}^{M} a_m e^{-jm\omega}|^2} \tag{19}$$

where P and a_m are M + 1 constants such that

$$P c_m = -a_m + \sum_{i=1}^{M-m} a_i a_{i+m}, \qquad 0 \leq m \leq M, \tag{20}$$

where we assumed that $a_0 = -1$.

These constants can be determined recursively from Levinson's algorithm. We state below the algorithm, omitting the well known proof [Burg, 1967]:

$$a_m^N = a_m^{N-1} - K_N a_{N-m}^N \qquad 1 \leq m \leq N-1 \tag{21a}$$

$$a_N^N = K_N \tag{21b}$$

$$P_{N-1} K_N = R[N] - \sum_{m=1}^{N-1} R[N-k] a_k^{N-1} \tag{21c}$$

The constants K_N (reflection coefficients) are determined recursively from Eq. (21c) and they are such that

$$|K_N| \leq 1. \tag{22}$$

The iteration starts with $P_0 = R[0]$, and for N = M it yields

$$P_M = P \tag{23}$$

$$a_m^M = a_m \tag{24}$$

where P and a_m are the M+1 constants in Eq. (19).

Equations (14) through (21) can be used to determine R[m] in terms of a set of constants K_m. If these constants satisfy Eq. (22), then the sequence R[m] so obtained is positive definite (p.d.); that is, the DFT $S(e^{j\omega})$ is positive.

Autoregressive processes (AR). From Eq. (19) it follows that

$$\mathbf{x}[n] - a_1\mathbf{x}[n-1] - \ldots - a_M\mathbf{x}[n-M] = \boldsymbol{\varepsilon}[n] \tag{25}$$

where $\varepsilon[n]$ is white noise and

$$P = E\{\varepsilon^2[n]\} . \tag{26}$$

From this it follows readily that the sum

$$\hat{\mathbf{x}}[n] = a_1\mathbf{x}[n-1] + \dots + a_M\mathbf{x}[n-M] \tag{27}$$

is the linear mean square estimate of $\mathbf{x}[n]$ in terms of its entire past and P is the resulting mean square error. Hence [see Eqs. (11) and (16)]

$$\frac{\ln P}{2} = \overline{H}(\mathbf{x}) \tag{28}$$

where $\overline{H}(\mathbf{x})$ is the entropy rate of the process $\mathbf{x}[n]$.

Nonconsecutive data. Suppose, next, that D is an arbitrary set of integers. In this case, the ME spectrum is again given by Eq. (19); however, now the coefficients c_m are such that

$$c_m = 0 , \qquad m \in \overline{D} , \tag{29}$$

where $\overline{D}$ is the set of integers in the interval (0,M) that are not in D. To determine $S(e^{j\omega})$, it suffices, therefore, to search only among all admissible AR spectra, that is, spectra of the form (19) satisfying the constraints (15). The class of these spectra and of the corresponding autocorrelations R[m] will be denoted by $\mathbf{C}_M$. Clearly, the ME spectrum is a member of this class. To determine it, we shall use a steepest ascent method based on Eqs. (17) and (18).

The entropy rate $\overline{H}(\mathbf{x})$ of the process $\mathbf{x}[n]$ is a function of the unspecified values of R[m], and, in the c_m space, it is maximum at the origin [see Eq. (29)]. Furthermore,

$$\frac{\partial \overline{H}}{\partial R[m]} = \frac{c_m}{2} , \qquad m \in \overline{D} . \tag{30}$$

This shows that, in the space with coordinates R[m], the gradient of the hyper-surface $\overline{H}$ = constant is a vector whose coordinates are proportional to c_m. This leads to the following iterative method for determining the ME spectrum.

Steepest ascent. At the ith iteration step, we obtain a sequence $R_i[m]$ of the class $\mathbf{C}_m$, the corresponding coefficients P_i and $a_{m,i}$, the reflection coefficients $K_{m,i}$, and the Fourier series coefficients $c_{m,i}$ of $1/S(e^{j\omega})$. The iteration proceeds as follows:

Suppose $R_{i-1}[m]$ is a sequence of the class $\mathbf{C}_M$. With δ_i a positive constant, we form the new sequence

$$R_i[m] = \begin{cases} R_{i-1}[m], & m \in D \\ R_{i-1}[m] - \delta_i c_{m,i-1}, & m \in \overline{D} \end{cases} \tag{31}$$

Inserting into Levinson's algorithm (21), we obtain the new constants

$$P_i,\ a_{m,i},\ K_{m,i}\ .$$

The coefficients $c_{m,i}$ are determined from Eq. (20):

$$P_i\, c_{m,i} = -a_{m,i} + \sum_{k=1}^{M-m} a_{k,i}\, a_{k+m,i}\ . \tag{32}$$

The choice of the constant δ_i is dictated by the following two requirements: (a) the sequence $R_i[m]$ obtained from Eq. (31) must be in the class $\mathbf{C}_m$; (b) the corresponding entropy rate $\overline{H}_i(\mathbf{x})$ must exceed the entropy rate $\overline{H}_{i-1}(\mathbf{x})$.

Both requirements are satisfied if δ_i is sufficiently small. In fact, we have a simple way of checking them without any additional computation. Indeed, as we noted, $R_i[m]$ is in the class $\mathbf{C}_M$ iff

$$|K_{m,i}| < 1\ . \tag{33}$$

Furthermore [see Eq. (28)],

$$\overline{H}_i(\mathbf{x}) > \overline{H}_{i-1}(\mathbf{x}) \ \text{ iff } \ P_i > P_{i-1}\ . \tag{34}$$

With δ_i so determined, we continue the process until

$$\sum_{m\in\overline{D}} c_{m,i}^2 \ll \sum_{m\in D} c_{m,i}^2\ . \tag{35}$$

It follows from Eq. (29) that $\overline{H}_i(\mathbf{x})$ is close to the maximum.

To start the iteration, we must find a sequence $R_0[m]$ belonging to the class $\mathbf{C}_M$. To do so, we use Levinson's algorithm Eqs. (14) to (21) to determine K_m in terms of the given data $R[m] = \alpha_m$ for every $m \in D$. For $m \in \overline{D}$, we use the algorithm to determine $R[m]$ in terms of an arbitrary set of reflection coefficients K_m. In fact, to remain at the center of each admissible p.d. interval of $R[m]$, we set

$$K_m = 0\,, \qquad m \in \overline{D}\ . \tag{36}$$

We have, thus, specified $R[m]$ and K_m for every m in the interval $(0,M)$. If the coefficients K_m are such that

$$|K_m| < 1 , \qquad 0 \leq m \leq M , \tag{37}$$

then the sequence R[m] is admissible and can, therefore, be used to start the iteration.

Modified data. Suppose, however, that Eq. (37) is not true. In this case, $|K_m| > 1$ for some $m \in D$; hence the sequence R[m] is no longer in the class $\mathbf{C}_M$. To find a member of the class $\mathbf{C}_M$, we modify the given data, adding a constant r to the given value $R[0] = \alpha_0$ of the sequence R[0] obtained as above. If r is sufficiently small, then the resulting sequence

$$\overline{R}[m] = \begin{cases} R[0] + r , & m = 0 \\ R[m] , & 0 < m \leq M \end{cases} \tag{38}$$

is in the class $\mathbf{C}_m$.

Starting the iteration with this sequence, we obtain an ME spectrum satisfying the modified constraints

$$\overline{R}[m] = \begin{cases} \alpha_0 + r , & m = 0 \\ \alpha_m , & 0 \neq m \in D . \end{cases} \tag{39}$$

This spectrum is not, of course, the solution to our problem because $\overline{R}[0]$ is not equal to α_0. To correct for this, we form a sequence $S_k(e^{j\omega})$ of ME spectra satisfying the constraints

$$R_k[m] = \begin{cases} \alpha_0 + r - \Delta_k , & m = 0 \\ \alpha_m , & m \neq 0 . \end{cases} \tag{40}$$

If the numbers Δ_k are sufficiently small, then each starting sequence is in the class $\mathbf{C}_M$. The corresponding ME spectrum $S_k(e^{j\omega})$ is then determined with the steepest ascent iteration discussed earlier. The above process terminates with $\overline{R}_k[0] = \alpha_0$, that is, when the original data are restored.

We have, thus, developed a steepest ascent method for obtaining the ME solution of the spectral estimation problem with nonconsecutive constraints. The method converges rapidly, and it utilizes the computational simplicity of Levinson's algorithm.

3. Entropy Rate and Linear Systems

If the input to a linear system is a stationary process $\mathbf{x}[n]$ with entropy rate $\overline{H}(\mathbf{x})$, then the resulting output $\mathbf{y}[n]$ is also stationary and its entropy rate equals

$$\overline{H}(\mathbf{y}) = \overline{H}(\mathbf{x}) + \frac{1}{2\pi}\int_{-\pi}^{\pi} \ln|L(e^{j\omega})|\, d\omega \tag{41}$$

where L(z) is the system function.

This result is based on the fact that the power spectrum $S_y(e^{j\omega})$ of $\mathbf{y}[n]$ is given by

$$S_y(e^{j\omega}) = S_x(e^{j\omega})|L(e^{j\omega})|^2 . \tag{42}$$

To prove Eq. (41), we assume, first, that $\mathbf{x}[n]$ is a normal process. In this case, $\mathbf{y}[n]$ is also normal; hence [see Eqs. (11) and (42)],

$$\overline{H}(\mathbf{y}) = \frac{1}{4\pi}\int_{-\pi}^{\pi} \ln[S_x(\omega)|L(e^{j\omega})|^2]\, d\omega , \tag{43}$$

and Eq. (41) results.

Suppose, next, that $\mathbf{x}[n]$ is an arbitrary process. As we know, if $\mathbf{x}_i$ and $\mathbf{y}_i$ are two sets of linearly dependent random variables, then the corresponding joint entropies are such that [Papoulis, 1984]

$$H(\mathbf{y}_1, \ldots, \mathbf{y}_n) = H(\mathbf{x}_1, \ldots, \mathbf{x}_n) + \ln|\Delta| \tag{44}$$

where Δ is the determinant of the transformation that maps $\mathbf{x}_i$ into $\mathbf{y}_i$. Extending this to infinitely many variables, we conclude that

$$\overline{H}(\mathbf{y}) = \overline{H}(\mathbf{x}) + K \tag{45}$$

where K is a constant that depends only on the system L(z). As we have just shown, if $\mathbf{x}[n]$ is a normal process, then K equals the integral in Eq. (41). And since K down not depend on the statistics of $\mathbf{x}[n]$, it follows from Eq. (45) that Eq. (41) holds for any $\mathbf{x}[n]$.

Output constraints. We shall use Eq. (41) to give a simple solution to the following problem [Ihara, 1982].

We are given a linear system with input $\mathbf{x}[n]$ and output $\mathbf{y}[n]$. We wish to estimate the power spectrum $S_x(e^{j\omega})$ of $\mathbf{x}[n]$ under the constraint that the first M+1 values

$$R_y[m] = \beta_m , \qquad 0 \leq m \leq M , \tag{46}$$

of the autocorrelation $R_y[m]$ of $\mathbf{y}[n]$ are specified.

We shall solve this problem using again the ME method. Our objective is, thus, the minimization of the entropy rate $\overline{H}(\mathbf{x})$ of $\mathbf{x}[n]$ subject to the output constraints (46). Since the system function L(z) is specified, it follows from Eq. (41) that, to maximize $\overline{H}(\mathbf{x})$, it suffices to maximize $\overline{H}(\mathbf{y})$ subject, again, to the constraints (46).

Using Eq. (41), we have, thus, reduced the above problem to the familiar problem of determining the ME spectrum $S_y(e^{j\omega})$ of $\mathbf{y}[n]$. As we see from Eq. (19), this spectrum is the all-pole function

$$S_y(e^{j\omega}) = \frac{P}{\left|1 - \sum_{m=1}^{M} a_m e^{-jm\omega}\right|^2}, \tag{47}$$

where the constant P and the M coefficients a_m are determined from Eqs. (14) to (21) in terms of the given data β_m. Inserting into Eq. (42), we conclude that the ME solution of our problem is the function

$$S_x(\omega) = \frac{P\left|L(e^{j\omega})\right|^2}{\left|1 - \sum_{m=1}^{M} a_m e^{-jm\omega}\right|^2}. \tag{48}$$

Autoregressive Moving Average (ARMA). We note in particular that, if

$$L(z) = b_0 + b_1 z^{-1} + \ldots + b_M z^{-m}, \tag{49}$$

then

$$S_x(e^{j\omega}) = P \frac{\left|\sum_{m=1}^{M} b_m e^{-jm\omega}\right|^2}{\left|1 - \sum_{m=1}^{M} a_m e^{-jm\omega}\right|^2}. \tag{50}$$

This shows that the corresponding $\mathbf{x}[n]$ is an ARMA process.

4. Acknowledgment

This work was supported by the Joint Services Technical Advising Committee under Contract F49620-80-C-0077.

5. References

Burg, J. P. (1967), "Maximum entropy spectral analysis," in Proc. 37th Ann. Int. Mtg. Soc. Explor. Geophysicists. Reprinted (1978) in Modern Spectrum Analysis, D. Childers, ed., IEEE Press, New York.

Ihara, S. (1982), "Maximum entropy spectral analysis," Transactions of the Ninth Prague Conference on Information Theory.

Papoulis, A. (1977), Signal Analysis, McGraw-Hill, New York.

Papoulis, A. (1981), "Maximum entropy spectral estimation: a review," IEEE Trans. Acoust., Speech, Signal Processing 29, pp. 1176-1186.

Papoulis, A. (1984), Probability, Random Variables, and Stochastic Processes, McGraw-Hill, New York.

STATE SPACES AND INITIAL ESTIMATES IN MINIMUM RELATIVE-ENTROPY INVERSION WITH APPLICATION TO SPECTRUM ANALYSIS AND IMAGE ENHANCEMENT*

John E. Shore**

Naval Research Laboratory, Washington, DC 20375

The principles of maximum entropy (ME) and minimum relative entropy (MRE) are information-theoretic methods for estimating unknown probability distributions based on information about their expected values [Elsasser, 1937; Jaynes, 1957; Kullback, 1959; Csiszár, 1975; Shore and Johnson, 1980]. MRE differs from ME by taking into account an initial estimate of the unknown distribution. Both ME and MRE are used in a variety of successful applications, but widespread use has been hindered by some unresolved issues. One issue is the choice of state space in which to express problems and their solutions. Another issue is the choice and interpretation of initial estimates for MRE. A third issue is the identity of the "correct" expression to use for entropy when ME is applied to spectrum analysis and image enhancement. This paper shows that these issues are interrelated, and presents results that help to resolve them.

The principles of ME and MRE provide information-theoretic means for inverting the equations

$$\int s_r(x)\, q(x)\, dx = \overline{s_r} \qquad (r = 0,\ldots,M) \tag{1}$$

to find q(x) when the $s_r(x)$ are known functions and the $\overline{s_r}$ are known values. In general, Eq. (1) has no unique solution for q(x). The constraints (1) and

$$\int dx\, q(x) = 1 \tag{2}$$

*The full version of this paper has been submitted to the IEEE Transactions on Information Theory.

**Present address: Entropic Processing, Inc., Washington Research Laboratory, 600 Pennsylvania Ave. S.E., Suite 202, Washington, DC 20003.

C. R. Smith and G. J. Erickson (eds.),
Maximum-Entropy and Bayesian Spectral Analysis and Estimation Problems, 51–56.

are satisfied by a convex constraint set of densities that contains q; the constraints themselves do not distinguish any particular density in the constraint set. Both ME and MRE provide an estimate of q(x) by solving a functional extremalization problem within the constraint set. In the case of ME [Elsasser, 1937; Jaynes, 1957], the estimate is determined by minimizing the entropy

$$H(q) = -\int dx\, q(x) \log q(x) . \tag{3}$$

When an initial estimate of the unknown q is available in addition to the constraints (1) and (2), MRE applies. Given such a strictly positive initial estimate p(x), the MRE final estimate is chosen by minimizing the relative-entropy (cross-entropy, discrimination information, directed divergence, I-divergence, Kullback-Leibler number),

$$H(q,p) = \int dx\, q(x) \log[q(x)/p(x)] . \tag{4}$$

ME and MRE have been applied to the power spectrum analysis of stationary time series and to the enhancement of two-dimensional images. These applications proceed by expressing the spectrum estimation and image enhancement problems in terms that require inversion of the equation

$$\sum_{k=1}^{N} s_{rk} O_k = D_r \qquad (r = 0,\ldots,M) , \tag{5}$$

where D_r and s_{rk} are known, and where $M \leq N$. In spectrum estimation,

D_r = autocorrelation value at lag t_r
O_k = power at frequency f_k
s_{rk} = Fourier functions = $2 \cos(2\pi f_k t_r)$

In image enhancement,

D_r = image intensity at point r
O_k = object intensity at point k
s_{rk} = point spread function

When M = N, Eq. (5) can be written in matrix form as $\mathbf{s \cdot O = D}$, which can in principle be solved by

$$\mathbf{O} = \mathbf{s}^{-1} \cdot \mathbf{D} . \tag{6}$$

In spectrum analysis applications, M < N often holds; in this case, Eq. (5) has no unique solution. A large class of spectrum estimation methods proceeds by extrapolating D_r so as to take on reasonable values in the region $M < r \leq N$, and then solving for the power spectrum by Eq. (6). Perhaps the best known extrapolation method is Burg's maximum entropy spectral analysis (MESA) [Burg, 1967, 1975; Shore, 1981], in which the power spectrum S(f) is estimated by maximizing

$$\sum_{k=1}^{N} \log O_k \tag{7}$$

subject to the constraints (5). The result is

$$O_k = \frac{1}{\sum_{r=0}^{M} \beta_r s_{rk}} , \tag{8}$$

where the β_r are chosen so that Eq. (8) satisfies Eq. (5). For equally spaced autocorrelation lags, $t_r = r\Delta t$, Eq. (8) takes on the familiar form

$$O_k = \frac{\sigma^2}{|\sum_{r=0}^{M} a_r z^{-r}|^2} \tag{9}$$

where $z = \exp(-2\pi i f_k \Delta t)$, a_r are inverse filter sample coefficients, and σ^2 is a gain [Shore, 1981]. This is the well known all-pole, autoregressive, or linear prediction form, which can also be derived by various equivalent formulations [VanDenBos, 1971; Markel and Gray, 1976; Kay and Marple, 1981; Papoulis, 1981]. It has become a widely used spectrum analysis technique in geophysical data processing [Lacoss, 1971; Smylie et al., 1973; Ulrych and Bishop, 1975; Robinson; 1982] and speech processing [Markel and Gray, 1976; Gray et al., 1981].

In image enhancement, when one wishes to estimate the object intensities with the same resolution that the image intensities are detected, $M = N$ holds and Eq. (5) can be solved in principle by Eq. (6). But D_r and s_{rk} usually are not known exactly, which causes the solution (6) to be ill-behaved or misleading [Frieden, 1980]. For this reason, "maximum entropy spectrum analysis" is also used in image restoration, but the phrase refers in this case not only to successful estimates produced by maximizing Eq. (7) subject to Eq. (5) [Wernecke and D'Addario, 1977; Ables, 1974; Wernecke, 1977], but also to estimates produced by maximizing

$$-\sum_{k=1}^{N} O_k \log O_k \tag{10}$$

subject to Eq. (5) [Gordon and Herman, 1971; Frieden, 1972; Gull and Daniell, 1978; Skilling and Gull, 1985]. The result has the form

$$O_k = \exp\left[-1 - \sum_{r=0}^{M} \beta_r s_{rk}\right] . \tag{11}$$

As was the case with Eq. (8), the β_r are chosen so that Eq. (11) satisfies Eq. (5). Spectrum estimates based on the maximization of Eq. (10) have also been studied for ARMA, meteorological, and speech time series [Nadeu et al., 1981; Ortigueira et al., 1981; Shore and Johnson, 1984].

In previous work [Shore and Johnson, 1984], we showed that the results of maximizing both Eqs. (7) and (10) subject to Eq. (5) are equivalent to relative-entropy minimization with a uniform initial estimate of **O**, the difference stemming from a different choice for the underlying state space. In particular, in one case we treated O_k as the expected value of the power at frequency f_k; in the other, we normalized the O_k and treated them as probabilities. In one sense these results were satisfying, since they showed that the two estimators arise from the same underlying principle: relative-entropy minimization. But we didn't answer the question of whether Eq. (7) or Eq. (10) is the correct expression for entropy; we replaced the question with another one, namely, which state space is correct?

Here we treat the question from a different viewpoint. We derive both estimators by treating O_k as an expected value. For the spectrum estimator, O_k is the expected power at f_k (as in Shore and Johnson [1984]); for the image estimator, O_k is the expected number of photons received at point k. We show that the difference between the estimators (8) and (11) corresponds to different choices for the initial estimate—a Gaussian initial estimate, which is appropriate for time series, leads to Eq. (8), while a Poisson initial estimate, which is appropriate for images, leads to Eq. (11). This unified derivation of spectrum and image estimators is also shown to extend to cases involving multiple signals (for example, additive noise), weighted initial estimates, and uncertain measurements.

These results show that the spectrum and image estimators arise from MRE analyses in the same state space, but with different initial estimates. But the results leave unclear the relationship between initial estimates and state-space choices. To clarify this relationship, we show that, if a nonuniform initial estimate for MRE is interpreted as the result of aggregating a uniform initial estimate from some "larger" state space, then MRE with the nonuniform initial estimate is equivalent to ME in that larger space. This result sheds further light on the relationship between the spectrum and image estimators, namely that both can be viewed as arising from ME in appropriate spaces. If the amplitudes in a time series are thought of as arising from the sum of many uniformly distributed increments, then the Gaussian initial estimate can be thought of as resulting from an aggregation transformation. Similarly, the Poisson initial estimate can be thought of as the aggregate probability of x total photons arriving during a standard time interval, given (in the limit) a uniform probability for the time instant at which any individual photon might arrive.

The paper closes with a general discussion of choosing initial estimates in MRE problems.

Acknowledgment

I thank Rodney W. Johnson for many helpful discussions.

References

Ables, J. G. (1974), "Maximum entropy spectral analysis," Astron. Astrophys. Supp. 15, pp. 383-393.

Burg, J. P. (1967), "Maximum Entropy Spectral Analysis," presented at the 37th Annual Meeting, Society of Exploration Geophysicists, Oklahoma City, Okla.

Burg, J. P. (1975), "Maximum Entropy Spectral Analysis," Ph.D. Dissertation, Stanford University, Stanford, Calif. (University Microfilms No. AAD75-25,499).

Csiszár, I. (1975), "I-divergence geometry of probability distributions and minimization problems," Ann. Prob. 3, pp. 146-158.

Elsasser, W. M. (1937), "On quantum measurements and the role of the uncertainty relations in statistical mechanics," Phys. Rev. 52, pp. 987-999.

Frieden, B. R. (1972), "Restoring with maximum likelihood and maximum entropy," J. Opt. Soc. Am. 62, pp. 511-518.

Frieden, B. R. (1980), "Statistical models for the image restoration problem," Comp. Graphics Image Processing 12, pp. 40-59.

Gordon, R., and G. T. Herman (1971), "Reconstruction of pictures from their projections," Quart. Bull. Center for Theor. Biol. 4, pp. 71-151.

Gray, R. M., A. H. Gray, Jr., G. Rebolledo, and J. E. Shore (1981), "Rate-distortion speech coding with a minimum discrimination information distortion measure," IEEE Trans. Inf. Theory IT-27, pp. 708-721.

Gull, S. F., and G. J. Daniell (1978), "Image reconstruction from incomplete and noisy data," Nature 272, pp. 686-690.

Jaynes, E. T. (1957), "Information theory and statistical mechanics I," Phys. Rev. 106, pp. 620-630.

Kay, S. M., and S. L. Marple, Jr. (1981), "Spectrum analysis—a modern perspective," Proc. IEEE 69, pp. 1380-1419.

Kullback, S. (1959), Information Theory and Statistics, Wiley, New York (Dover, New York, 1968).

Lacoss, R. T. (1971), "Data adaptive spectral analysis methods," Geophys. 36, pp. 661-675.

Markel, J. D, and A. H. Gray, Jr. (1976), Linear Prediction of Speech, Springer-Verlag, New York.

Nadeu, C., E. Sanvicente, and M. Bertran (1981), "A new algorithm for spectral estimation," Proc. International Conf. on DSP, pp. 463-470.

Ortigueira, M. D., R. Garcia-Gomez, and J. M. Tribolet (1981), "An iterative algorithm for maximum flatness spectral analysis," Proc. International Conf. on DSP, pp. 810-818.

Papoulis, A. (1981), "Maximum entropy and spectral estimation: a review," IEEE Trans. Acoust., Speech, Sig. Processing ASSP-29, pp. 1176-1186.

Robinson, E. A. (1982), "A historical perspective of spectrum estimation," Proc. IEEE 76, pp. 885-907.

Shore, J. E. (1981), "Minimum cross-entropy spectral analysis," IEEE Trans. Acoust, Speech, Sig. Processing ASSP-29, pp. 230-237.

Shore, J. E., and R. W. Johnson (1980), "Axiomatic derivation of the principle of maximum entropy and the principle of minimum cross-entropy," IEEE Trans. Inf. Theory IT-26, pp. 26-37. (See also comments and corrections in IEEE Trans. Inf. Theory IT-29, p. 942, 1983.)

Shore, J. E., and R. W. Johnson (1984), "Which is the better entropy expression for speech processing: -S log S or log S?" IEEE Trans. Acoust., Speech, Sig. Processing ASSP-32, pp. 129-137.

Skilling, J., and S. F. Gull (1985), "Algorithms and applications," in Maximum-Entropy and Bayesian Methods in Inverse Problems, C. Ray Smith and W. T. Grandy, eds., D. Reidel, Dordrecht, pp. 83-132.

Smylie, D. E., G. K. C. Clarke, and T. J. Ulrych (1973), "Analysis of irregularities in the earth's rotation," Meth. Comp. Phys. 13, pp. 391-431.

Ulrych, T. J., and T. N. Bishop (1975), "Maximum entropy spectral analysis and autoregressive decomposition," Rev. Geophys. Space Phys. 43, pp. 183-200.

VanDenBos, A. (1971), "Alternative interpretation of maximum entropy spectral analysis," IEEE Trans. Inf. Theory IT-17, pp. 493-494.

Wernecke, S. J. (1977), "Two-dimensional maximum entropy reconstruction of radio brightness," Radio Sci. 12, pp. 831-844.

Wernecke, S. J., and L. D'Addario (1977), "Maximum entropy image reconstruction," IEEE Trans. Computers C-26, pp. 351-364.

RELATIVE-ENTROPY MINIMIZATION WITH UNCERTAIN CONSTRAINTS: THEORY AND APPLICATION TO SPECTRUM ANALYSIS

Rodney W. Johnson*

Computer Science and Systems Branch, Naval Research Laboratory, Washington, DC 20375

*Present address: Entropic Processing, Inc., Washington Research Laboratory, 600 Pennsylvania Ave. S.E., Suite 202, Washington, DC 20003.

C. R. Smith and G. J. Erickson (eds.),
Maximum-Entropy and Bayesian Spectral Analysis and Estimation Problems, 57–73

1. Introduction

The relative-entropy principle (REP) is a general, information-theoretic method for inference when information about an unknown probability density $q^\dagger$ consists of an initial estimate p and additional constraint information that restricts $q^\dagger$ to a specified convex set of probability densities. Typically the constraint information consists of linear-equality constraints, that is, expected values

$$\overline{f}_r = \int f_r(x)\, q^\dagger(x)\, dx \tag{1}$$

for known $f_r(x)$ and $\overline{f}_r$, $r = 0,1,\ldots,M$. The principle states that one should choose the final estimate q that satisfies

$$H(q,p) = \min_{q'} H(q',p) , \tag{2}$$

where H is the relative entropy (cross entropy, discrimination information, directed divergence, I-divergence, K-L number, etc.),

$$H(q,p) = \int q(x) \log \frac{q(x)}{p(x)}\, dx , \tag{3}$$

and where q' varies over the set of densities that satisfy the constraints. When these are linear-equality constraints [Eq. (1)], the final estimate has the form

$$q(x) = p(x) \exp \left[-\alpha - \sum_r \beta_r f_r(x)\right] , \tag{4}$$

where the β_r and α are Lagrangian multipliers determined by Eq. (1) (with $q^\dagger$ replaced by q) and by the normalization constraint

$$\int q(x)\, dx = 1 . \tag{5}$$

Properties of REP solutions and conditions for their existence are discussed by Csiszár [1975] and by Shore and Johnson [1981]. Expressed in terms of the expected values and the Lagrangian multipliers, the relative entropy at the minimum is given by

$$H(q,p) = -\alpha - \sum_r \beta_r \overline{f}_r . \tag{6}$$

The normalization multiplier α is given by

$$\alpha = \log \int p(x) \exp [-\sum_r \beta_r f_r(x)] \, dx . \tag{7}$$

The quantity $Z = e^{\alpha}$ is often referred to as the partition function. If the partition function can be evaluated analytically, that is, if the integral in Eq. (7) can be performed, then the relations

$$-\frac{\partial \alpha}{\partial \beta_r} = \overline{f}_r \tag{8}$$

can sometimes be solved to express the β_r as functions of the expected values $\overline{f}_r$. If not, various computational methods can be used to find the values for the α and β_r in Eq. (4) that satisfy Eqs. (1) and (5) [Johnson, 1979a]. As a general method of statistical inference, the REP was first introduced by Kullback [1959], has been advocated in various forms by others [Good, 1963; Jaynes, 1968; Johnson, 1979b], and has been applied in a variety of fields [for a list of references, see Shore and Johnson, 1980].

Informally speaking, of the densities that satisfy the constraints, the REP selects the one that is closest to p in the sense measured by relative entropy. In more formal terms, the REP can be justified on the basis of the information-theoretic properties of relative entropy [Kullback, 1959], or on the basis of consistency axioms for logical inference [Shore and Johnson, 1980]. In applications of the REP, the known expected values $\overline{f}_r$ in Eq. (1) frequently correspond to physical measurements. Such measurements usually are subject to error, so strict equality in Eq. (1) is unrealistic.

In Section 2 we discuss the REP with "uncertain constraints," a form of the principle appropriate for applications with uncertainty in the expected values. In Section 3, relative-entropy minimization with uncertain constraints is applied to spectrum analysis; a relative-entropy spectrum estimate from uncertain autocorrelations is derived. Sections 4 and 5 are devoted to a numerical example and a concluding discussion, respectively.

2. Relative-Entropy Minimization with Uncertain Constraints

In this section we extend the results on the REP with linear-equality constraints to incorporate uncertainty about the values of the $\overline{f}_r$ in Eq. (1). We define an error vector **v** with components

$$v_r = \int f_r(x)\, q^\dagger(x)\, dx - \overline{f}_r. \tag{9}$$

A simple generalization would be to replace the set of constraints (1) with a bound on the magnitude of **v**:

$$\sum_r \left[\int f_r(x)\, q^\dagger(x)\, dx - \overline{f}_r \right]^2 \leq \varepsilon^2 . \tag{10}$$

However, not all components v_r may have equal uncertainty, and different components may be correlated. We therefore replace Eq. (10) with the more general constraint

$$\sum_{rs} M_{rs}\, v_r\, v_s \leq \varepsilon^2 . \tag{11}$$

In matrix notation this is

$$\mathbf{v} \cdot \mathbf{M} \mathbf{v} \leq \varepsilon^2 , \tag{12}$$

where **M** is any positive-definite matrix.

We assume we are given an initial estimate p of $q^\dagger$, measured values $\overline{f}_r$ of the expectations (1) of functions f_r for a finite set of indices r, and an error estimate ε. We first derive the form of the final estimate q under the assumption that the constraint has the form (10) and that the $\overline{f}_r$ are zero; that is, we assume a constraint

$$\sum_r \left[\int f_r(x)\, q^\dagger(x)\, dx \right]^2 \leq \varepsilon^2 . \tag{13}$$

Next we show how to reduce the more general constraint (11) to this case. We conclude this section with a remark on the relation between the result with $\varepsilon > 0$ and that for "exact constraints" ($\varepsilon = 0$).

Our problem is to minimize the relative entropy H(q,p) subject to the constraint (13) (with q in place of $q^\dagger$) and the normalization constraint (5). If the initial estimate satisfies the constraint (that is, if Eq. (13) holds with p in place of q), then setting q = p gives the minimum. Otherwise, equality holds in Eq. (13) and the criterion for a minimum is that the variation of

$$\int q(x) \log \frac{q(x)}{p(x)} dx + \lambda \sum_r \left[\int f_r(x)\, q(x)\, dx \right]^2 + (\alpha-1) \int q(x)\, dx \qquad (14)$$

with respect to q(x) is zero for some Lagrange multipliers $\lambda > 0$, corresponding to Eq. (13), and $\alpha-1$, corresponding to Eq. (5). (We write $\alpha-1$ instead of α for later convenience.) With $\lambda > 0$, the criterion intuitively implies that a small change δq in q that leaves $\int q(x)\, dx$ fixed and decreases H(q,p) must increase the error term $\Sigma_r[\int f_r(x)\, q(x)\, dx]^2$.

Equating the variation of expression (14) to zero gives

$$\log \frac{q(x)}{p(x)} + \alpha + \lambda \sum_r 2f_r(x) \int f_r(x')\, q(x')\, dx' = 0 . \qquad (15)$$

Therefore q satisfies

$$q(x) = p(x) \exp [-\alpha - \sum_r \beta_r f_r(x)] , \qquad (16)$$

where

$$\beta_r = 2\lambda \int f_r(x)\, q(x)\, dx . \qquad (17)$$

Conversely, if q has the form (16), and if α, λ, and the β_r are chosen so that Eq. (17), the constraint (13), and the normalization condition (5) hold, then q is a solution to the minimization problem. But if Eq. (17) holds, the constraint with equality is equivalent to

$$\sum_r \left[\frac{\beta_r}{2\lambda} \right]^2 = \varepsilon^2 \qquad (18)$$

or to

$$\lambda = \frac{1}{2\varepsilon} \|\boldsymbol{\beta}\| , \qquad (19)$$

where we have written $\|\boldsymbol{\beta}\|$ for the Euclidean norm $(\Sigma_r \beta_r^2)^{1/2}$. Thus if we choose α and β_r in Eq. (16) so that Eq. (5) and

$$\varepsilon \frac{\beta_r}{\|\boldsymbol{\beta}\|} = \int f_r(x)\, q(x)\, dx \qquad (20)$$

hold, then the constraint (13) will be satisfied, and we can ensure that Eq. (17) holds by the choice of λ.

Next, assume a constraint of the general form (11), (9), with a symmetric, positive-definite matrix **M**. Then there is a matrix **A**, not in general unique, such that $\mathbf{A}^t\mathbf{A} = \mathbf{M}$. Now

$$\mathbf{v}\cdot\mathbf{M}\mathbf{v} = \mathbf{v}\cdot\mathbf{A}^t\mathbf{A}\mathbf{v} = (\mathbf{A}\mathbf{v})\cdot(\mathbf{A}\mathbf{v}) = \sum_r \left[\sum_s A_{rs} v_s\right]^2 , \tag{21}$$

and so the constraint assumes the form

$$\sum_r u_r^2 \lesssim \varepsilon^2 , \tag{22}$$

where

$$u_r = \sum_s A_{rs} v_s . \tag{23}$$

In view of Eq. (5), we may rewrite Eq. (9) as

$$v_r = \int [f_r(x) - \overline{f}_r]\, q(x)\, dx \tag{24}$$

and obtain

$$u_r = \int \sum_s A_{rs}[f_s(x) - \overline{f}_s]\, q(x)\, dx . \tag{25}$$

Defining

$$g_r(x) = \sum_s A_{rs}[f_s(x) - \overline{f}_s] , \tag{26}$$

we obtain

$$\sum_r \left[\int g_r(x)\, q(x)\, dx\right]^2 \lesssim \varepsilon^2 \tag{27}$$

from Eq. (22). Thus, constraints of the general form (11) can be transformed to (27), which is of the same form as (13).

Note that Eq. (16) is identical to Eq. (4): that is, the functional form of the solution with uncertain constraints is the same as that for exact constraints. The difference is that, for uncertain constraints, the conditions that determine the β_r have the general form (20). These conditions reduce to the exact-constraint case for $\varepsilon = 0$. One way of viewing this identity of form for the solutions of the two problems is to note that every solution q of an uncertain-constraint problem is simultaneously a solution of an exact-constraint problem with the same functions f_k and appropriately modified values for the $\overline{f}_k$.

The relative entropy at the minimum may be computed by substituting Eq. (16) into Eq. (3), which leads to

$$H(q,p) = -\alpha - \sum_r \beta_r \int f_r \, q(x) \, dx \, . \tag{28}$$

In the case of nonzero expected values, $\overline{f}_r \neq 0$, Eq. (20) becomes

$$\varepsilon \frac{\beta_r}{\|\boldsymbol{\beta}\|} = \int f_r(x) \, q(x) \, dx - \overline{f}_r \, . \tag{29}$$

(For simplicity we take **M** to be the identity.) Substituting Eq. (29) into Eq. (28) yields

$$H(q,p) = -\alpha - \sum_r \beta_r \overline{f}_r - \varepsilon \|\boldsymbol{\beta}\| \, , \tag{30}$$

which is the generalization of Eq. (6) in the case of uncertain constraints. The normalization multiplier α has the same functional form as in the exact-constraint case, Eq. (7); the generalization of Eq. (8) therefore results from differentiating Eq. (7), which yields

$$-\frac{\partial \alpha}{\partial \beta_r} = \int f_r(x) \, q(x) \, dx \, , \tag{31}$$

and then substituting Eq. (29), which yields

$$-\frac{\partial \alpha}{\partial \beta_r} = \overline{f}_r + \varepsilon \frac{\beta_r}{\|\boldsymbol{\beta}\|} \, . \tag{32}$$

Note that Eqs. (30) and (32) reduce respectively to Eqs. (6) and (8) when $\varepsilon = 0$.

3. Application to Spectrum Analysis

Relative-entropy spectrum analysis (RESA), introduced by Shore [1981], is an extension of Burg's [1967, 1975] maximum-entropy spectral analysis (MESA). Like MESA, RESA estimates a spectrum from values of the autocorrelation function. Unlike MESA, however, it also takes into account prior information in the form of an initial estimate of the spectrum. Multisignal RESA (MRESA), introduced by Johnson and Shore [1983], simultaneously estimates the power spectra of several signals when an initial estimate for each spectrum is available and new information is obtained in the form of values of the autocorrelation function of the sum. The resulting final estimates are the solution of a constrained minimization problem: they are consistent with the autocorrelation information and otherwise as similar as possible to the respective initial estimates in a precisely defined information-theoretic sense. MRESA has recently been extended by Johnson, Shore, and Burg [1984] to incorporate weighting factors associated with each initial spectrum estimate to allow for the fact that initial estimates may not be uniformly reliable.

The autocorrelation values were treated by Shore [1981], Johnson and Shore [1983], and Johnson, Shore, and Burg [1984] as exactly given. Usually, however, these are estimated or measured values, subject to error. By basing a derivation on the REP with uncertain constraints, we will show how to incorporate an error bound to allow for uncertainty in autocorrelation values.

MRESA assumes the existence of L independent signals with power spectra $S_i(f)$ and autocorrelations

$$R_{ir} = \int C_r(f)\, S_i(f)\, df\,, \tag{33}$$

where

$$C_r(f) = \cos 2\pi t_r f\,. \tag{34}$$

Given initial estimates $P_i(f)$ of the power spectrum of each signal S_i, and autocorrelation measurements on the sum of the signals, MRESA provides final estimates for the S_i. In particular, if the measurements R_r^{tot} satisfy

$$R_r^{tot} = \sum_{i=1}^{L} \int C_r(f)\, Q_i(f)\, df\,, \tag{35}$$

for lags $r = 0,\ldots,M$, the resulting final estimates are

$$Q_i(f) = \frac{1}{\frac{1}{P_i(f)} + \sum_r \beta_r C_r(f)} , \tag{36}$$

where the B_r are chosen so that the Q_i satisfy the autocorrelation constraints (35) [Johnson and Shore, 1983]. Since some initial estimates may be more reliable than others, these results have been extended recently to include a frequency-dependent weight $w_i(f)$ for each initial estimate $P_i(f)$ [Johnson, Shore, and Burg, 1984]. The larger the value of $w_i(f)$, the more reliable the initial estimate $P_i(f)$ is considered to be. With the weights included, result (36) becomes

$$Q_i(f) = \frac{1}{\frac{1}{P_i(f)} + \frac{1}{w_i(f)} \sum_r \beta_r C_r(f)} . \tag{37}$$

Before generalizing MRESA to include uncertain constraints, we review some notation and results from Johnson and Shore [1983] and Shore [1979]. In Johnson and Shore [1983], for each of the L signals, we used a discrete-spectrum approximation

$$s_i(t) = \sum_{k=1}^{N} (a_{ik} \cos 2\pi f_k t + b_{ik} \sin 2\pi f_k t) , \qquad i = 1,\ldots,L, \tag{38}$$

with nonzero frequencies f_k, not necessarily uniformly spaced. The a_{ik} and b_{ik} were random variables with independent, zero-mean, Gaussian initial distributions. We defined random variables

$$x_{ik} = \tfrac{1}{2} (a_{ik}^2 + b_{ik}^2) , \tag{39}$$

representing the power of process s_i at frequency f_k, and we described the collection of signals in terms of their joint probability density $q^\dagger(\mathbf{x})$, where $\mathbf{x} = (x_1,\ldots,x_L)$ and $\mathbf{x}_i = (x_{i1},\ldots,x_{iN})$. We expressed the power spectrum S as an expectation

$$S_i(f_k) = \int x_{ik}\, q^\dagger(\mathbf{x})\, d\mathbf{x} . \tag{40}$$

In terms of initial estimates $P_{ik} = P_i(f_k)$ of $S_i(f_k)$, we wrote initial estimates p of $q^\dagger$ in the form

$$p(\mathbf{x}) = \prod_{i=1}^{L} \prod_{k=1}^{N} p_{ik}(x_{ik}) \tag{41}$$

where

$$p_{ik}(x_{ik}) = \frac{1}{P_{ik}} \exp \frac{-x_{ik}}{P_{ik}} \; . \tag{42}$$

The assumed Gaussian form of the initial distribution of a_{ik} and b_{ik} is equivalent to this exponential form for $p_{ik}(x_{ik})$; the coefficients were chosen to make the expectation of x_{ik} equal to P_{ik}. Using Eq. (40), we wrote a discrete-frequency form of Eq. (35) as linear constraints

$$R_r^{tot} = \sum_{i=1}^{L} \sum_{k=1}^{N} \int c_{rk} \, x_{ik} \, q^\dagger(\mathbf{x}) \, d\mathbf{x} \tag{43}$$

on expectation values of $q^\dagger$, where

$$c_{rk} = C_r(f_k) \; . \tag{44}$$

We obtained a final estimate q of $q^\dagger$ by minimizing the relative entropy

$$H(q,p) = \int q(\mathbf{x}) \log \frac{q(\mathbf{x})}{p(\mathbf{x})} \, d\mathbf{x} \tag{45}$$

subject to the constraints [Eq. (43) with q in place of $q^\dagger$] and the normalization condition

$$\int q(\mathbf{x}) \, d\mathbf{x} = 1 \; ; \tag{46}$$

the result had the form

$$q(\mathbf{x}) = \prod_{i=1}^{L} \prod_{k=1}^{N} q_{ik}(x_{ik}) \, , \tag{47}$$

where the q_{ik} were related to the final estimates

$$Q_{ik} = Q_i(f_k) = \int x_{ik} \, q(\mathbf{x}) \, d\mathbf{x} \tag{48}$$

of the power spectra of the s_i by

$$q_{ik}(x_{ik}) = \frac{1}{Q_{ik}} \exp \frac{-x_{ik}}{Q_{ik}} \ . \tag{49}$$

This led to a discrete-frequency version of Eq. (36)

$$Q_{ik} = \frac{1}{\dfrac{1}{P_{ik}} + \sum_{r=0}^{M} \beta_r c_{rk}} , \tag{50}$$

where the β_r had to be chosen so that

$$\sum_{i=1}^{L} \sum_{k=1}^{N} c_{rk} Q_{ik} = R_r^{tot} \tag{51}$$

was satisfied.

To handle uncertain constraints, we first replace Eq. (35) with a bound

$$\sum_r \left[\sum_i \int C_r(f)\, Q_i(f)\, df - R_r^{tot} \right]^2 \leq \varepsilon^2 \tag{52}$$

on the Euclidean norm of the error vector $\mathbf{v}$ given by

$$v_r = \sum_i \int C_r(f)\, Q_i(f)\, df - R_r^{tot} \ . \tag{53}$$

We write a discrete-frequency form of Eq. (52) in terms of expectations of q:

$$\sum_{r=0}^{M} \left[\sum_{i=1}^{L} \sum_{k=1}^{N} \int c_{rk}\, x_{ik}\, q(\mathbf{x})\, d\mathbf{x} - R_r^{tot} \right]^2 \leq \varepsilon^2 \ . \tag{54}$$

This has the form (36); by Eq. (16), minimizing relative entropy subject to these constraints gives

$$q(\mathbf{x}) = p(\mathbf{x}) \exp \left[-\alpha - \sum_{r=0}^{M} \beta_r \sum_{i=1}^{L} \sum_{k=1}^{N} c_{rk}\, x_{ik} \right] , \tag{55}$$

where the β_r are to be determined so that

$$\varepsilon \frac{\beta_r}{\|\boldsymbol{\beta}\|} = \sum_{i=1}^{L} \sum_{k=1}^{N} \int c_{rk}\, x_{ik}\, q(\mathbf{x})\, d\mathbf{x} - R_r^{tot} \tag{56}$$

[compare Eq. (20)]. Using Eq. (42), we find that q has the form (47), where $q_{ik}(x_{ik})$ is proportional to

$$\exp\left[\frac{-x_{ik}}{Q_{ik}} - \sum_{r=0}^{M} \beta_r \sum_{i=1}^{L} \sum_{k=1}^{N} c_{rk}\, x_{ik}\right] . \tag{57}$$

Consequently q_{ik} is given by Eq. (49), where Q_{ik} is given by Eq. (50). Rewriting Eq. (56) in terms of Q_{ik} and passing from discrete to continuous frequencies gives

$$Q_i(f) = \frac{1}{\frac{1}{P_i(f)} + \sum_r \beta_r C_r(f)} , \tag{58}$$

where the β_r are to be determined so that

$$\varepsilon \frac{\beta_r}{\|\boldsymbol{\beta}\|} = \sum_{i=1}^{L} \int C_r(f)\, Q_i(f)\, df - R_r^{tot} . \tag{59}$$

The functional form (58) of the solution with uncertain constraints is the same as the form (36) for exact constraints; the difference is in the conditions that determine the β_r: Eq. (35) for exact constraints and Eq. (59) for uncertain constraints. This is a consequence of the analogous result for probability-density estimation, noted in Section 2.

In the case of the more general constraint form

$$\sum_{rs} M_{rs}\, v_r\, v_s \leq \varepsilon^2 , \tag{60}$$

with the error vector **v** as in Eq. (53), it is convenient to carry the matrix through the derivation rather than transforming the constraint functions as in Eq. (26). The result is that the final estimates again have the form (36), while the conditions (59) on the β_r are replaced by

$$\varepsilon \frac{\beta'_r}{(\beta' \cdot M \beta')^{1/2}} = \sum_{i=1}^{L} \int C_r(f)\, Q_i(f)\, df - R_r^{tot} , \qquad (61)$$

where

$$\beta' = M^{-1}\beta . \qquad (62)$$

In the uncertain-constraint case, when we include weights $w_i(f)$ as in Johnson, Shore, and Burg [1984], the functional form of the solution becomes generalized to Eq. (37); the conditions that determine the β_r, Eq. (59) or (61), remain the same.

4. Example

We shall use a numerical example from Johnson and Shore [1983] and Johnson, Shore, and Burg [1984]. We define a pair of spectra, S_B and S_S, which we think of as a known "background" component and an unknown "signal" component of a total spectrum. Both are symmetric and defined in the frequency band from -0.5 to +0.5, though we plot only their positive-frequency parts. S_B is the sum of white noise with total power 5 and a peak at frequency 0.215 corresponding to a single sinusoid with total power 2. S_S consists of a peak at frequency 0.165 corresponding to a sinusoid of total power 2. Figure 1 shows a discrete-frequency approximation to the sum S_B+S_S, using 100 equi-spaced frequencies. From the sum, six autocorrelations were computed exactly. As the initial estimate P_B of S_B, we used S_B itself; that is, P_B was Fig. 1 without the left-hand peak. For P_S we used a uniform (flat) spectrum with the same total power as P_B.

Figure 2 shows unweighted MRESA final estimates Q_B and Q_S [Johnson and Shore, 1983]. The signal peak shows up primarily in Q_S, but some evidence of it is in Q_B as well. This is reasonable since P_B, although exactly correct, is treated as an initial estimate subject to change by the data. The signal peak can be suppressed from Q_B and enhanced in Q_S by weighting the background estimate P_B heavily [Johnson, Shore, and Burg, 1984].

Figure 3 shows final estimates for uncertain constraints with an error bound of ε = 1. The Euclidean distance [a constraint of the form (52)], was used. The estimates were obtained with Newton-Raphson algorithms similar to those developed by Johnson [1983]. Both final estimates in Fig. 3 are closer to the corresponding initial estimates than is the case in Fig. 2, since the sum of the final estimates is no longer constrained to satisfy the autocorrelations.

Figure 4 shows results for ε = 3; these final estimates are even closer to the initial estimates. Because the example was constructed with exactly known autocorrelations, it is not surprising that the exactly constrained final estimates are better than those in Figs. 3 and 4, which illustrate the more conservative deviation from initial estimates that results from incorporating the uncertain constraints.

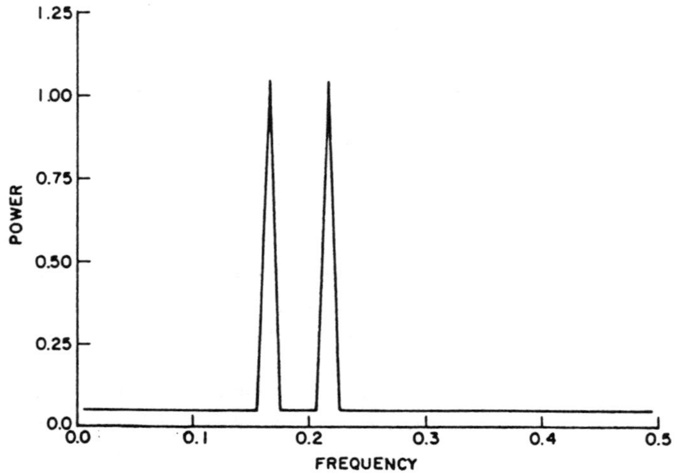

Figure 1. Sum S_B+S_S of original spectra.

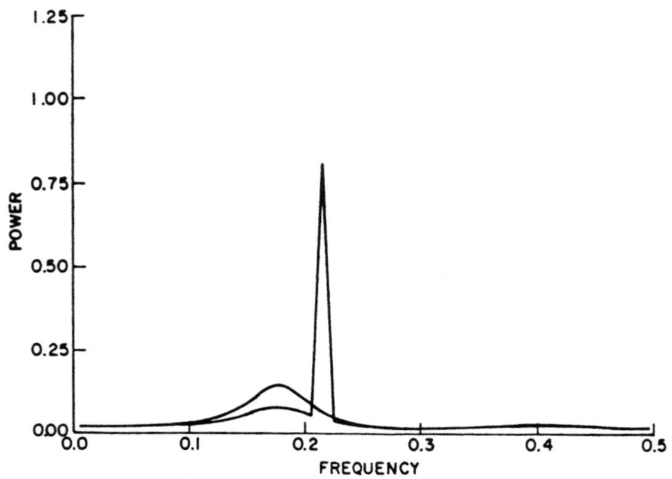

Figure 2. MRESA final estimates Q_B and Q_S.

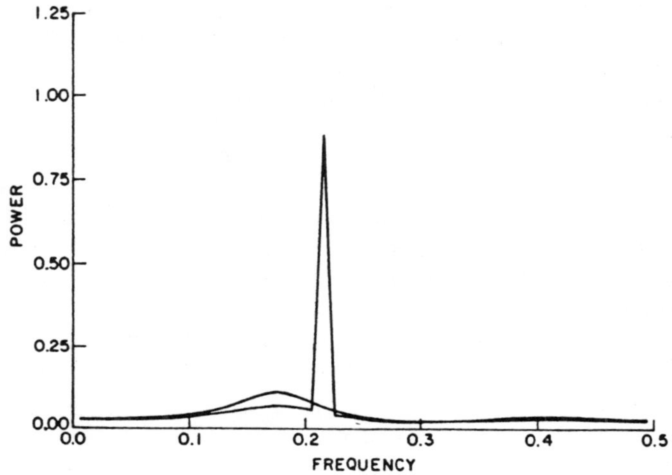

Figure 3. Final estimates Q_B and Q_S with $\varepsilon = 1$.

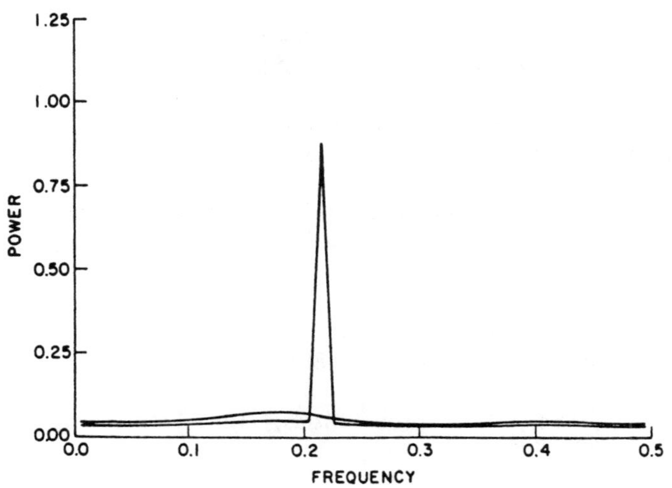

Figure 4. Final estimates Q_B and Q_S with $\varepsilon = 3$.

5. Discussion

A pleasant property of the new estimator, both in its general probability-density form and in the power-spectrum form, is that it has the same functional form as that for exact constraints. In the case of the power spectrum estimator, this means that resulting final estimates are still all-pole spectra whenever the initial estimates are all-pole and the weights are frequency independent.

It appears that Ables [1974] was the first to suggest using an uncertain constraint of the Euclidean form (52) in MESA. The use of this and a weighted Euclidean constraint in MESA was studied by Newman [1977, 1981]. This corresponds to a diagonal matrix **M** in Eq. (12). The generalization to general matrix constraints has been studied by Schott and McClellan [1983], who offer advice on how to choose **M** appropriately. The results presented herein differ in two main respects: treatment of the multisignal case and inclusion of initial estimates. Uncertain constraints have also been used in applying maximum entropy to image processing [Gull and Daniell, 1978; Skilling and Gull, 1981] although with a different entropy expression [Shore, 1983].

6. References

Ables, J. G. (1974), "Maximum entropy spectral analysis," Astron. Astrophys. Suppl. 15, pp. 383-393.

Burg, J. P. (1967), "Maximum entropy spectral analysis," presented at the 37th Annual Meeting, Society of Exploration Geophysicists, Oklahoma City, Okla.

Burg, J. P. (1975), "Maximum Entropy Spectral Analysis," Ph.D. dissertation, Stanford University (University Microfilms No. AAD75-25,499).

Csiszár, I. (1975), "I-divergence geometry of probability distributions and minimization problems," Ann Prob. 3, pp. 146-158.

Good, I. J. (1963), "Maximum entropy for hypothesis formulation, especially for multidimensional contingency tables," Ann. Math. Stat. 34, pp. 911-934.

Gull, S. F., and G. J. Daniell (1978), "Image reconstruction from incomplete and noisy data," Nature 272, pp. 686-690.

Jaynes, E. T. (1968), "Prior probabilities," IEEE Trans. Syst. Sci. Cybernet. SSC-4, pp. 227-241.

Johnson, R. W. (1979a), "Determining probability distributions by maximum entropy and minimum cross-entropy," APL79 Conference Proceedings, pp. 24-29, ACM 0-89791-005 (May 1979).

Johnson, R. W. (1979b), "Axiomatic characterization of the directed divergences and their linear combinations," IEEE Trans. Inf. Theory IT-25, pp. 709-716.

Johnson, R. W. (1983), "Algorithms for single-signal and multisignal minimum-cross-entropy spectral analysis," NRL Rept. 8667, Naval Research Laboratory, Washington, DC (AD-A132 400).

Johnson, R. W., and J. E. Shore (1983), "Minimum-cross-entropy spectral analysis of multiple signals," IEEE Trans. Acoust. Speech Signal Process. ASSP-31, pp. 574-582; also see NRL MR 4492 (AD-A097531).

Johnson, R. W., J. E. Shore, and J. P. Burg (1984), "Multisignal minimum-cross-entropy spectrum analysis with weighted initial estimates," IEEE Trans. Acoust. Speech Signal Process. ASSP-32, pp. 531-539.

Kullback, S. (1959), Information Theory and Statistics, Wiley, New York; reprinted Dover, New York (1968).

Newman, W. I. (1977), "Extension to the maximum entropy method," IEEE Inf. Theory IT-23, pp. 89-93.

Newman, W. I. (1981), "Extension to the maximum entropy method III," in S. Haykin, ed., Proc. First ASSP Workshop on Spectral Estimation, Aug. 1981, McMaster University, pp. 3.2.1-3.2.7.

Schott, J. P., and J. H. McClellan (1983), "Maximum entropy power spectrum estimation with uncertainty in correlation measurements," Proc. ICASSP 83, pp. 1068-1071, IEEE.

Shore, J. E. (1979), "Minimum cross-entropy spectral analysis," NRL Memorandum Rept. 3921, Naval Research Laboratory, Washington, DC 20375 (AD-A064183).

Shore, J. E. (1981), "Minimum cross-entropy spectral analysis," IEEE Trans. Acoust. Speech Signal Process. ASSP-29, pp. 230-237.

Shore, J. E. (1983), "Inversion as logical inference—theory and applications of maximum entropy and minimum cross-entropy," SIAM-AMS Proc. Vol. 14, Am. Math. Soc., Providence, RI, pp. 139-149.

Shore, J. E., and R. W. Johnson (1980), "Axiomatic derivation of the principle of maximum entropy and the principle of minimum cross-entropy," IEEE Trans. Inf. Theory IT-26, pp. 26-37; see also comments and corrections in IEEE Trans. Inf. Theory IT-29, p. 942 (1983).

Shore, J. E., and R. W. Johnson (1981), "Properties of cross-entropy minimization," IEEE Trans. Inf. Theory IT-27, pp. 472-482.

Skilling, J., and S. F. Gull (1981), "Algorithms and applications," in C. Ray Smith and W. T. Grandy, Jr., eds. (1985), Maximum-Entropy and Bayesian Methods in Inverse Problems, D. Reidel Publ. Co., Dordrecht (Holland), pp. 83-132.

A PROOF OF BURG'S THEOREM*

B. S. Choi

Department of Applied Statistics, Yonsei University, Seoul, Korea

Thomas M. Cover

Departments of Statistics and Electrical Engineering, Stanford University, Stanford, CA 94305

There are now many proofs that the maximum entropy stationary stochastic process, subject to a finite number of autocorrelation constraints, is the Gauss Markov process of appropriate order. The associated spectrum is Burg's maximum entropy spectral density. We pose a somewhat broader entropy maximization problem, in which stationarity, for example, is not assumed, and shift the burden of proof from the previous focus on the calculus of variations and time series techniques to a string of information theoretic inequalities. This results in a simple proof.

*Expanded version of a paper published originally in Proceedings of the IEEE 72, pp. 1094-1095 (1984).

C. R. Smith and G. J. Erickson (eds.),
Maximum-Entropy and Bayesian Spectral Analysis and Estimation Problems, 75–84

1. Preliminaries

We shall give some necessary definitions and go directly to a proof of the characterization of the maximum entropy stochastic process given covariance constraints. Section 5 has the history. In the concluding section, we mention a conditional limiting characterization of Gauss Markov processes.

Let $\{X_i\}_{i=1}^{\infty}$ be a stochastic process specified by its marginal probability density functions $f(x_1,x_2,\ldots,x_n)$, $n = 1,2\ldots$. Then the <u>(differential) entropy</u> of the n-sequence $X_1,X_2,\ldots,X_n$ is defined by

$$h(X_1,X_2,\ldots,X_n) = -\int f(x_1,\ldots,x_n) \ln f(x_1,\ldots,x_n)\, dx_1\, dx_2 \ldots dx_n$$

$$= h(f) \,. \tag{1}$$

The stochastic process $\{X_i\}$ will be said to have an <u>entropy rate</u>

$$h = \lim_{n\to\infty} \frac{h(X_1,X_2,\ldots,X_n)}{n} \tag{2}$$

if the limit exists. It is known that the limit always exists for stationary processes.

2. The Proof

We prove the following theorem:

<u>Theorem 1</u>: The stochastic process $\{X_i\}_{i=1}^{\infty}$ that maximizes the differential entropy rate h subject to the autocorrelation constraints

$$E\, X_i\, X_{i+k} = \alpha_k\,, \quad k = 0,1,2,\ldots,p, \quad i = 1,2,\ldots, \tag{3}$$

is the minimal order Gauss Markov process satisfying these constraints.

<u>Remark</u>: This pth order Gauss Markov process simultaneously solves the maximization problems

$$\max \frac{h(X_1,X_2,\ldots,X_n)}{n}\,, \quad n = 1,2,\ldots, \tag{4}$$

subject to the above autocorrelation constraints.

<u>Proof</u>: Let $X_1,X_2,\ldots,X_n$ be any collection of random variables satisfying Eq. (3). Let $Z_1,Z_2,\ldots,Z_n$ be zero mean multivariate normal with a covariance matrix given by the correlation matrix of $X_1,X_2,\ldots,X_n$. And let $Z_1',Z_2',\ldots,Z_n'$ be the pth order Gauss Markov process with covariance specified in Eq. (3). Then, for $n \geq p$,

$$h(X_1,\ldots,X_n) \leq h(Z_1,Z_2,\ldots,Z_n) \tag{5a}$$

$$= h(Z_1,Z_2,\ldots,Z_p) + \sum_{k=p+1}^{n} h(Z_k \mid Z_{k-1},\ldots,Z_1) \tag{5b}$$

$$\leq h(Z_1,Z_2,\ldots,Z_p) + \sum_{k=p+1}^{n} h(Z_k \mid Z_{k-1},Z_{k-2},\ldots,Z_{k-p}) \tag{5c}$$

$$= h(Z_1',Z_2',\ldots,Z_p') + \sum_{k=p+1}^{n} h(Z_k' \mid Z_{k-1}',\ldots,Z_{k-p}') \tag{5d}$$

$$= h(Z_1',Z_2',\ldots,Z_n') \,. \tag{5e}$$

Here inequality (b) is the chain rule for entropy, and inequality (c) follows from $h(A \mid B,C) \leq h(A \mid B)$. [See standard texts like Ash, 1965, and Gallager, 1968.] Inequality (a) follows from the information inequality, as shown in Section 3. Thus the pth order Gauss Markov process $Z_1',Z_2',\ldots,Z_n'$ with covariances $\alpha_0,\alpha_1,\ldots,\alpha_p$ has higher entropy $h(Z_1',Z_2',\ldots,Z_n')$ than any other process satisfying the autocorrelation constraints $\alpha_0,\alpha_1,\ldots,\alpha_p$. Consequently,

$$\lim_{n\to\infty} \frac{1}{n} h(X_1,\ldots,X_n) \leq \lim_{n\to\infty} \frac{1}{n} h(Z_1',\ldots,Z_n') = h \,, \tag{6}$$

for all stochastic processes $\{X_i\}$ satisfying the covariance constraints, thus proving the theorem.

3. Comments on the Proof

For completeness, we provide a proof of the well known inequality (a) in the proof of Theorem 1. [See for example Berger, 1971.] Let $f(x_1,\ldots,x_n)$ be a probability density function, and let

$$\phi(\mathbf{x}) = \frac{1}{(2\pi)^{n/2} \, |K|^{1/2}} \, e^{-\mathbf{x}^t K^{-1} \mathbf{x}/2} \tag{7}$$

be the n-variate normal probability density with covariance matrix

$$K = \int \mathbf{x}\,\mathbf{x}^t f(\mathbf{x})\, d\mathbf{x} \,. \tag{8}$$

Thus, ϕ and f have the same correlation matrix K.

Let

$$D(f\|g) = \int f \ln \frac{f}{g} \tag{9}$$

denote the Kullback-Leibler information number for f relative to g. It is known from Jensen's inequality that $D(f\|g) \geq 0$ for any probability densities f and g. Thus,

$$0 \leq D(f\|\phi) \overset{\Delta}{=} \int f(\mathbf{x}) \ln \frac{f(\mathbf{x})}{\phi(\mathbf{x})}\, d\mathbf{x}$$

$$= \int f \ln f - \int f \ln \phi\,. \tag{10}$$

But

$$\int f \ln \phi = \int \phi \ln \phi \tag{11}$$

because both are expectations of quadratic forms in $\mathbf{x}$. These expected quadratic forms are completely determined by Eq. (3), and are thus equal. Substituting Eq. (11) into Eq. (10) and using Eq. (1), we have

$$0 \leq -h(f) - \int f \ln \phi$$

$$= -h(f) - \int \phi \ln \phi$$

$$= -h(f) + h(\phi) \tag{12}$$

and

$$h(f) \leq h(\phi)\,, \tag{13}$$

as desired. This completes the proof of inequality (5a).

Remark: A pleasing byproduct of the proof is that the solutions to all of the finite-dimensional maximization problems, and therefore of the (limiting) entropy rate maximization problem, are given by the finite dimensional marginal densities $f(x_1,x_2,\ldots,x_n)$, $n = 1,2,\ldots$, of a single stochastic process: the Gauss Markov process of order p.

4. Equivalent Characterizations of the Solutions

Now that the maximum entropy process has been characterized, it is simple to provide an equivalent characterization.

We shall give the autoregressive characterization of the maximum entropy process by means of the Yule-Walker equations. If the p × p symmetric Toeplitz matrix whose (i,j)th element is $\alpha_{|i-j|}$ is positive definite, then there exists a unique solution set $\{a_1,\ldots,a_p\}$ of the Yule-Walker equations

$$\sum_{i=0}^{p} a_i \, \alpha_{|\ell-i|} = 0 , \qquad \ell = 1,\ldots,p, \tag{14}$$

where $a_0 = 1$. And then it can be proved [Choi, 1983] that $\Sigma_{i=0}^{p} a_i \alpha_i$ is positive. Thus, we can define

$$\sigma^2 = \sum_{i=0}^{p} a_i \, \alpha_i \, . \tag{15}$$

Consider the corresponding autoregressive process $\{X_n\}$ of order p,

$$X_n = -\sum_{i=1}^{p} a_i \, X_{n-i} + Z_n \, , \tag{16}$$

where $Z_1, Z_2, \ldots$ are independent and identically distributed normal random variables with mean 0 and variance σ^2. Inspection of Eqs. (3) and (14) yields the remaining autocovariance values

$$\alpha_\ell = \sum_{j=1}^{p} a_j \, \alpha_{|\ell-j|} \, , \qquad \ell \geq p+1 \, . \tag{17}$$

Thus, as was observed by Burg, the maximum entropy stochastic process is not obtained by setting the unspecified covariance terms equal to zero, but instead is given by letting the pth order autoregressive process "run" according to the Yule-Walker equations.

Finally, taking the Fourier transform of $\alpha_0, \alpha_1, \ldots$ given in Eqs. (3) and (17), yields the spectral density $S(\lambda)$:

$$S(\lambda) = \frac{1}{2\pi} \sum_{\ell=-\infty}^{\infty} \alpha_\ell e^{-i\lambda\ell}$$

$$= \frac{\sigma^2}{2\pi} \frac{1}{|\sum_{j=0}^{p} a_j e^{i\lambda j}|^2} . \tag{18}$$

This is Burg's maximum entropy spectral density subject to the covariance constraints $\alpha_0, \alpha_1, \ldots, \alpha_p$.

The resulting maximum entropy rate is

$$h = \lim_{n\to\infty} \frac{1}{n} \left[h(X_1, \ldots, X_p) + \sum_{\ell=p+1}^{n} h(X_\ell \mid X_{\ell-1}, \ldots, X_{\ell-p}) \right]$$

$$= h(X_{p+1} \mid X_p, X_{p-1}, \ldots, X_1)$$

$$= \frac{1}{2} \ln (2\pi e \sigma^2) , \tag{19}$$

where σ^2 is given in Eq. (15). Incidentally, the maximum entropy process will be less than pth order, although still determined by Eqs. (14), (15), (16), if Eq. (3) is not strictly positive definite. The true order of the process is the largest k for which $[\alpha_{|i-j|}]_{1\leq i,j\leq k}$ is positive definite.

5. History

Burg (1967) introduced the maximum entropy spectral density among Gaussian stochastic processes by exhibiting the solution to the problem of maximizing the entropy rate

$$h = \frac{1}{2} \ln(2\pi e) + \frac{1}{4\pi} \int_{-\pi}^{\pi} \ln [2\pi S(\lambda)] \, d\lambda \tag{20}$$

where

$$S(\lambda) = \frac{1}{2\pi} \sum_{\ell=-\infty}^{\infty} \sigma(\ell) e^{-i\lambda\ell} , \tag{21}$$

and $\{\sigma(\ell)\}_{\ell=-\infty}^{\infty}$ is an arbitrary autocovariance function satisfying the constraints

$$\sigma(0) = \alpha_0 , \quad \sigma(1) = \alpha_1 , \quad \ldots , \quad \sigma(p) = \alpha_p . \tag{22}$$

Proof that the pth order Gaussian autoregressive process spectral density is the maximum entropy spectral density has been established by variational methods by Smylie, Clarke, and Ulrych [1973, pp. 402-419], using the Lagrange multiplier method, and independently by Edward and Fitelson [1973]. Burg [1975], Ulrych and Bishop [1975], Haykin and Kesler [1979, pp. 16-21], and Robinson [1982] follow Smylie's method. Ulrych and Ooe [1979] and McDonough [1979] use Edward's method. See also Grandell et al. [1980].

The calculus of variations necessary to show that

$$S^*(\lambda) = \frac{\sigma^2}{2\pi} \frac{1}{\left|\sum_{j=0}^{p} a_j e^{i\lambda j}\right|^2} \tag{23}$$

is the solution to Eq. (20) is tricky. Smylie et al. [1973] show that the first variation about $S^*(\lambda)$ is zero. Further considerations establish S^* as a maximum.

Van den Bos [1971] maximizes the entropy $h(x_1,x_2,\ldots,x_{p+1})$ subject to the constraints (22) by differential calculus, but further argument is required to extend his solution to the maximization of $h(X_1,\ldots,X_n)$, $n > p+1$. Feder and Weinstein [1984] have carried this out.

Akaike [1977] maximizes another form of the entropy rate h, that is,

$$h = \tfrac{1}{2}\log(2\pi e) + \tfrac{1}{2}\operatorname{Var}(\varepsilon_t), \tag{24}$$

where ε_t is the prediction error of the best linear predictor of X_t in terms of all the past $X_{t-1}, X_{t-2}, \ldots$. Of course, Eq. (24) holds only if the process is Gaussian. Equation (24) can be derived from Eq. (20) through Kolmogorov's equality [1941]:

$$\operatorname{Var}(\varepsilon_t) = 2\pi \exp\left[\frac{1}{2\pi}\int_{-\pi}^{\pi} \ln S_X(\lambda)\, d\lambda\right]. \tag{25}$$

Using prediction theory, one can show that $\operatorname{Var}(\varepsilon_t)$ has its maximum if

$$\varepsilon_t = X_t + a_1X_{t-1} + a_2X_{t-2} + \cdots + a_pX_{t-p}, \tag{26}$$

where $a_1, a_2, \ldots, a_p$ are given in Eq. (14). For details, see Priestley [1981, pp. 604-606].

More details of proofs in this section can be found in Choi [1983].

With hindsight, we see that all of the maximization can be captured in the information theoretic string of inequalities in Eq. (5) of Theorem 1, and that the global maximality of $S^*(\lambda)$ follows automatically from verifying that $S^*(\lambda)$ is the spectrum of the process specified by the theorem.

6. Conclusions

A bare bones summary of the proof is that the entropy of a finite segment of a stochastic process is bounded above by the entropy of a segment of a Gaussian random process with the same covariance structure. This entropy is in turn bounded above by the entropy of the minimal order Gauss Markov process satisfying the given covariance constraints. Such a process exists and has a convenient characterization by means of the Yule-Walker equations. Thus the maximum entropy stochastic process is obtained.

We mention that the maximum entropy spectrum actually arises as the answer to a certain "physical" question. Suppose $X_1, X_2, \ldots$ are independent identically distributed uniform random variables. Suppose also that the following empirical covariance constraints are observed:

$$\frac{1}{n} \sum_{i=1}^{n} X_i X_{i+k} = \alpha_k, \quad k = 0,1,\ldots,p. \tag{27}$$

What is the conditional distribution on $(X_1, X_2, \ldots, X_m)$? It is shown in Choi and Cover [1987] that the limit, as $n \to \infty$, of the conditional probability densities given the empirical constraint (27) tends to the unconditional probability density function of the maximum entropy process specified in Theorem 1. Thus, an independent uniform process conditioned on empirical correlations looks like a Gauss Markov process.

7. Acknowledgments

This work was partially supported by National Science Foundation Grant ECS82-11568 and Joint Services Electronics Program DAAG29-81-K-0057.

A shortened version of this paper appears as a Letter in the Proceedings of the IEEE [Choi and Cover, 1984].

8. References

Akaike, H. (1977), "An entropy maximization principle," in P. Krishnaiah, ed., Proceedings of the Symposium on Applied Statistics, North-Holland, Amsterdam.

Ash, R. (1965), Information Theory, Wiley Interscience, New York.

Berger, T. (1971), Rate Distortion Theory, A Mathematical Basis for Data Compression, Prentice-Hall, N.J.

Burg, J. P. (1967), "Maximum entropy spectral analysis," presented at the 37th Meeting of the Society of Exploration Geophysicists; reprinted in D. G. Childers, ed. (1978), Modern Spectrum Analysis, IEEE Press, pp. 34-41.

Burg, J. P. (1975), "Maximum Entropy Spectral Analysis," Ph.D. dissertation, Department of Geophysics, Stanford University, Stanford, Calif.

Choi, B. S. (1983), "A Conditional Limit Characterization of the Maximum Entropy Spectral Density in Time Series Analysis," Ph.D. dissertation, Statistics Department, Stanford University.

Choi, B. S., and T. M. Cover (1984), "An information-theoretic proof of Burg's maximum entropy spectrum" (letter), Proc. IEEE 72, pp. 1094-1095.

Choi, B. S., and T. M. Cover (1987), "A conditional limit characterization of Gauss Markov processes," submitted to JASA.

Edward, J. A., and M. M. Fitelson (1973), "Notes on maximum-entropy processing," IEEE Trans. Inf. Theory IT-19, pp. 232-234; reprinted in D. G. Childers, ed. (1978), Modern Spectrum Analysis, IEEE Press, pp. 94-96.

Feder, M., and E. Weinstein (1984), "On the finite maximum entropy extrapolation," Proc. IEEE 72, pp. 1660-1662.

Gallager, R. (1968), Information Theory and Reliable Communication, Wiley, New York.

Grandell, J., M. Hamrud, and P. Toll (1980), "A remark on the correspondence between the maximum entropy method and the autoregressive model," IEEE Trans. Inf. Theory IT-26, pp. 750-751.

Haykin, S., and S. Kesler (1979), "Prediction-error filtering and maximum entropy spectral estimation," in S. Haykin, ed., Nonlinear Methods of Spectral Analysis, Springer, New York, pp. 9-72.

Kolmogorov, A. N. (1941), "Interpolation und Extrapolation von Stationären Zufälligen Folgen," Bull. Acad. Sci. URSS, Ser. Math. 5, pp. 3-41.

McDonough, R. N. (1979), "Application of the maximum-likelihood method and the maximum entropy method to array processing," in S. Haykin, ed., Nonlinear Methods of Spectral Analysis, Springer, New York, pp. 181-244.

Priestley, M. B. (1981), Spectral Analysis and Time Series, Vol. 1, Academic Press, New York.

Robinson, E. A. (1982), "A historical perspective of spectrum estimation," Proc. IEEE 70, pp. 885-907.

Smylie, D. G., G. K. C. Clarke, and T. J. Ulrych (1973), "Analysis of irregularities in the earth's rotation," Meth. Comp. Phys. 13, pp. 391-430.

Ulrych, T., and T. Bishop (1975), "Maximum entropy spectral analysis and autoregressive decomposition," Rev. Geophys. and Space Phys., 13, pp. 183-200; reprinted in D. G. Childers, ed. (1978), Modern Spectrum Analysis, IEEE Press, pp. 54-71.

Ulrych, T., and M. Ooe (1979), "Autoregressive and mixed autoregressive-moving average models and spectra," in S. Haykin, ed., Nonlinear Methods of Spectral Analysis, Springer, New York, pp. 73-126.

Van den Bos, A. (1971), "Alternative interpretation of maximum entropy spectral analysis," IEEE Trans. Inf. Theory IT-17, pp. 493-494; reprinted in D. G. Childers, ed. (1978), Modern Spectrum Analysis, IEEE Press, pp. 92-93.

A BAYESIAN APPROACH TO ROBUST LOCAL FACET ESTIMATION

Robert M. Haralick

Virginia Polytechnic Institute and State University, Blacksburg, VA 24061

C. R. Smith and G. J. Erickson (eds.),
Maximum-Entropy and Bayesian Spectral Analysis and Estimation Problems, 85–97.

1. Introduction

The facet model for image processing takes the observed pixel values to be a noisy discretized sampling of an underlying gray tone intensity surface that in each neighborhood of the image is simple. To process the image requires the estimation of this simple underlying gray tone intensity surface in each neighborhood of the image. Prewitt (1970), Haralick and Watson (1981), and Haralick (1980, 1982, 1983, 1984) all use a least squares estimation procedure. In this note we discuss a Bayesian approach to this estimation problem. The method makes full use of prior probabilities. In addition, it is robust in the sense that it is less sensitive to small numbers of pixel values that might deviate highly from the character of the other pixels in the neighborhood.

Two probability distributions define the model. The first distribution specifies the conditional probability density of observing a pixel value, given the true underlying gray tone intensity surface. The second distribution specifies the conditional probability density of observing a neighborhood having a given underlying gray tone intensity surface.

To motivate the representations we choose, and to help make clear what underlying gray tone intensity surface means, consider the following thought experiment. Suppose we have a noiseless image that is digitized to some arbitrary precision. Suppose, for the moment, we take simple underlying gray tone intensity surface to mean a constant surface in each neighborhood. Now begin moving a fixed and reasonable sized neighborhood window through the image. Most neighborhoods (probably all of them) will not have constant values. Many would be constant except for illumination shading or texture effects; those neighborhoods are nearly constant. Some have an object edge passing through; these are not constant.

The nearly constant neighborhoods can be thought of as having arisen from small perturbations of a constant neighborhood. The perturbation is due, not to sensor noise, but to the difference between the idealization of the model (perfectly constant neighborhoods) and the observed perfect reality. In this case, we take the underlying gray tone intensity surface to be a constant, the value of which is representative of the values in the observed nearly constant neighborhood.

What does it mean to determine a value that is representative of the values in the neighborhood? Does it mean an equally weighted average, for example? To answer this question, fix attention on the center pixel of the neighborhood. We expect that the neighbors of the center pixel would have a value close to the value of the center pixel. The neighbors of these neighbors, the second neighbors, would have values that could deviate more from the value of the center pixel than the first neighbors. This expectation—that the closer a pixel is to the center pixel, the less the deviation is likely to be from the center pixel—should find a way to be incorporated into the model explicitly. Under these conditions, the representative gray tone intensity of the underlying gray tone intensity surface in the neighborhood can be estimated as an unequally weighted average of the pixel values in the neighborhood, those pixels farther away from the center pixel getting less weight.

We have neglected the neighborhoods having an edge or a line passing through them. These neighborhoods do not satisfy the spirit of a model that is "constant in each neighborhood." This suggests that we need to be examining models in which the spatial distribution of gray tones in the neighborhood is more complex than constant. An appropriate model, for example, may be one in which the ideal gray tone intensity surface is a low order polynomial of row and column positions.

Now suppose that our model is that the underlying gray tone intensity surface in each neighborhood is a bivariate cubic polynomial. Again take our hypothetical noiseless perfect image and pass a neighborhood window through it. As before, there probably will be no neighborhoods that fit a cubic precisely, but this time most neighborhoods will nearly or almost nearly fit. The cubic model can represent constants, slopes, edges, and lines.

Fix attention on one of the neighborhoods. Suppose it is mostly constant, especially near its center, with a small portion of a line or edge at its boundary. Instead of thinking of the polynomial underlying gray tone intensity surface as representative, in the sense of fitting, of the entire neighborhood, think of it as containing the values of all partial derivatives of order 3 or less evaluated at the center pixel. Since the area around the center pixel is nearly constant, we should expect all partial derivatives of order 1 to order 3 to be small or zero, despite some significant disturbance at the boundary of the neighborhood and despite the fact that a least squares fit of the pixel values in the neighborhood would certainly not produce near-zero partial derivatives.

At this point we begin to uncover a few concepts about which deeper understanding is needed. The first is the difference between estimating the derivatives at the center pixel of a neighborhood and least squares fitting an entire neighborhood. The second is the notion of neighborhood size. The larger the neighborhood, the more different things are likely to happen near and around its boundary and the more we will want to ignore the values around the boundary in estimating the partial derivatives at the neighborhood center. At the same time, should the pixel values near and around the boundary of the neighborhood fit in with the spatial distribution at the center of the neighborhood, we would definitely want to have the estimation procedure utilize these neighborhood boundary pixels in a supportive way.

The conclusion we can draw from this perspective is that we can expect the underlying polynomial gray tone intensity surface to be more representative of what is happening at the center of the neighborhood than at its periphery. That is, the observed values at the periphery of the neighborhood are likely to deviate more from the corresponding values of the underlying gray tone intensity surface than are the observed values at the center of the neighborhood. Furthermore, we need to pay careful attention to the similarity or dissimilarity of pixels at the periphery of the neighborhood so that their values can be used in a supportive way.

In section 2 we discuss a model and estimation procedure that makes these ideas precise.

2. The Model

Let $\sum_{\substack{ij \\ i+j\leq 3}} \alpha_{ij} r^i c^j$ represent the underlying gray tone intensity surface of a neighborhood, and let $J(r,c)$ represent the observed gray tone values in the neighborhood. At each pixel (r,c) the squared deviation between the representative underlying gray tone intensity surface and the observed image is given by $[\sum_{\substack{ij \\ i+j\leq 3}} \alpha_{ij} r^i c^j - J(r,c)]^2$. The expected value of this squared deviation is the variance of $J(r,c)$ around $\sum_{\substack{ij \\ i+j\leq 3}} \alpha_{ij} r^i c^j$. It is a function of (r,c), and our perspective has suggested that it increases as a monotonic function of the distance between (r,c) and $(0,0)$. We can express this by writing

$$E[\sum_{\substack{ij \\ i+j\leq 3}} \alpha_{ij} r^i c^j - J(r,c)]^2 = \sigma^2[1 + k(r^2+c^2)^P] . \tag{1}$$

To help make our notation compact, we rewrite this in vector notation. Let J be the vector of observed pixel values in a neighborhood. Let α be the vector of coefficients for the underlying gray tone intensity surface. Let F be a matrix whose columns constitute the discretized polynomial basis. Thus the column corresponding to the basis function $r^i c^j$ has component values that are $r^i c^j$ evaluated at all pixel positions in the neighborhood.

Assuming an ellipsoidally symmetric distribution for the deviations between the observed pixel values and the underlying gray tone intensity surface, we have

$$P(J|F\alpha) = h[(J-F\alpha)' \, {\textstyle\not{\Sigma}}_J^{-1}(J-F\alpha)] , \tag{2}$$

where Σ_J is the covariance matrix of the deviations of the observed values J from the ideal values $F\alpha$.

For the prior distribution of α we likewise take the deviations between the neighborhood α and an α_0 representative of the distribution of α's over all neighborhoods to be distributed in an ellipsoidally symmetric form (typically $\alpha_0 = 0$):

$$P(\alpha) = h[(\alpha-\alpha_0)' \Sigma_\alpha^{-1}(\alpha-\alpha_0)] . \tag{3}$$

From a Bayesian point of view, having observed J we wish to estimate an α that maximizes the probability of α given J. Now,

$$P(\alpha|J) = \frac{P(J|\alpha)P(\alpha)}{P(J)} . \tag{4}$$

Maximizing $P(\alpha|J)$ is equivalent to maximizing $P(J|\alpha)P(\alpha)$, and this is equivalent to maximizing $\log P(J|\alpha) + \log P(\alpha)$. The necessary condition is for the partial derivative of $\log P(J|\alpha) + \log P(\alpha)$ with respect to each component of α to be equal to zero. This yields

$$(-2)\frac{h'[(J-F\alpha)' \Sigma_J^{-1}(J-F\alpha)]}{h[(J-F\alpha)' \Sigma_J^{-1}(J-F\alpha)]}[F' \Sigma_J^{-1}(J-F\alpha)]$$

$$+ (-2)\frac{h'[(\alpha-\alpha_0)' \Sigma_\alpha^{-1}(\alpha-\alpha_0)]}{h[(\alpha-\alpha_0)' \Sigma_\alpha^{-1}(\alpha-\alpha_0)]} \Sigma_\alpha^{-1}(\alpha_0-\alpha) = 0 . \tag{5}$$

In the case where h is the multivariate normal density,

$$h(x^2) = Ae^{-x^2/2} . \tag{6}$$

Or, with a simple argument μ replacing x^2,

$$h(\mu) = Ae^{-\mu/2} . \tag{7}$$

Hence,

$$-2\,\frac{h'(\mu)}{h(\mu)} = -2\,\frac{Ae^{-\mu/2}(-1/2)}{Ae^{-\mu/2}} = 1\,. \tag{8}$$

In the multivariate normal case, the equation simplifies to

$$-F' \Sigma_J^{-1}(J-F\alpha) + \Sigma_\alpha^{-1}(\alpha-\alpha_0) = 0 \tag{9}$$

or

$$[F' \Sigma_J^{-1} F + \Sigma_\alpha^{-1}]\alpha = F' \Sigma_J^{-1} J + \Sigma_\alpha^{-1} \alpha_0\,. \tag{10}$$

To relate this to standard least squares, take $\Sigma_J^{-1} = \sigma^2 I$ and $\Sigma_\alpha^{-1} = 0$, in which case we have $F'F\alpha = F'J$, which is the usual normal equation. $\Sigma_\alpha^{-1} = 0$ means that the variance of α is very large. In essence, it says that nothing is known about α. $\Sigma_J^{-1} = \sigma^2 I$ means that the deviations of the observed from the ideal are uncorrelated and that the expected squared deviations are identical throughout the neighborhood rather than increasing for pixels closer to the periphery as suggested earlier.

Now let us move on to a nonnormal case, in which the tails of the distribution are much fatter than the normal distribution. One such distribution is the slash distribution, which arises from a normal (0,1) variate being divided by a uniform (0,1) variate. Another such distribution is the Cauchy distribution.

The slash density function has the form

$$s(x) = \frac{1 - e^{-x^2/2}}{2\pi x^2}\,. \tag{11}$$

Because we have squared the argument before the evaluation, we have

$$s(\mu) = \frac{1 - e^{-\mu/2}}{2\pi\mu}\,, \qquad \mu \geq 0\,. \tag{12}$$

Thus,

$$- \frac{2s'(\mu)}{s(\mu)} = 2 \, \frac{1 - (1 + \mu/2)e^{-\mu/2}}{\mu^2} , \tag{13}$$

a function that is always positive, having largest magnitude for small μ and a monotonically decreasing magnitude for larger μ.

The Cauchy distribution has the form

$$c(x) = \frac{1}{\pi(1+x^2)} . \tag{14}$$

Because we have squared the argument before evaluation, we have

$$c(x) = \frac{1}{\pi(1+\mu)} , \quad \mu \geq 0 . \tag{15}$$

Thus,

$$- \frac{c'(\mu)}{c(\mu)} = \frac{1}{1 + \mu} , \tag{16}$$

a function that is always positive, having largest magnitude for small μ and a monotonically decreasing magnitude for larger μ.

On the basis of the behavior of h'/h for slash and Cauchy distributions, we can discuss the meanings of h'/h in Eq. (5). Simply, if the fit $F\alpha$ to J is relatively good compared to our prior uncertainty about α, then the estimated α is determined mostly by the least squares fit and hardly at all by the prior information we have about α. If the fit $F\alpha$ to J is comparable in uncertainty to our prior uncertainty about α, then the estimated α is determined in equal measure by the least squares fit and by the prior information. If the fit $F\alpha$ to J has more error than our prior uncertainty about α, then the estimated α is determined more by the prior information than by the fit.

To see how this works more precisely, let

$$\lambda_J(\alpha) = \frac{h'[(J-F\alpha)' \not{\sum}_J^{-1} (J-F\alpha)]}{h[(J-F\alpha)' \not{\sum}_J^{-1} (J-F\alpha)]} \tag{17}$$

and

$$\lambda_\alpha(\alpha) = \frac{h'[(\alpha-\alpha_0)' \Sigma_\alpha^{-1}(\alpha-\alpha_0)]}{h[(\alpha-\alpha_0)' \Sigma_\alpha^{-1}(\alpha-\alpha_0)]} \; . \tag{18}$$

Equation (5) becomes

$$[\lambda_J(\alpha)F' \Sigma_J^{-1} F + \lambda_\alpha(\alpha) \Sigma_\alpha^{-1}]\alpha$$

$$= \lambda_J(\alpha)F' \Sigma_J^{-1} J + \lambda_\alpha(\alpha) \Sigma_\alpha^{-1} \alpha_0 \; . \tag{19}$$

We can solve this equation iteratively. Let α^n be the value of the estimated α at the nth iteration. Take the initial $\alpha^{(1)}$ to satisfy Eq. (10). Suppose $\alpha^{(n)}$ has been determined. Substitute $\alpha^{(n)}$ into Eqs. (17) and (18) to obtain $\lambda_J(\alpha^{(n)})$ and $\lambda_\alpha(\alpha^{(n)})$. Then substitute these values for $\lambda_J(\alpha^n)$ and $\lambda_\alpha(\alpha^n)$ into Eq. (19) to determine $\alpha^{(n+1)}$.

3. The Independence Assumption

An alternative model for the distributions would be for the deviations of the observed values from the values of the underlying gray tone intensity surface to be assumed independent. In this case,

$$P(J/\alpha) = \prod_{(r,c)} P_{rc}(J(r,c)|\alpha)$$

$$= \prod_{(r,c)}^{N} h\left[\left(\frac{J(r,c)^2 - \sum_{n=1} \alpha_n f_n(r,c)}{\sigma_J(r,c)}\right)^2\right] \tag{20}$$

and

$$P(\alpha) = \prod_{n=1}^{N} P_n(\alpha_n | \alpha_{n0})$$

$$= \prod_{n=1}^{N} h\left[\left(\frac{\alpha_n - \alpha_{n0}}{\sigma_{an}}\right)^2\right], \quad (21)$$

where $\alpha' = (\alpha_1,\ldots,\alpha_N)$ and $\alpha_0' = (\alpha_{10},\ldots,\alpha N_0)$.

Proceeding as before, we obtain that the maximizing α must satisfy

$$\left[F' \Lambda_J \sum_J{}^{-1} F + \Lambda_\alpha \sum_\alpha{}^{-1}\right]\alpha = \left[F' \Lambda_J \sum_J{}^{-1} J + \Lambda_\alpha \sum_\alpha{}^{-1}\right]\alpha_0, \quad (22)$$

where $\sum_J$, $\sum_\alpha$, Λ_J, and Λ_α are diagonal matrices

$$\sum_J = \begin{pmatrix} \ddots & & 0 \\ & \sigma_J(r,c)^2 & \\ 0 & & \ddots \end{pmatrix} \quad (23)$$

$$\sum_\alpha = \begin{pmatrix} \ddots & & 0 \\ & \sigma_\alpha(n)^2 & \\ 0 & & \ddots \end{pmatrix} \quad (24)$$

$$\Lambda_J = \begin{pmatrix} \ddots & & 0 \\ & \lambda_J(r,c) & \\ 0 & & \ddots \end{pmatrix} \quad (25)$$

$$\Lambda_\alpha = \begin{pmatrix} \ddots & & 0 \\ & \lambda_\alpha(n) & \\ 0 & & \ddots \end{pmatrix} \quad (26)$$

and the diagonal entries of Λ_J and Λ_α are given by

$$\lambda_J(r,c) = \frac{h'\left[\left(\dfrac{J(r,c) - \sum_{n=1}^{N} \alpha_n f_n(r,c)}{\sigma_J(r,c)}\right)^2\right]}{h\left[\left(\dfrac{J(r,c) - \sum_{n=1}^{N} \alpha_n f_n(r,c)}{\sigma_J(r,c)}\right)^2\right]} \tag{27}$$

$$\lambda_\alpha(n) = \frac{h'\left[\left(\dfrac{\alpha_n - \alpha_{n0}}{\sigma_{\alpha n}}\right)^2\right]}{h\left[\left(\dfrac{\alpha_n - \alpha_{n0}}{\sigma_{\alpha n}}\right)^2\right]} \; . \tag{28}$$

The solution for α can be obtained iteratively. Take the first Λ_J and Λ_α to be the corresponding identity matrices. Solve Eq. (22) for α. Then substitute into Eqs. (27) and (28) for the next Λ_J and Λ_α.

Because the solution for α is iterative, it is not necessary to take the required $NK(1+N+K) + 2N + N^3$ operations to solve Eq. (22) exactly. (The vector J is $K \times 1$ and the vector α is $N \times 1$.) There is a quicker computation procedure. Suppose the basis that is the columns of F satisfies

$$F' \mathop{\not{\Sigma}}_J{}^{-1} F = I \; . \tag{29}$$

This means that the basis vectors are discretely orthonormal with respect to the weights that are the diagonal entries of the diagonal matrix $\mathop{\not{\Sigma}}_J{}^{-1}$. In this case, Eq. (22) holds if and only if

$$\left(F' \Lambda_J \mathop{\not{\Sigma}}_J{}^{-1} F + \Lambda_\alpha \mathop{\not{\Sigma}}_\alpha{}^{-1} + I - F' \mathop{\not{\Sigma}}_J{}^{-1} F\right)\alpha$$

$$= F' \Lambda_J \mathop{\not{\Sigma}}_J{}^{-1} J + \Lambda_\alpha \mathop{\not{\Sigma}}_\alpha{}^{-1} \alpha \; . \tag{30}$$

Rewriting this equation, we have

$$J(I + \Lambda_\alpha \Sigma_\alpha^{-1})\alpha$$

$$= F'[\Lambda_J \Sigma_J^{-1} J + (I - \Lambda_J) \Sigma_J^{-1} F\alpha + \Sigma_J^{-1} F \Lambda_\alpha \Sigma_\alpha^{-1} \alpha_0] \; . \tag{31}$$

This equation suggests the following iterative procedure for the determination of α:

Take

$$\alpha^{(1)} = (I + \Lambda_\alpha \Sigma_\alpha^{-1})^{-1} (F' \Sigma_J^{-1} J + \Sigma_\alpha^{-1} \alpha_0) \; . \tag{32}$$

Suppose $\alpha^{(n)}$ has already been determined. Define

$$\alpha^{(n+1)} = (I + \Lambda_\alpha \Sigma_\alpha^{-1})^{-1}$$

$$\cdot F' \Sigma_J^{-1} [\Lambda_J J + (I - \Lambda_J) F\alpha^{(n)} + F \Lambda_\alpha \Sigma_\alpha^{-1} \alpha_0] \; . \tag{33}$$

Each iteration of Eq. (33) requires 3KN+4K+3N operations, and only two to four iterations are necessary to get a reasonably close answer.

4. Robustness

The model assuming the independence of the deviations between the observed values and the underlying gray tone intensity surface is robust. If there are some pixel positions in which $J(r,c)$ deviates greatly from the corresponding value $\sum_{n=1}^{N} \alpha_n f_n(r,c)$ of the underlying gray tone intensity surface, then since $\lambda(r,c)$ is defined by Eq. (27), that is,

$$\lambda_J(r,c) = \frac{h'\left[\left(\dfrac{J(r,c) - \sum_{n=1}^{N} \alpha_n f_n(r,c)}{\sigma_J(r,c)}\right)^2\right]}{h\left[\left(\dfrac{J(r,c) - \sum_{n=1}^{N} \alpha_n f_n(r,c)}{\sigma_J(r,c)}\right)^2\right]} \tag{27}$$

and $-h'/h$ is small for large arguments, $\lambda_J(r,c)$ will be small. To understand the effect of a small $\lambda_J(r,c)$, examine Eq. (33). On the right-hand side of that equation is the expression $\Lambda_J J + (I - \Lambda_J)F\alpha$, which consists of a generalized convex combination of J, a term depending on the observed data, and $F\alpha$, a term depending on the fit to the data. In those components where $\lambda_J(r,c)$ is small, the generalized convex combination tends to ignore $J(r,c)$ and, in effect, to substitute for it the fit $\sum_{n=1}^{N} \alpha_n f_n(r,c)$. Thus small values of $\lambda_J(r,c)$ substitute the fitted values for the observed values. Values of the weight $\lambda_J(r,c)$ close to 1 tend to make the procedure ignore the fitted values and use only the observed values.

The technique is inherently robust. Any observed value that deviates greatly from the fitted value is in a sense ignored and replaced with a fitted value interpolated on the basis of the other pixel values.

To do the λ_J computation, a density function h is required. As we have seen, a normal distributional assumption leads to each λ_J being the same identical constant. Distributional assumptions such as slash or Cauchy lead to λ_J being some monotonically decreasing function of the squared difference between the observed and fitted values. The monotonically decreasing function depends on the distributional assumption being made.

One way to avoid the distributional assumption is to use a form λ_J that has proved to work well over several different kinds of distributions. One such form is Tukey's bisquare function, used in computing the biweight:

$$\lambda_J(r,c) = \begin{cases} [1 - s^2(r,c)]^2, & \text{if } s^2(r,c) < 1 \\ 0, & \text{otherwise} \end{cases} \tag{34}$$

where

$$s^2(r,c) = \frac{[J(r,c) - \sum_{n=1}^{N} \alpha_n f_n(r,c)]^2}{C\,\sigma_J^2(r,c)} \tag{35}$$

and C is a constant with value between 6 and 9. In this case, the estimated coefficients $\alpha_1,\ldots,\alpha_N$ are generalizations of the biweight, and the computational procedure discussed in section 2.1 corresponds to Tukey's iterative reweighted least squares regression procedure [Mosteller and Tukey, 1977].

5. References

Haralick, R. M. (1980), "Edge and region analysis for digital image data," Comput. Graphics and Image Processing 12, pp. 113-129.

Haralick, R. M. (1982), "Zero-crossing of second directional derivative edge operator," Proceedings of the SPIE Technical Symposium East, Arlington, Va., May 3-7, 1982, 336, p. 23.

Haralick, R. M. (1983), "Ridges and valleys on digital images," Comput. Vision Graphics and Image Processing 22, pp. 28-38.

Haralick, R. M. (1984), "Digital step edges from zero-crossing of second directional derivative," IEEE Trans. Pattern Analysis and Machine Intelligence PAMI-6, No. 1, pp. 58-68.

Haralick, R. M., and Layne Watson (1981), "A facet model for image data," Comput. Graphics and Image Processing 15, pp. 113-129.

Mosteller, Frederick, and John Tukey (1977), Data Analysis and Regression, Addison-Wesley, Reading, Mass., pp. 356-358.

Prewitt, Judy (1970), "Object enhancement and extraction," in Picture Processing and Psychopictorics, B. Lipkin and A. Rosenfeld, eds., Academic Press, New York, pp. 75-149.

THE MAXIMUM ENTROPY METHOD: THE PROBLEM OF MISSING DATA

William I. Newman

Department of Earth and Space Sciences and Department of Astronomy, University of California, Los Angeles, CA 90024

The maximum entropy method is reviewed and adapted to treat the problems posed by noisy and by missing data. Examples of the use of the extended maximum entropy method are presented.

C. R. Smith and G. J. Erickson (eds.),
Maximum-Entropy and Bayesian Spectral Analysis and Estimation Problems, 99–125.

1. Introduction

In this paper we explore the problem of spectral estimation and of linear prediction of time series and correlation function data in the presence of noise and of missing information. We begin by reviewing some fundamental concepts of spectral analysis and information theory. We then explore an idealized problem utilizing the maximum entropy method (MEM) or maximum entropy spectral analysis (MESA) method and its relation to prediction by means of autoregressive methods (ARM). Then we consider extensions to real-world problems, including missing data and noise in the correlation function, and present some examples.

The above methods are often expressed in the form of a problem in constrained optimization, a feature that is not esthetically appealing from an information theoretic viewpoint. We explore an ad hoc formulation of the problem that is Bayesian in nature and that results in an unconstrained optimization problem. Finally, we explore methods of extending time series data containing gaps, and develop a variational approach, predicated on a principle first advanced by Burg [1968], for accomplishing this objective. We discuss the relationship of this extension of Burg's method to existing techniques. Illustrative examples for the use of this method are presented.

2. Review of Spectral Analysis and Information Theory

Consider a band-limited power spectrum $S(\nu)$ such that

$$S(\nu) = 0, \qquad |\nu| > \nu_N . \tag{1}$$

If $\rho(t)$ is the correlation function corresponding to $S(\nu)$, then we have the Fourier transform pair

$$\begin{aligned} S(\nu) &= \int_{-\infty}^{\infty} \rho(t) \exp(2\pi i\nu t)\, dt \\ \rho(t) &= \int_{-\infty}^{\infty} S(\nu) \exp(-2\pi i\nu t)\, d\nu \end{aligned} \tag{2}$$

From Eqs. (2) it follows that, if the correlation function $\rho(t)$ is exactly known, the spectrum $S(\nu)$ is also exactly known. But this rarely happens; indeed, in real-world applications it never happens. So, it is important to understand precisely what limits our ability to determine the spectrum.

Recalling that the process we are considering is band-limited, there must exist a Fourier series representation for the spectrum, assuming that the correlation function is not pathological in some sense. This Fourier series can be expressed in terms of coefficients ρ_n (whose nature is yet to be determined) and has the form

$$S(\nu) = \begin{cases} \Delta t \sum_{-\infty}^{\infty} \rho_n \exp(2\pi i n \nu \Delta t), & |\nu| \leq \nu_N \\ 0, & |\nu| > \nu_N \end{cases} \quad (3)$$

where we define the time interval Δt according to

$$\Delta t = (2\nu_N)^{-1} . \quad (4)$$

We observe that we can write an expression of the form (3) if and only if $\Delta t \lesssim (2\nu_N)^{-1}$. Since the coefficients ρ_n have yet to be determined, we introduce this latter expression into Eq. (2) for the correlation function and obtain the so-called Whittaker interpolation formula, that is,

$$\rho(t) = \sum_{-\infty}^{\infty} \rho_n \frac{\sin[2\pi\nu_N(n\Delta T - t)]}{2\pi\nu_N(n\Delta t - t)} . \quad (5)$$

Owing to the presence of the sinc function [the sin(x)/x term] in this expression, we observe that, when t is an integral multiple of the time interval Δt, we can make the identification

$$\rho_n \equiv \rho(n\Delta t) . \quad (6)$$

Therefore, the coefficients ρ_n are sampled values of the correlation function, a result that is known as the sampling theorem. As long as we sample at a rate $(\Delta t)^{-1}$ that is greater than the bandwidth, namely $2\nu_N$, and as long as our set of samples is complete, that is, we know ρ_n for all n, both the spectrum and the correlation function are completely known and can be evaluated by using Eqs. (3) and (5).

This, therefore, elucidates the fundamental obstacle encountered in real situations: sampled data are available for only a finite duration of time. Given ρ_n for $n = 0,1,\ldots,N$, we know by virtue of Eqs. (2) that $\rho_{-n} = \rho_n^*$, where the asterisk denotes complex conjugation and ρ_n is said to be Hermitian. Therefore, we know ρ_n for $n = -N,\ldots,0,\ldots,N$, and we are faced with the problem of somehow extrapolating ρ_n for $n \to \pm\infty$.

A minimal requirement that must be met, in developing a method for extrapolating the correlation function, is that the estimate of the correlation function must correspond to a nonnegative spectrum, whereupon we say that the correlation function is Toeplitz. If we were simply to assume that the correlation function vanishes beyond the range of measured values, this positivity requirement almost certainly would be violated. Alternatively, we could multiply the measured correlation values by a quantity that smoothly goes to zero as $n \to \pm N$. This process, which is referred to as tapering, corresponds to introducing a convolutional filter or "window function" in the frequency domain, for example, the Bartlett, Hamming, and Hanning filters [Blackman and Tukey, 1958]. Although this procedure does produce a positive spectrum, it suffers from the fact that we have prejudiced the information content resident in our data; the spectrum that results from this windowing will not reproduce the original data. We prefer to obtain a spectrum that does not bias the information at hand and that makes the fewest assumptions about the missing information. To be parsimonious with respect to estimating missing information, we wish to develop a procedure that minimizes the assumed information content residing in the extrapolated correlation function or, conversely, to maximize the entropy implicit in the process.

From a probabilistic standpoint, the entropy H of a process with probability distribution function P(x) is given by

$$H = -\int_{-\infty}^{\infty} P(x) \ln[P(x)]\, dx = -\langle \ln(P) \rangle . \tag{7}$$

For example, if P(x) described a Gaussian process with zero mean and variance σ^2, the entropy for the process would be $H = \ln[2\pi e \sigma^2]^{1/2}$. Suppose that the correlation function and the spectrum are the outcome of a stationary, Gaussian process. It is convenient to think of this process as the outcome of an ensemble of independent sinusoidal oscillators, each at a different frequency. The variance σ^2 of the process at a given frequency ν can be associated with the spectrum $S(\nu)$, and in loose terms we can think of $\ln[2\pi e S(\nu)]^{1/2}$ as a measure of the entropy associated with the sinusoidal oscillator at that frequency [Papoulis, 1981]. The entropy for the process can be shown to be [Bartlett, 1955]

$$H = (2\nu_N)^{-1} \int_{-\nu_N}^{\nu_N} \ln[2\pi e S(\nu)]\, d\nu$$

$$= (4\nu_N)^{-1} \int_{-\nu_N}^{\nu_N} \ln[S(\nu)]\, d\nu + \frac{1}{2} \ln[2\pi e] . \tag{8}$$

Therefore, we wish to develop a mechanism for maximizing the entropy consistent with <u>all</u> available information.

3. The MEM and Its Relation to Linear Prediction

The maximum entropy method (MEM), or maximum entropy spectral analysis (MESA), is a nonlinear technique for estimating power spectra with increased resolution. The method was developed independently by Burg [1967] and by Parzen [1969]. Edward and Fitelson [1973] formally proved Burg's conjecture that the MEM provides a spectrum that maximizes the entropy of a stationary random process consistent with the first N samples or "lags" of the correlation function. A detailed discussion of the method is found in the review by Ulrych and Bishop [1975]. Briefly, the problem posed by Burg, Parzen, and others is to maximize the entropy

$$H = (4\nu_N)^{-1} \int_{-\nu_N}^{\nu_N} \ln[S(\nu)]\, d\nu , \tag{9}$$

where we have dropped the additive constant term contained in the second of Eqs. (8), subject to the constraints imposed by the correlation function lag data

$$\rho_n = \int_{-\nu_N}^{\nu_N} S(\nu) \exp(-2\pi i n \nu \Delta t)\, d\nu , \qquad n = -N,\ldots,0,\ldots,N . \tag{10}$$

This procedure parallels that employed in statistical mechanics, as has been pointed out by Jaynes [1957a,b] and others. In that application, one maximizes the entropy of the distribution function describing a gas subject to mass conservation and to temperature constraints. The outcome of this latter process is that the kinetic equations of statistical mechanics reduce to the laws of equilibrium thermodynamics. This methodological link provides an appealing rationale for employing an entropy-maximizing procedure.

In extrapolating the correlation function, there are an infinity of estimates for the unknown coefficients that preserve the Toeplitz property, that is, correspond to a nonnegative power spectrum. Of these solutions, the MEM solution is "closest" to the greatest concentration of solutions by virtue of being an extremum in a variational problem and is, in some sense, the statistically most probable solution.

There are two formally equivalent realizations possible to the problem of rendering the entropy an extremum subject to the constraints posed by the data. We can either (1) vary the unmeasured correlation function lags to maximize the entropy, or (2) perform a variation on the power spectrum to make the entropy an extremum. We now consider these two approaches in turn.

3.1. Implicit Constraint Approach

Recall that we can express the power spectrum in the form

$$S(\nu) = \Delta t \sum_{-\infty}^{\infty} \rho_n \exp(2\pi i n \nu \Delta t) , \tag{11}$$

where the ρ_n, $n = -N,\ldots,0,\ldots,N$, are known. We vary the unmeasured ρ_n so that $\partial H/\partial \rho_n = 0$ for $|n| > N$. In this way, we are implicitly preserving the measured values of ρ_n. As a consequence, we obtain

$$\int_{-\nu_N}^{\nu_N} \frac{\exp(2\pi i n \nu \Delta t)}{S(\nu)} \, d\nu = 0 , \qquad |n| > N . \tag{12}$$

This implies that the power spectrum is the reciprocal of a truncated Fourier series. In other words, the Fourier series representation for $S^{-1}(\nu)$ has coefficients that vanish if the corresponding correlation function lag vanishes. Thus, we can write

$$S(\nu) = \left\{ \sum_{-\infty}^{\infty} b_n^* \exp(-2\pi i n \nu \Delta t) \right\}^{-1} . \tag{13}$$

It is possible to generalize our conclusion concerning the character of the b_n coefficients: b_n vanishes for <u>any</u> value of n where ρ_n is unmeasured. Thus, if certain intermediate (that is, $|n| \lesssim N$) correlation lags are unmeasured, the corresponding b_n coefficients are zero. A case in point occurs in applications to sonar arrays, where elements of the array are sometimes inoperative [see, for example, Newman, 1979]. However, we point out that this formulation of the MEM is incapable of treating situations where correlation data are contaminated by noise, a problem that we address in the next section.

3.2. Explicit Constraint Approach

In this approach, we employ a Lagrangian function L and we employ coefficients b_n as Lagrange multipliers. Thus, following Edward and Fitelson [1973], we define L by

$$L = -\sum_{-N}^{N} b_n^* \left\{ \int_{-\nu_N}^{\nu_N} S(\nu) \exp(-2\pi i n \nu \Delta t) \, d\nu - \rho_n \right\} / 4\nu_N . \tag{14}$$

In addition, we recall the definition of the entropy, H, apart from an additive constant, namely [Eq. (9)]:

$$H = (4\nu_N)^{-1} \int_{-\nu_N}^{\nu_N} \ln[S(\nu)]\, d\nu \,. \tag{15}$$

We perform a variation on the power spectrum $S(\nu)$ so that

$$\delta(H + L) = 0 \tag{16}$$

and, at the same time, require that each term in braces in Eq. (14) vanish. Therefore, we obtain, as before,

$$S(\nu) = \left\{ \sum_{-N}^{N} b_n^* \exp(-2\pi i n \nu \Delta t) \right\}^{-1} . \tag{17}$$

This procedure may be generalized readily by adopting a more appropriate constraint in the Lagrangian function. For noisy problems, we might consider that the weighted sum of residuals is constrained, that is,

$$\sum_{-N}^{N} W_N \left| \int_{-\nu_N}^{\nu_N} S(\nu) \exp(-2\pi i n \nu \Delta t)\, d\nu - \rho_n \right|^2 \leq \sigma^2 . \tag{18}$$

(In this approach, we also may consider cases where the noise between the different lags is correlated.) In this case, the constraint (18) can be adapted to a form appropriate to a Lagrangian function, namely,

$$L = \alpha \left\{ \sigma^2 - \sum_{-N}^{N} W_n \left| \int_{-\nu_N}^{\nu_N} S(\nu) \exp(-2\pi i n \nu \Delta t)\, d\nu - \rho_n \right|^2 \right\} . \tag{19}$$

Again, we perform a variation in the power spectrum $S(\nu)$ so that Eq. (16) holds, and we seek an extremum in the entropy, with L vanishing. Newman [1977] gives a more complete description of this problem. We shall show that this approach can accommodate unmeasured data as well.

3.3. Relation to the Autoregressive Method

The spectral estimate produced by the MEM is algebraically equivalent to that produced by the autoregressive method (ARM), an equivalence demonstrated by van den Bos [1971]. One way to show this is to employ Wald's method [Edward and Fitelson, 1973], recalling that

$$S(\nu) = \left\{ \sum_{-N}^{N} b_n^* \exp(-2\pi i n \nu \Delta t) \right\}^{-1} \geq 0 . \tag{20}$$

Since $S(\nu)$ is nonnegative and real, the coefficients b_n are Hermitian. By employing the fundamental theorem of algebra and the distribution of roots in ν to the latter, we can show that the spectrum may be factorized as

$$S(\nu) = \Delta t [A(\nu)\, A(\nu)^*]^{-1} , \tag{21}$$

where

$$A(\nu) = \sum_{n=0}^{N} \gamma_n^* \exp(2\pi i n \nu \Delta t) \tag{22}$$

is chosen to be minimum phase, that is, $A(\nu)^{-1}$ is analytic in the upper half ν-plane. A simple relation between the b_n coefficients and the γ_n coefficients can be obtained by employing Eqs. (20), (21), and (22).

Van den Bos [1971] and Edward and Fitelson [1973] employed contour integration techniques and the analytic properties of $A(\nu)$ to obtain a simple relationship between the γ_n coefficients and the measured correlation function lags ρ_n. They found that the γ_n coefficients satisfied the Yule-Walker or normal equations [Jenkins and Watts, 1968]

$$\begin{bmatrix} \rho_0 & \rho_1 & \rho_2 & \cdots & \rho_N \\ \rho_{-1} & \rho_0 & \rho_1 & \cdots & \rho_{N-1} \\ \rho_{-2} & \rho_{-1} & \rho_0 & \cdots & \cdot \\ \cdot & \cdot & \cdot & \cdots & \cdot \\ \rho_{-N} & \rho_{1-N} & \cdot & \cdots & \rho_0 \end{bmatrix} \begin{bmatrix} \gamma_0 \\ \gamma_1 \\ \gamma_2 \\ \cdot \\ \gamma_N \end{bmatrix} = \begin{bmatrix} 1/\gamma_0^* \\ 0 \\ 0 \\ \cdot \\ 0 \end{bmatrix} . \tag{23}$$

Van den Bos [1971] observed that the γ_n coefficients, when properly normalized, corresponded directly to the N+1 point noise prediction filter or prediction error filter obtained when autoregressive methods are used to estimate the power spectrum. This establishes the correspondence between the spectrum estimated by the MEM and by prediction or autoregressive methods.

4. Problems Posed by Noise and Missing Intermediate Data

In this section we extend the MEM in order to estimate the spectrum, given a set of correlation function samples that are contaminated by noise and that contain gaps in the information available. As a secondary objective, we estimate the unmeasured or missing lags, "correct" the measured lags for the effect of contaminating noise, and extrapolate the correlation function to infinity in a statistically unbiased manner. As a case in point where both of these complications occur, the ionospheric radar facility at Arecibo Observatory maintains a 1008-channel digital correlator. Contamination of the correlation function estimates by noise due to short observation times and by digitization is frequently significant (although rarely apparent during the course of observations). Not infrequently, one or more channels of the device fail, leading to missing intermediate data.

By using a combination of the two methods described earlier, it is possible to overcome both of these obstacles simultaneously [Newman, 1981]. In many ways the principal computational problem emerges from the routine "bookkeeping" that is required. To overcome this obstacle we develop a systematic notation. We consider correlation lags ρ_n for $n = -N,\ldots,0,\ldots,N$. Some of these are unmeasured; the rest are contaminated by noise. We assume that M positive lags are missing, and we designate these by the index q_k, for $k = \pm1,\pm2,\ldots,\pm M$. The remaining N-M lags are presumed to be known but contaminated by noise, and we designate these by the index p_ℓ, for $\ell = 0,\pm1,\pm2,\ldots,\pm(N-M)$. We assume that ρ_0 is known, albeit in the presence of noise, or at least has a known upper bound. Otherwise the maximum entropy method requires that it be infinite. If $\rho_{\pm N}$ is unmeasured, it is computationally preferable to reduce N by 1. For convenience, we assume that the indices p_ℓ and q_k are antisymmetric in their indices and are distinct, that is,

$$q_k = -q_{-k} \qquad p_\ell = -p_\ell \qquad q_k \neq p_\ell \qquad q_k \neq 0 \qquad q_k \neq \pm N \tag{24}$$

$$k = \pm1,\pm2,\ldots,\pm M \qquad \ell = 0,\pm1,\pm2,\ldots,\pm(N-M)$$

We now employ this notation as an expression of confidence in the measured but noisy lags. Extending the explicit constraint approach associated with Eq. (18), we assume that we have "weight factors" W_{p_ℓ} that provide a measure of confidence in each of the measured lags and that a weighted sum of residuals is bounded by some threshold. Hence, we write

$$\sum_{\ell=M-N}^{N-M} W_{p_\ell} \left| \int_{-\nu_N}^{\nu_N} S(\nu)\exp(-2\pi i p_\ell \nu \Delta t)\, d\nu - \rho_{p_\ell} \right|^2 \leq \sigma^2 . \tag{25}$$

This expression implies that the noise effects in different correlation function lags are uncorrelated and that the measured lags are confined within a hyperellipsoid centered around the true correlation function lags. It can be shown, in the limit that $\sigma \to 0$ and $M = 0$, that maximizing the entropy (9) subject to Eq. (25) reproduces the standard maximum entropy result for the spectrum.

We can be assured that a solution to the problem of maximizing the entropy subject to the latter constraint exists if the weights and confidence thresholds are properly chosen and if the instrumentation that provides the correlation function is working properly. In practical situations, this may not always be the case, so we wish to develop a computational procedure that is capable of testing for the existence of solutions and that must be able to accommodate data sequences that are non-Toeplitz owing to contaminating noise.

We can also show that the solution to this problem is unique since the entropy (9) is a concave functional over the convex space of power spectra defined by Eq. (25). However, we can show [Newman, 1977, 1981] that the entropy is maximized on the surface of the hyperellipsoid, that is, that equality in the constraint holds. This feature reduces the intuitive appeal of the method since its justification is no longer exclusively probabilistic. We consider an ad hoc resolution to this problem in Sec. 5.

We now wish to develop a variational approach to the problem of spectral information in the presence of missing information and noise. As before, we define a Lagrange multiplier α and a Lagrangian function

$$L = \alpha \left\{ \sigma^2 - \sum_{\ell=M-N}^{N-M} W_{p_\ell} \left| \int_{-\nu_N}^{\nu_N} S(\nu) \exp(-2\pi i p_\ell \nu \Delta t)\, d\nu - \rho_{p_\ell} \right|^2 \right\}, \tag{26}$$

with the entropy H defined as in Eq. (9). Then, we perform a variation on the power spectrum $S(\nu)$ so that

$$\delta(H + L) = 0 . \tag{27}$$

We then determine α so that equality in the constraint (25) is obtained, that is, $L = 0$. The result of the variation is that

$$S(\nu) = \left\{ \sum b_n^* \exp(-2\pi i n \nu \Delta t) \right\}^{-1} \tag{28}$$

where

$$\begin{aligned} b_{q_k} &= 0 , && k = \pm 1, \pm 2, \ldots, \pm M \\ b_{p_\ell} &\propto \int_{-\nu_N}^{\nu_N} S(\nu) \exp(-2\pi i p_\ell \nu \Delta t)\, d\nu - \rho_{p_\ell} , && \ell = 0, \pm 1, \ldots, \pm(N-M) \end{aligned} \tag{29}$$

The first of Eqs. (29) corresponds to the unmeasured lag problem described in the implicit constraint approach [Newman, 1979]; the second of Eqs. (29) corresponds to the noisy lag problem described in the explicit constraint approach [Newman, 1977]. The proportionality factor that emerges in the latter is an algebraic combination of σ and of the weight factors W_{p_ℓ}.

Wald's spectral factorization method [Edward and Fitelson, 1973] can be employed here as proposed by Newman [1981].

The formal solution to this problem can be expressed in the following way. For convenience, we set $W_{q_k} = 1$ for $k = \pm1, \pm2, \ldots, \pm M$. We recall that the b_n coefficients may be directly obtained from the γ_n coefficients according to Eqs. (20), (21), and (22). Thus, we must determine γ_n coefficients (and corresponding b_n coefficients) as well as obtain values for ρ_{q_k} so that the following nonlinear modification of the Yule-Walker equations is satisfied:

$$\sum_{n=0}^{N} \gamma_n \left\{ \rho_{n-k} + \frac{\sigma b_{n-k}}{W_{n-k}} \left[\sum_{m=0}^{N} \frac{b_m b_m^*}{W_m} \right]^{-1/2} \right\}$$

$$= [\gamma_0^*]^{-1} \delta_{k,0} , \qquad k = 0,1,\ldots,N . \tag{30}$$

We observe that Eq. (30) corresponds exactly to the Yule-Walker equations associated with a "derived" correlation function $\boldsymbol{\rho}_n$ defined by

$$\boldsymbol{\rho}_n = \rho_n + \frac{\sigma\, b_n}{W_n} \left[\sum_{m=-N}^{N} \frac{b_m\, b_m^*}{W_m} \right]^{-1/2}$$

$$= \int_{-\nu_N}^{\nu_N} S(\nu) \exp(-2\pi i n \nu \Delta t)\, d\nu, \qquad n = 0,1,\ldots,N . \tag{31}$$

In that case, the modified Yule-Walker equations (30) become

$$\sum_{n=0}^{N} \gamma_n \boldsymbol{\rho}_{n-k} = [\gamma_0^*]^{-1} \delta_{k,0} , \qquad k = 0,1,\ldots,N , \tag{32}$$

a form that exactly parallels the usual form, that is, Eq. (23). In some sense, we can regard the derived correlation function $\boldsymbol{\rho}_n$ as a form of maximum entropy method "fit" in the presence of contaminating noise [for $n = p_\ell$, $\ell = 0, \pm1, \pm2, \ldots, \pm(N-M)$], as a maximum entropy method interpolant for missing intermediate lags [for $n = q_k$, $k = \pm1, \pm2, \ldots, \pm M$], and as a maximum entropy method extrapolation for the unmeasured lags [$n > N$ and $n < -N$].

We wish to develop a practical computational procedure for solving this problem. To do this, we perform a set of iterations that increase the entropy consistent with the constraint at each step. Since the maximum entropy state or solution is unique, we can invoke the uniqueness of the solution and recognize that with each iteration we are "closer" to the desired result. Each iteration consists of two sets of steps:

> "Fill holes." We vary the ρ_{q_k}, for $k = \pm 1, \pm 2, \ldots, \pm M$, employing a multidimensional Newton method. Newman [1979] showed that this step was absolutely convergent and was asymptotically quadratic.
>
> "Correct for noise." We vary the γ_n's, employing an iterative relaxation or invariant imbedding procedure. Newman [1977] showed that this step generally converges although special care is sometimes needed in difficult problems.

In each iteration, these two steps are performed in succession. We stop the iterations once self-consistency is achieved. When the initial estimates of the correlation function are not Toeplitz, we employ an artificially elevated value of ρ_0 until convergence of the above iterative steps emerges. Then ρ_0 is relaxed and the steps are repeated until convergence is obtained. In especially difficult cases, σ is also varied in a relaxation scheme until convergence occurs. When variation of ρ_0 and σ in this manner does not permit their relaxation to their known values, we can be assured that no solution exists, owing to instrument failure or faulty weight or variance estimates.

Applications of this extended maximum entropy technique abound, particularly in problems associated with radar and sonar. This approach avoids line splitting or broadening and, because it is consistent with all available information, it provides an unbiased estimate of the solution. As an illustration, we consider the ionospheric backscatter studies performed at Arecibo Observatory, which sample data at rates too high and for times too long to permit the storage of a time series and the application of the Burg algorithm. Instead, researchers there have constructed a digital correlator that produces single-bit lag products in 1008 correlator registers or channels. The correlation function estimates that emerge are unbiased but are contaminated by the effect of single-bit digitization and often suffer from improperly functioning channels (usually providing no direct indication of which channels are providing misleading information). As a simple representation for the kind of data provided by this device, we construct a time series x_n and an approximate correlation function ρ_n defined by

$$\rho_n = K^{-1} \sum_{m=1}^{K} x_{m+n}\, x_m^* \,. \tag{33}$$

For illustration, we select the time series to correspond to two unit-amplitude sinusoids plus 0.5 amplitude rms white noise, and we take $N = 9$ and $K = 150$.

In Fig. 1, we consider cases free from noise where measurements for $M = 3$ intermediate lags are not available, as in Newman [1977]. [There, the correlation function was defined to be the expectation value of the quantity given in Eq. (33).] In Figs. 1(a) and (b), we take $q_1 = 3$, $q_2 = 6$,

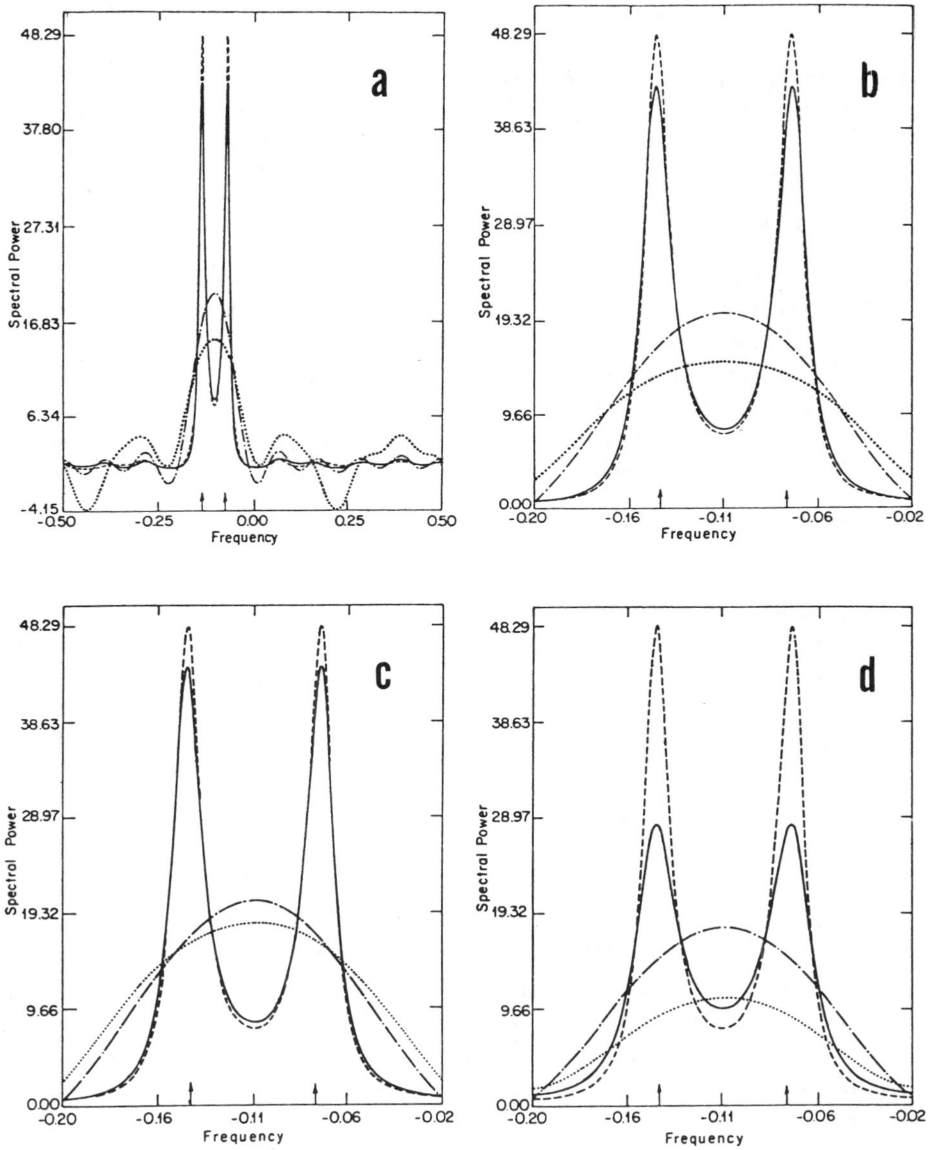

Figure 1. Comparison of spectral techniques for missing lags problem with no noise in lags; N = 9. (a) Full passband; $q_1 = 3$, $q_2 = 6$, $q_3 = 7$. (b) Limited passband; $q_1 = 3$, $q_2 = 6$, $q_3 = 7$. (c) Limited passband; $q_1 = 6$, $q_2 = 7$, $q_3 = 8$. (d) Limited passband; $q_1 = 2$, $q_2 = 3$, $q_3 = 4$.

MEM spectrum with MEM estimates in missing lags (——). Standard MEM spectrum with true values in missing lags (— — —). Truncated correlation-function-derived spectrum with zeroes in missing lags (• • •) and with MEM estimates in missing lags (— • — •). Arrowheads indicate true frequencies of spectral peaks.

and $q_3 = 7$; in 1(c) we take $q_1 = 6$, $q_2 = 7$, and $q_3 = 8$; and in 1(d) we take $q_1 = 2$, $q_2 = 3$, and $q_3 = 4$. [Figure 1(b) is an enlargement of the frequency region in Fig. 1(a) containing the spectral peaks. Succeeding figures include the full passband as well as an enlargement of the spectral peak region.] The choice of q_k's was motivated so that we can better appreciate the effect of displacement from the origin of the missing information on our ability to resolve spectral features. The computational procedure described above for "filling holes" was employed. The four curves represent—

(1) the MEM spectrum that accommodates the effects of missing data,

(2) the MEM spectrum that would result if all intermediate values of the correlation function were known,

(3) the truncated correlation-function-derived spectrum where the missing lag values were set to zero,

(4) the truncated correlation-function-derived spectrum where the missing lag values were estimated by maximizing the entropy.

Arrowheads indicate the true frequencies of spectral peaks. We observe that truncated correlation-function-derived spectra are characterized by slowly decaying oscillations with regions of negative power and poor resolution, properties common to unwindowed periodograms. Neither periodogram resolves the two peaks. The MEM-associated spectra, however, resolve the peaks and give accurate determinations of their frequencies. The effect of unmeasured data upon the maximum entropy estimates of the spectrum is to reduce the degree of resolution obtained. In addition, we observe that the more distant the unmeasured data are from the correlation function time origin, the less deleterious the reduction in resolving power. This gives rise to the ad hoc rule that information content in correlation functions is inversely related to distance from the time origin.

In Fig. 2 we consider cases where all intermediate correlation function values are known according to Eq. (33) but where contamination by noise has occurred, as in Newman [1979]. The four curves represent—

(1) the MEM spectrum that accommodates the effects of correlator noise and missing data,

(2) the MEM spectrum that overcorrects for the influence of correlator noise (that is, the estimate for σ is replaced by 4σ),

(3) the MEM spectrum that does not correct for noise (that is, $\sigma = 0$),

(4) the truncated correlation-function-derived spectrum, or unwindowed periodogram.

The computational procedure described above where we "correct for noise" was employed. As before, the truncated correlation-function-derived spectrum is characterized by slowly decaying oscillations with regions of negative power and poor resolution, properties common to unwindowed periodograms. Away from the spectral peaks, the three MEM-associated spectra are indistinguishable. In Fig. 2(b) we see that the periodogram fails to resolve the two peaks, while the MEM-associated spectra resolve the

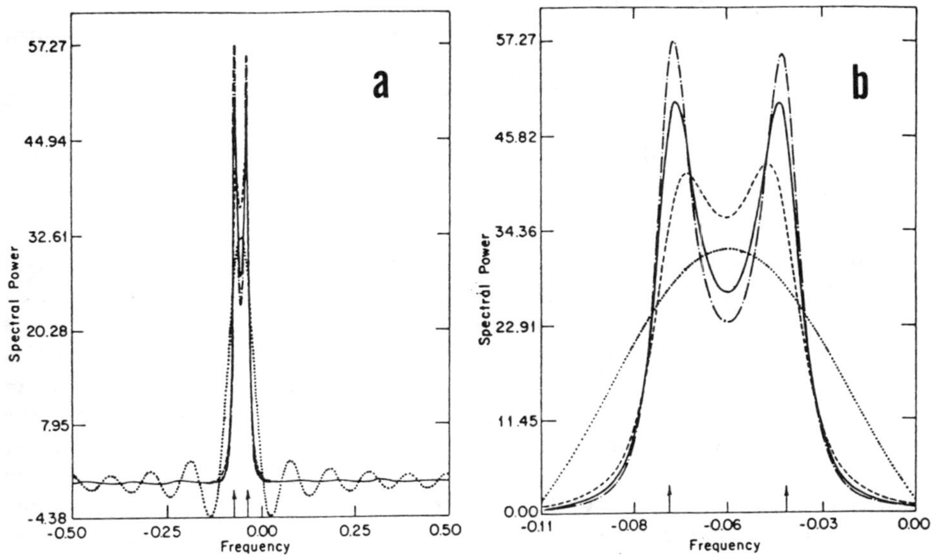

Figure 2. Comparison of spectral techniques for noisy lags problem with no unmeasured lags. (a) Full passband. (b) Limited passband.

Corrected MEM spectrum (——). Overcorrected MEM spectrum (— — —). Standard MEM spectrum (— • — •). Truncated correlation-function-derived spectrum (• • •). Arrowheads indicate true frequencies of spectral peaks.

features. There we observe that the peaks for the noise-corrected MEM spectra are not as tall as those of the standard MEM, but not significantly broader. The error in the noise-corrected MEM estimate of the peak frequencies is only 0.2% of the bandwidth. The overcorrected spectrum has very poor resolution. This gives rise to the ad hoc rule that, as the entropy (a monotonically nondecreasing function of σ) increases, the associated information content of the data decreases and the resolution of the spectrum is reduced. The relative amplitudes of the two peaks and the relative areas under them are not correctly estimated by the standard MEM. (We note that, if σ were larger, the correlation function estimates would not be Toeplitz and the standard MEM could not be used.) The noise-corrected spectrum rectifies this situation, while the overcorrected spectrum inverts the relative heights of the peaks obtained by the application of the standard MEM.

In Fig. 3 we consider the case of noisy measured correlation data as well as missing intermediate lags, namely $M = 3$, $q_1 = 3$, $q_2 = 6$, and $q_3 = 7$, as in Newman [1981]. The computational procedure described above where we iteratively "fill holes" and "correct for noise" was employed, including the variation of trial values for ρ_0. The four curves represent—

(1) the MEM spectrum that accommodates the effects of correlator noise and missing data,
(2) the MEM spectrum that overcorrects for the influence of correlator noise (that is, the estimate of σ was replaced by 8σ),
(3) the truncated correlation-function-derived spectrum, where the unmeasured lag values were set to zero,
(4) the truncated correlation-function-derived spectrum, where the missing lag values were estimated by the extended MEM described above.

Because the estimated correlation function lags were not Toeplitz, a conventional MEM spectrum cannot be determined. In Fig. 3(a) the truncated correlation-function-derived spectra are characterized by slowly decaying oscillations with regions of negative power and poor resolution, properties we had identified earlier. Away from the spectral peaks, the two MEM-associated spectra are indistinguishable. In 3(b), neither periodogram resolves the two peaks, but the MEM-associated spectra resolve the features. As in our second case [Newman, 1977], the noise-corrected maximum entropy spectral estimate with the correct value of σ determined the peak frequencies with an error of only 0.2% of the bandwidth and accurately reproduced the relative heights of the other two features. On the other hand, the overcorrected MEM spectrum displayed reduced resolution, displaced the

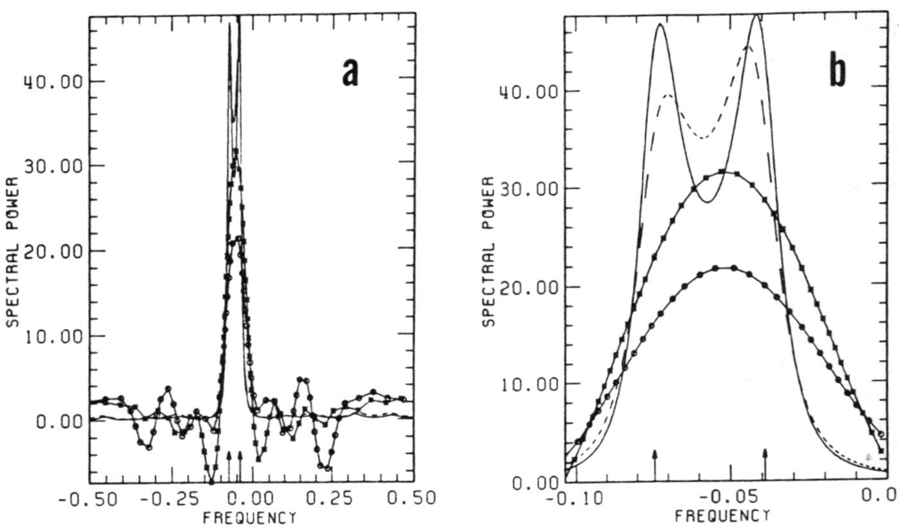

Figure 3. Comparison of spectral techniques for noisy lags with some unmeasured lags. (a) Full passband. (b) Limited passband.

Corrected MEM spectrum (——). Overcorrected MEM spectrum (— — —). Truncated correlation-function-derived spectrum with zeroes in missing lags (—o—o—) and with MEM estimates in missing lags (—*—*—).

peak frequencies (they would ultimately merge as σ was increased), and introduced a bias into the relative amplitude of the peaks. Added insight into the influence of contaminating noise and missing data may be inferred from our previous case studies, where we considered these effects individually.

Finally, in comparing the above extensions of the MEM with other techniques of spectral analysis, it is useful to think of the MEM estimate of $S(\nu)$ as the Fourier transform of the derived correlation function. The derived correlation function provides no information that is not already included in the given correlation function estimate. Indeed, the derived correlation function fits the imprecisely measured lags, provides an interpolatory estimate of the missing data, provides a stable extrapolation for the correlation function in the time domain, and provides a power spectral density that neither loses nor assumes any added information. However, in accomplishing these ends we have inadvertently introduced a firm constraint into the problem: that the derived correlation function lies on a hyperellipsoid instead of being characterized by some distribution function. We discuss this problem next.

5. Ad Hoc Bayesian Approach to Unconstrained Optimization

Recall that the MEM yields a spectral estimate that is "closest" in some sense to the greatest concentration of solutions and that this estimate is the "statistically most probable solution" for the power spectrum. From the asymptotic equipartition theorem, we have that e^{-H} is the probability of the process x, where $H = H(x)$. A nonrigorous way of seeing how this result emerges is to recall [Eq. (7)] that $H \equiv -\langle \ln P \rangle$. In an ad hoc way, we can regard the probability of obtaining the derived correlation function $\boldsymbol{\rho}_n$, for $n = 0,1,2,\ldots,N$, as being approximately $\exp[-H(\boldsymbol{\rho})]$. We call this probability $f(\boldsymbol{\rho})$ and regard the $\boldsymbol{\rho}_n$ as the "true" correlation function.

At the same time, we have measurements of the correlation function, namely ρ_n, for $n = 0,1,2,\ldots,N$. We wish to obtain the Bayesian probability that we have measured the ρ_n, given the "true" correlation function $\boldsymbol{\rho}_n$, and we denote this probability as $f_B(\rho|\boldsymbol{\rho})$. The desired result is given by the Wishart distribution, which we can approximate by

$$f_B(\rho|\boldsymbol{\rho}) \propto \exp\left\{-\sigma^{-2} \sum_{\ell=N-M}^{M-N} W_{p_\ell} \left| \int_{-\nu_N}^{\nu_N} S(\nu) \exp(-2\pi i p_\ell \nu \Delta t)\, d\nu - \rho_{p_\ell} \right|^2 \right\}, \tag{34}$$

where we identify the integral with the derived correlation function $\boldsymbol{\rho}_{p_\ell}$. Recalling Eq. (15) for the entropy H, we can express the probability of the derived correlation function as

$$f(\boldsymbol{\rho}) = \exp[-H(\boldsymbol{\rho})] = \exp\left[-(4\nu_N)^{-1}\int_{-\nu_N}^{\nu_N} \ln S(\nu)\, d\nu\right] . \tag{35}$$

Using conventional Bayesian arguments, the joint probability distribution $f(\rho,\boldsymbol{\rho})$ must be given by

$$f(\rho,\boldsymbol{\rho}) = f_B(\rho|\boldsymbol{\rho})\, f(\boldsymbol{\rho}) . \tag{36}$$

We now define $f(\rho)$ as the probability that a correlation function ρ_n, for n = 0,1,...,N, is measured. Although we have no direct way of calculating this probability, knowledge of the precise nature of $f(\rho)$ is not essential to our purpose. Further application of Bayesian arguments yields, therefore,

$$f_B(\boldsymbol{\rho}|\rho) = \frac{f(\rho,\boldsymbol{\rho})}{f(\rho)} = f_B(\rho|\boldsymbol{\rho})\, \frac{f(\boldsymbol{\rho})}{f(\rho)} . \tag{37}$$

We are now in a position to employ a maximum likelihood approach. We wish to find the derived correlation function $\boldsymbol{\rho}$ so that $f_B(\boldsymbol{\rho}|\rho)$ is a maximum. Although we do not know $f(\rho)$, we may regard the measured correlation function as given, and seek values of the derived correlation function that renders the right-hand side of Eq. (37) an extremum. This occurs when $f_B(\rho|\boldsymbol{\rho})f(\boldsymbol{\rho})$ is a maximum with respect to $\boldsymbol{\rho}$.

To obtain the maximum likelihood solution to this problem, we must maximize the quantity

$$(4\nu_N)^{-1}\int_{-\nu_N}^{\nu_N} \ln S(\nu)\, d\nu$$

$$+\ \sigma^{-2} \sum_{\ell=N-M}^{M-N} W_{p_\ell} \left| \int_{-\nu_N}^{\nu_N} S(\nu)\exp(-2\pi i p_\ell \nu \Delta t)\, d\nu - \rho_{p_\ell} \right|^2 . \tag{38}$$

After performing the necessary algebraic manipulations, we observe that the outcome of this variation is formally equivalent to the constrained optimization problem if the Lagrange multiplier α is set to 1. This shows that the constrained variational method is a useful approximation to this ad hoc maximum likelihood estimate of the power spectrum.

The methods described in the preceding sections for extending the MEM to accommodate unmeasured data and contaminating noise can also be adapted to the method of cross entropy maximization [Johnson and Shore, 1984]. In addition, the ad hoc approach developed here can also be adapted to the cross entropy problem to show that the constrained cross entropy calculation is a useful approximation to the maximum likelihood estimate of the power spectrum.

6. Time Series Data with Gaps

We now develop methods for adapting a maximum-entropy-like principle to the spectral analysis of time series with gaps in the data. In many instances, researchers compute an approximate spectrum for each time series and then compute an appropriately weighted mean. We explore the idea that, by knowing the length of the gaps between time series segments, we have additional information that can be incorporated into the calculation of the spectrum. We consider an ad hoc approach that has been employed in the past, and provide a rigorous foundation for this method. In addition, we develop some computational devices to make this approach more robust and more generally applicable.

We begin by recalling the purpose of the MEM for providing a smooth extrapolation of data into a domain where no measurements are available. At the same time, we recall the formal equivalence of the MEM and the autoregressive method. We wish to exploit the methodological link between these two methods of spectral analysis in application to time series with gaps. To do this, we first review some of the basic features of the autoregressive method for determining prediction filters for contiguous time series.

In the manner suggested by Wiener, we consider a signal x_i at time $t_0+i\Delta t$ to consist of a message s_i and noise n_i, that is,

$$x_i = s_i + n_i , \qquad i = 1,\ldots,N. \tag{39}$$

Further, we assume that the noise at time interval i can be calculated from a linear combination of earlier signals. In this way, we define a forward (causal) filter with M points $\Gamma_j^{(M)}$, for $j = 0,1,\ldots,M-1$, which we refer to as an M-point noise prediction filter or prediction error filter. From Eq. (39), it is appropriate to normalize $\Gamma_0^{(M)} = 1$. We assume, therefore, that our estimate for the noise obtained using the forward filter, namely $n_i^{(f)}$, can be expressed as

$$n_i^{(f)} = \sum_{j=0}^{M-1} x_{i-j}\, \Gamma_j^{(M)*} , \qquad i = M,\ldots,N. \tag{40}$$

Similarly, Burg [1975] showed that we can estimate the noise by using the same filter in the backward direction. In analogy to Eq. (40), we write

$$n_i^{(b)} = \sum_{j=0}^{M-1} x_{i+j}\, \Gamma_j^{(M)} , \qquad i = M,\ldots,N+1-M. \tag{41}$$

In Wiener's original formulation of the prediction theory, the optimal choice of the Γ coefficients was that which minimized the estimate of the power resident P in the noise, that is, minimized the quantity

$$P = \langle n_i n_i^* \rangle \tag{42}$$

where we assume that the noise is uncorrelated; that is,

$$\langle n_i n_j^* \rangle = 0 , \qquad \text{for } i \neq j. \tag{43}$$

In a data set with finite numbers of samples, Burg [1968] proposed that the expectation values employed by Wiener be replaced by arithmetic averages. Moreover, as we have a choice of whether to employ $n_i^{(b)}$ or $n_i^{(f)}$, Burg proposed that we use both, to minimize the effect of running off the "end" of our data. As a practical measure following Burg, we replace Eq. (42) with

$$P_M = \frac{1}{2} \frac{1}{N+1-M} \left[\sum_{i=M}^{N} n_i^{(f)} n_i^{(f)*} + \sum_{i=1}^{N+1-M} n_i^{(b)} n_i^{(b)*} \right] . \tag{44}$$

This measure of the noise power is positive-definite and has no built-in bias toward the beginning or end of the data set. Minimization of this latter estimate of the noise power proceeds using the Burg-Levinson recursion scheme [Burg, 1968]. This scheme produces a filter that is minimum phase (that is, whose z-transform has all of its zeroes confined to within the unit circle) and is equivalent to a correlation function estimate that is Toeplitz. The Burg-Levinson scheme, however, does not necessarily produce a filter of minimum phase that at the same time minimizes the noise power in Eq. (44). Although computationally expedient, the Burg-Levinson procedure sometimes results in spectral features or lines that are artificially split. Fougere et al. [1976] developed a scheme that provides a minimum phase filter that minimizes the noise power P_M and overcomes the problem of line splitting, albeit at the cost of significantly more exhaustive computations. Our first objective here is to adapt Burg's approach, as epitomized in Eq. (44), to the problem of gaps.

The simplest approach to the problem can be met by modifying the range of summation for P_M to include only "meaningful" estimates of $n_i^{(f)}$ and $n_i^{(b)}$, that is, including in P_M estimates of the noise power coming from all contiguous segments of data. This approach was first suggested by Burg [1975]. Although a useful first estimate for the filter coefficients can be obtained this way, this approach ignores the information present in the continuity of the data and our knowing the lengths of the gaps. To see how information can be resident in the gaps, consider the paradigm of measuring the local gravitational acceleration by obtaining an accurate representation for the period of a pendulum that oscillates in a vacuum. Suppose we note the time that the pendulum passes over a well defined point (with an absolute error of ±0.1 sec) along its arc and we initiate a clock at 0 sec. We observe the first five swings to be at 1.1, 2.1, 3.0, 4.0, and 4.9 sec, respectively, and observe later swings at 7.1, 8.0, 10.1, 12.1, 15.0, 20.0, 25.1, 29.9, 38.1, 50.1, 59.9, 70.1, 85.1, 99.9 sec. From the first five swings, we conclude that the period of the swing is 0.98 ± 0.08 sec. Although we have obviously missed the sixth swing, we can readily associate the next three recorded measurements with the seventh, eighth, and tenth swings. Although we are now missing swings more and more frequently, our absolute error of measurements translates into a lower and lower relative error in the frequency of the pendulum. With the last recorded swing of the pendulum, we know we have witnessed the execution of its hundredth oscillation and that the period of the pendulum is 0.999 ± 0.001 sec. In time series analysis, we are often concerned with the problem of identifying the frequencies that characterize a given phenomenon. The example above shows that "gaps" in the data can be employed to help reduce the uncertainty in these frequencies when one extrapolates across the gaps. This opens up the possibility of obtaining spectral resolution characteristic of the overall length of the time series and also estimate the missing or unmeasured datum.

There is, however, a limitation to this idea. Suppose in the above "experiment" that the measurements at 7.1 and 8.0 sec were not available. Then we could not conclude with absolute certainty that the measurement at 10.1 sec corresponded to the tenth swing. Indeed, without the measurements at 7.1 and 8.0 sec, we could equally well conclude that either 9 or 11 swings had taken place. This uncertainty would then propagate through the rest of the data set, and we would obtain several distinct estimates for the frequency of the pendulum. From the above example, we see that spectral peaks must be separated by approximately $(N_{max}\Delta t)^{-1}$, where N_{max} is the length of the longest contiguous set of data, if they are to be resolved. Assuming that all pairs of spectral features can be resolved, the smallest uncertainty with which we can identify the frequencies of the spectral peaks is approximately $(N\Delta t)^{-1}$, where $N\Delta t$ is the time elapsed from the first to last sample (including intervening gaps). Moreover, we see that, by using an appropriate sampling strategy, we can sample data less and less frequently. In particular, we can introduce gaps whose durations are proportional to the time elapsed since the first datum was measured. Hence, it follows that the effective sampling rate, as time passes, can decrease as quickly as $1/N_{elapsed}$ (where $N_{elapsed}$ is the number of time intervals Δt that have passed since the first datum was recorded). Therefore, we need only obtain the order of $\ln(N)$ data to obtain frequency estimates with accuracy of order approximately $(N\Delta t)^{-1}$. Thus, this particular sampling strategy is an example of time decimation.

These comments on indeterminacy are also appropriate to situations where the gaps are not organized in some fashion. The variational problem for the prediction error filter without gaps implicit to Eqs. (42) and (44) has a unique solution. When gaps are present in practical problems, the order of the prediction filter should not be taken to be longer than N_{max} since the longest contiguous set of data determines the maximum number of sinusoids present in the time series that one can reasonably expect to find. The potential for uncertainty that exists if the gaps are too long, or, equivalently, if the longest contiguous set of measurements is too short, will manifest in nonuniqueness in the extension to Burg's criterion that we now present.

As a natural extension to Burg's scheme (44), we write

$$P'_M = \frac{1}{2}\,\frac{1}{N+1-M}\left\{\sum_{i=M}^{N}\left|\sum_{j=0}^{M-1} x_{i-j}\,\Gamma_j(M)^*\right|^2 + \sum_{i=1}^{N+1-M}\left|\sum_{j=0}^{M-1} x_{i+j}\Gamma_j\right|^2\right\}\,, \tag{45}$$

where $\Gamma_0(M) \equiv 1$ and where we vary the $\Gamma_j(M)$, for $j = 1,\ldots,M-1$, and the unknown x_i in order to minimize P'_M. As pointed out earlier, the solution is not necessarily unique since P'_M will be a concave functional only if sufficient quantities of data have been sampled. As a computational approach to solving this variational problem, we propose the following procedure, which reduces P'_M at each stage:

(1) "Fill the holes" by employing the conjugate gradient algorithm [Hestenes, 1980] to vary the unknown x_i. This is an absolutely convergent procedure and will perform in a manner that is independent of the character of the gap spacing.

(2) Calculate the revised filter coefficients, regarding the "fitted data" (estimates of unmeasured x_i in the above sense) as if those data had been measured. Here, employ the Burg-Levinson algorithm or, if line-splitting appears to be problematic, employ the Fougere et al. [1976] algorithm.

Repeat steps (1) and (2) until the missing data and the filter coefficients converge to some values. If P'_M has a unique minimum, this result will correspond to that minimum. Otherwise, this method will provide one of the minima, albeit not necessarily the global minimum. Upon reflection, we can show that filling the gaps in this variational scheme ties the various data segments together and can improve spectral resolution significantly (particularly if P'_M has a unique minimum) since the poles of the filter will migrate toward the unit circle.

The concept of employing the prediction error filter to estimate the unmeasured time series data was first advanced by Wiggins and Miller [1972]. However, they made no use of the information resident in the gaps to improve upon their filter estimate. Fahlman and Ulrych [1982], in an astronomical application to determining the periodicities of the variable star δ Delphini, recognized that they could improve upon their filter estimate by using the information present in the gap and then use the improved filter estimate to better estimate the missing datum. Their approach was an iterative one, in contrast to a variational principle, although they recognized its fundamental link to the maximum entropy principle. In practical terms, their approach is essentially equivalent to the one proposed here, but a critical operational difference emerges from the algorithm that they developed. In particular, the method they employed to "fill the gaps" required that the order of the filter be shorter than the shortest segment of contiguous data. Since large filter orders are needed to determine the character of complex spectra, they would necessarily ignore short data segments so that their numerical algorithm would work. The algorithm that we propose here has no intrinsic limitation on the length of the filter, and all data are employed. In practical terms, we limit the order of the filter to the length of the longest segment of contiguous data. Thus, this natural extension of Burg's method renders the maximum entropy method applicable to a much richer variety of problems. Moreover, if we know that the periods are sufficiently well separated, that is, that the distance between respective spectral peaks exceeds $(N_{max}\Delta t)^{-1}$, then we can obtain the frequencies of the spectral peaks with an uncertainty of $(N\Delta t)^{-1}$.

To illustrate this enhanced technique, we considered a variety of examples containing one or two unit-amplitude complex sinusoids together with $\sigma = 0.1$ amplitude Gaussian noise in order to test this method's sensitivity to the separation of spectral peaks and to the presence of harmonic structure. A sampling interval of $\Delta t = 1$ was assumed, the length of the data set was $N = 64$, and time series elements z_i for $i = 31,\ldots,50$ were assumed unmeasured although we had that information available for comparison purposes. The order of the assumed filter was taken to be $M = 5$.

In our first example, we have only one sinusoid present with a spectral peak at -5/64. We plot the real component of the time series in Fig. 4(a), with lines connecting each data point, and with a horizontal black line drawn along the portion of the x-axis where data are taken to be unavailable. The plotted time series elements that correspond to this range are the converged interpolants calculated by the above method. (In all cases illustrated below, time series interpolants agree with the original data with an accuracy corresponding to the noise level σ.) In Fig. 4(b), we illustrate the effect of harmonic structure by incorporating two sinusoids, at frequencies of -5/64 and -20/64, respectively. Although the interpolant is in excellent agreement with the original data and the spectral estimate of the frequencies of the spectral peaks is very accurate, the relative amplitude of the spectral peaks is approximately 3:1, which is not in accord with two unit-amplitude sinusoids. On the other hand, when the autoregressive method was applied to the original data set, including the data in the gap, the relative amplitude of the spectral peaks was approximately 5:2. The failure of the MEM or of the ARM to determine the correct amplitude of spectral peaks is a feature first noted by Lacoss [1971] although he observed that the <u>area</u> under each spectral peak was an accurate representation of the power associated with that spectral feature.

To test the resolving power of this method, we considered cases that we would expect to be immediately above and below the resolution limit, respectively. In the first case, the spectral peaks were at frequencies of

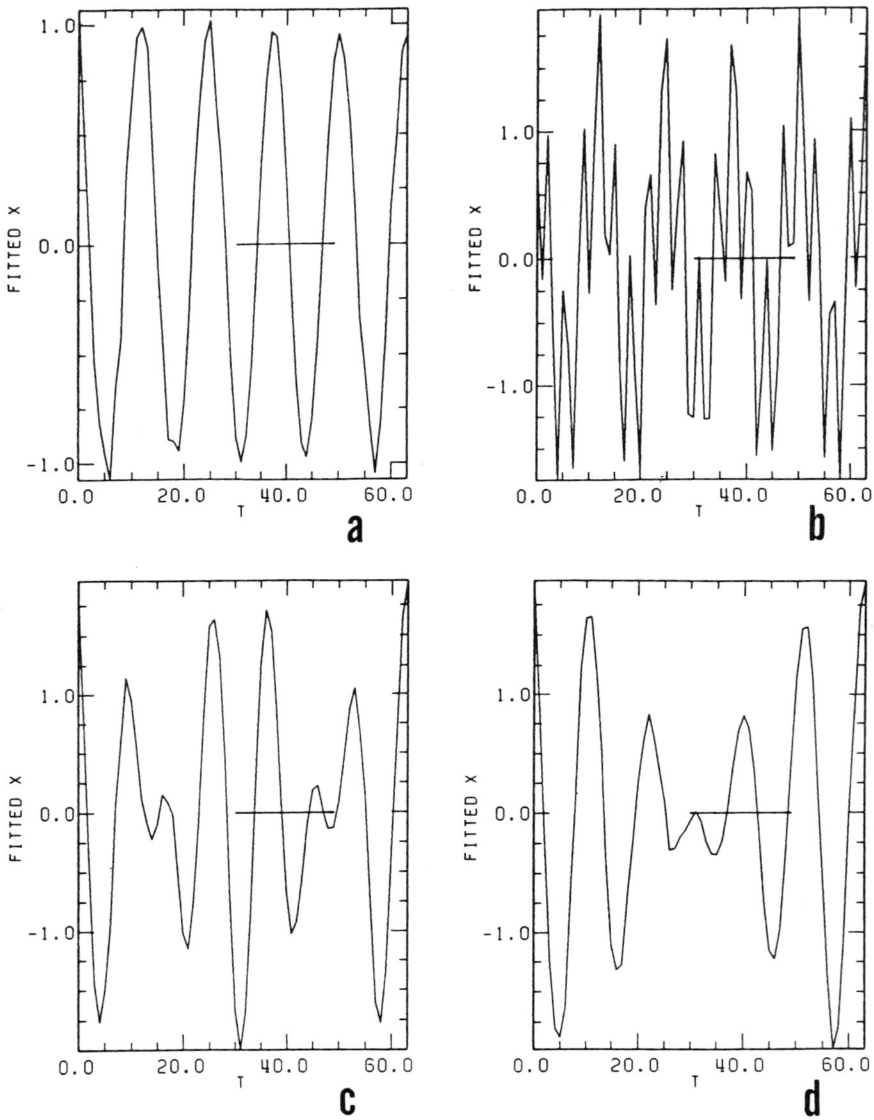

Figure 4. MEM interpolant for (a) time series with one spectral peak, and (b) time series with two spectral peaks where second peak is at fourth harmonic of first peak, (c) where second peak is barely resolved from first peak, and (d) where second peak is not resolved from first peak. In each case, the solid horizontal line indicates position of missing data.

-5/64 and -6/64, and Fig. 4(c) illustrates the excellent interpolant obtained. Figure 5(a) is the spectrum obtained for this case. The relative amplitude of the spectral peaks is in rough agreement with our expectations, and in excellent agreement with the original, ungapped data. Curiously, the spectral peaks associated with the interpolant are better resolved than the spectral peaks associated with the original data. Indeed, we must expect this (sometimes) desirable bias since the variational principle for P_M' seeks to minimize the prediction error and therefore can produce an interpolated estimate for the gapped data that is "smoother" than the true data, resulting in better separation in the associated spectral peaks. In Figs. 4(d) and 5(b) we consider a case immediately below the resolution threshold where the spectral peaks are at -5/64 and -6/64. The interpolated time series, Fig. 4(d), is in excellent agreement with the original data, as is the spectrum, Fig. 5(b). However, this spectral estimate is incapable of resolving the two features and, instead, reproduces a relatively broad feature whose width is an approximate measure of the spectral peak separation.

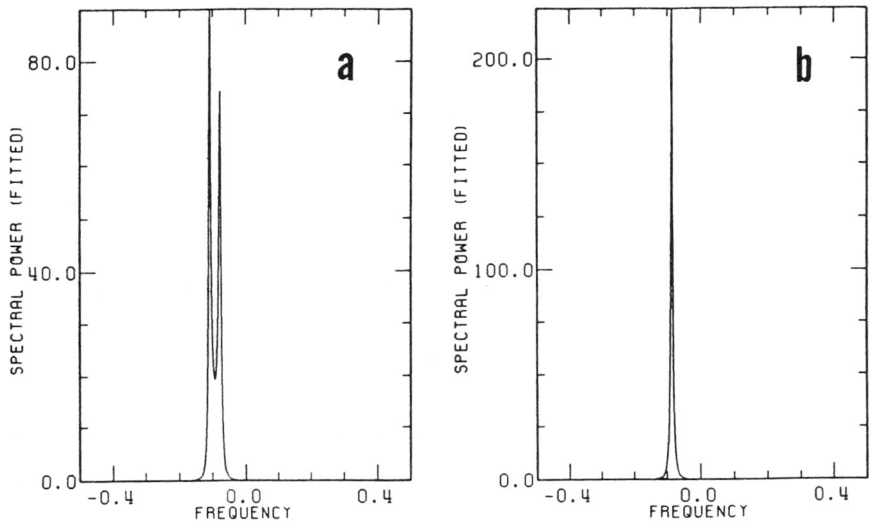

Figure 5. (a) Spectral estimate corresponding to Fig. 4(c); both features are resolvable. (b) Spectral estimate corresponding to Fig. 4(d); both features are not resolvable.

7. Acknowledgments

I wish to thank L. Knopoff, D. D. Jackson, T. J. Ulrych, and T. Fine for a number of stimulating discussions.

8. References

Bartlett, M. S. (1955), An Introduction to Stochastic Processes with Special Reference to Methods and Applications, Cambridge University Press, Cambridge, 214 pp.

Blackman, R. B., and J. W. Tukey (1958), The Measurement of Power Spectra, Dover, New York, pp. 98, 171.

Burg, J. P. (1967), "Maximum entropy spectral analysis," presented at the 37th Annual Meeting of the Society of Exploration Geophysicists, Oklahoma City.

Burg, J. P. (1968), "A new analysis technique for time series data," presented at the NATO Advanced Study Institute on Signal Processing with Emphasis on Underwater Acoustics, Enschede, Netherlands, Aug. 12-23, 1968.

Burg, J. P. (1975), "Maximum Entropy Spectral Analysis," Ph.D. Thesis in Geophysics, Stanford University.

Edward, J. A., and M. M. Fitelson (1973), "Notes on maximum-entropy processing," IEEE Trans. Inf. Theory IT-19, pp. 232-234.

Fahlman, G. G., and T. J. Ulrych (1982), "A new method for estimating the power spectrum of gapped data," Mon. Not. R. Astron. Soc. 199, pp. 53-65.

Fougere, P. F., E. J. Zawalick, and H. R. Radoski (1976), "Spontaneous line splitting in maximum entropy power spectrum analysis," Phys. Earth and Planet. Inter. 12, pp. 201-207.

Hestenes, M. R. (1980), Conjugate Direction Methods in Optimization, Springer-Verlag, New York, pp. 295-298.

Jaynes, E. T. (1957a), "Information theory and statistical mechanics," Phys. Rev. 106, pp. 620-630.

Jaynes, E. T. (1957b), "Information theory and statistical mechanics. II," Phys. Rev. 108, pp. 171-190.

Jenkins, G. M., and D. G. Watts (1968), Spectral Analysis and Its Applications, Holden-Day, San Francisco, pp. 105-107.

Johnson, R. W., and J. E. Shore (1984), "Power spectrum estimation by means of relative entropy minimization with uncertain constraints," in Proceedings of ICASSP 1984, IEEE International Conference on Acoustics, Speech, and Signal Processing, San Diego, Calif.

Lacoss, R. T. (1971), "Data adaptive spectral analysis methods," Geophysics 36, pp. 661-675.

Newman, W. I. (1977), "Extension to the maximum entropy method," IEEE Trans. Inf. Theory IT-23, pp. 89-93.

Newman, W. I. (1979), "Extension to the maximum entropy method II," IEEE Trans. Inf. Theory IT-25, pp. 705-708.

Newman, W. I. (1981), "Extension to the maximum entropy method III," in Proceedings of the First ASSP Workshop on Spectral Estimation, McMaster University, Aug. 17-18, 1981. (IEEE Acoustics, Speech and Signal Processing Society), pp. 1.7.1-1.7.6.

Papoulis, A. (1981), "Maximum entropy and spectral estimation: a review," IEEE Trans. Acoustics, Speech and Signal Processing ASSP-29, pp. 1176-1186.

Parzen, E. (1969), "Multiple time series modelling," in Multivariate Analysis II, P. R. Krishnaiah, ed., Academic, New York, pp. 389-410.

Ulrych, T. J., and T. N. Bishop (1975), "Maximum entropy spectral analysis and autoregressive decomposition," Rev. Geophys. and Space Phys. 13, pp. 183-200.

van den Bos, A. (1971), "Alternative interpretation of maximum entropy spectral analysis," IEEE Trans. Inf. Theory IT-17, pp. 493-494.

Wiggins, R. A., and S. P. Miller (1972), "New noise-reduction technique applied to long-period oscillations from the Alaskan earthquake," Bull. Seismolog. Soc. Am. 62, pp. 471-479.

ON THE ACCURACY OF SPECTRUM ANALYSIS OF RED NOISE PROCESSES USING MAXIMUM ENTROPY AND PERIODOGRAM METHODS: SIMULATION STUDIES AND APPLICATION TO GEOPHYSICAL DATA*

Paul F. Fougere

Air Force Geophysics Laboratory, Hanscom Air Force Base, MA 01731

Power spectra, estimated by the maximum entropy method and by a fast Fourier transform based periodogram method, are compared using simulated time series. The times series are computer generated by passing Gaussian white noise through low-pass filters with precisely defined magnitude response curves such that the output time series have power law spectra in a limited frequency range: $P(f) = Af^{-p}$, $f_1 \lesssim f \lesssim f_2$. Ten different values of p between 0.5 and 5.0 are used. Using 4000 independent realizations of these simulated time series, it is shown that maximum entropy results are superior (usually greatly superior) to the periodogram results even when end-matching or windowing or both are used before the power spectra are estimated. Without the use of end-matching or windowing or both, the periodogram results are useless at best and very misleading at worst. For an application to geophysical data, a 5-min section of ionospheric scintillation data from the MARISAT satellite was chosen because it illustrates a transition from low-level background noise to moderate scintillation and another transition to fully saturated scintillation. This section was broken into 60 sections, each 10 s long and overlapped by 5 s. Order 5 Burg-MEM spectra from the raw data are compared with periodograms computed from end-matched and windowed data. The superiority of Burg-MEM rests largely in the smoothness of the spectrum: real changes in spectral shape are not obscured by meaningless detail.

*Originally published in Journal of Geophysical Research 90, pp. 4355-4366 (May 1, 1985).

C. R. Smith and G. J. Erickson (eds.),
Maximum-Entropy and Bayesian Spectral Analysis and Estimation Problems, 127–148.

1. Introduction

A power law process is one whose power spectrum can be written as $P(f) = Af^{-p}$. In this paper, only real signals are discussed. Since a real signal has a power spectrum that is symmetric about zero frequency, only positive frequencies are used, to simplify the discussion. If such a spectrum is plotted as log P(f) versus log f, then the slope of the resulting straight line is -p, where p is the so-called spectral index. Many geophysical processes can be approximated as power law processes, including scintillation of radio waves passing through the ionosphere [Yeh and Liu, 1982], interplanetary scintillation used to measure solar wind properties [Armstrong and Coles, 1972], Landau damping of Alfvén waves [McKenzie, 1982], electron density fluctuations in the solar wind [Woo and Armstrong, 1979], turbulence in the upper troposphere and lower stratosphere [Larsen et al., 1982], and many others. In the case of ionospheric scintillation the spectral index then yields important information on the scale sizes of the ionospheric irregularities that produce the scintillation [Crane, 1976].

The power law approximation is merely convenient; that is, it is the simplest of the red noise type processes: it is linear on a log-log plot. A process is red if it exhibits appreciably more power at low frequencies than at high frequencies. Redness is a qualitative description rather than an exact specification. Nevertheless, it is quite useful to compare power spectra, obtained from different spectral analysis techniques, when applied to simulated processes with power spectra exactly known. We can then decide, for this general kind of process, that one technique yields results that are consistently superior to results of another technique for this class of spectra. It would then be a small intuitive leap to conclude that the superior technique is very likely to continue to be superior when applied to real, geophysical data if the spectra fall into the same general category, in this case, red noise: power-law-like. It is not necessary that the spectra of the real geophysical processes be accurate power law spectra but only that they be red and occasionally exhibit close to power law behavior. In particular, this recipe might exclude highly peaked spectra arising from nearly periodic processes and would certainly exclude processes that are blue, that is, those having more power at high frequencies than at low frequencies. But surely the underlying physical processes may exhibit broad small spectral peaks, which may even be caused by artifacts such as the ubiquitous 60-Hz (or 50-Hz) interference.

The motivation for examining in great detail processes with exactly known power law spectral shapes is that at least occasionally such processes really do occur in nature. A more immediate motivation for the present research is the ongoing analysis of ionospheric scintillation data as observed on the Wideband satellite and currently on the HILAT satellite. The kind of sampled periodogram referred to in this paper is the norm in the current processing of the HILAT data. The HILAT science team has decided that to average or smooth spectra to yield "consistent" power spectral estimates would cost too much in reduced time and frequency resolution, and thus such smoothing will not be done. In fact, it will be pointed out that to yield statistically consistent spectra, 100 periodograms would be needed to match the low variance of a single maximum entropy spectrum. However, if only the spectral index is required, fewer will do. In any case, a rational use of averaging of periodograms is recommended. Greatly superior results will be produced with averaging than without.

The above-mentioned authors estimate the power spectrum by a method based on the finite Fourier transform using the fast Fourier transform (FFT) algorithm. The finite Fourier transform of a set of real numbers is a complex sequence. When this complex sequence has its magnitude squared, the resulting sequence is called a periodogram, a term coined by Schuster [1898]. The periodogram is used as an estimator for the power spectrum; it has the disturbing property that the variance of the estimates increases as the number of original time series observations increases. Statistically, the periodogram provides what is called an "inconsistent" estimate of the power spectrum. A statistical estimator is called inconsistent if in the limit, as the sample size goes to infinity, the variance does not approach zero. For the periodogram of a white Gaussian noise process, the variance approaches the square of the power spectral density. Averaging is required to produce a consistent estimate of the power spectrum [Jenkins and Watts, 1968]. In an attempt to further reduce the variance of the periodogram, Welch [1967] suggests averaging many distinct periodograms obtained from overlapped data sections. These procedures reduce the variance of the spectral estimates by a factor proportional to the number of sections.

About 15 years ago, at two international meetings, Burg [1981a,b] introduced his new technique, which is related to the work of Shannon [Shannon and Weaver, 1949] and Jaynes [1982] on information theory. (The two Burg papers, as well as many others on modern spectrum estimation, are reprinted in the work of Childers [1981]; see also Burg [1975].) The method, called the maximum entropy method (MEM), has revitalized the subject of power spectral estimation. Hundreds of papers have appeared on the subject, but MEM is still not widely accepted and practiced throughout the geophysical community despite the fact that it has been shown to produce a smoother spectrum with higher resolution than the FFT-based techniques, including the Cooley-Tukey method and the slightly older Blackman-Tukey method (see, for example, Radoski et al. [1975, 1976]). For an excellent tutorial review on many of the modern spectral analysis techniques, including MEM and periodogram, see Kay and Marple [1981].

The MEM estimate will always be a smooth estimate, but as expected, its resolution decreases as the signal-to-noise ratio decreases. The MEM, as applied here, consists of two independent steps. The first step is the so-called Burg technique, which determines a prediction error filter with m weights. When run in both time directions over the n data points ($n > m$), this technique minimizes the mean square prediction error. The m weights are the coefficients in an mth-order autoregressive (AR) model of a linear process that could have operated on a white noise input to produce an output with the statistical properties of the observations. The second step uses the mth-order autoregressive process model to estimate the power spectral density. The two-step method is called the Burg-MEM for short. The parameter that controls the smoothness or complexity of the resultant spectrum is the order m of the process. For small m the spectrum will be smooth and simple; as m increases, the resolution and complexity of the spectrum will increase; for very large $m \to n$ the spectral appearance approaches that of the periodogram. But note that the MEM spectrum is a true power spectral density, a continuous function of frequency that may be evaluated over an arbitrary grid of any density; the periodogram is based on the finite Fourier transform, and hence it is essentially a sampled or discrete estimate of the power spectrum.

Section 2 of this paper compares the unaveraged periodogram and Burg-MEM techniques for power law processes using spectral indices ranging from 0.5 to 5.0.

2. Simulation of the Power Law Process

For any linear, time-invariant filter, the output power spectrum is equal to the input power spectrum times the square of the magnitude of the frequency response of the filter (see, for example, Rabiner and Gold [1975, p. 414]). If the input is Gaussian white noise, its power spectrum is a constant, the variance of the noise; the output power spectrum is simply a constant times the magnitude-squared frequency response of the filter. If a filter can be designed to have a frequency response in the form f^{-p}, then its output power spectrum when excited by white noise with unit variance will be

$$P(f) = f^{-2p} . \tag{1}$$

A program written by McClellan et al. [1979] and available on tape from the Institute of Electrical and Electronics Engineers (see also Digital Signal Processing Committee [1979]) can be used to design a finite input response (FIR) filter with any desired magnitude frequency response. It is only necessary to write a simple subroutine to define the desired frequency response. Since the frequency response of a power law filter with negative slope is infinite at zero frequency, this one point must be excluded from the calculations. In practice, a band is chosen such that the frequency response follows a power law from f_0 to f_1, where $f_0 > 0$ and $f_1 \lesssim 0.5$, where f is a frequency normalized by the sampling frequency and runs from 0 to 0.5. The sampled output from the FIR filter is a discrete time moving average (MA) process. An infinite impulse response (IIR) filter (all pole) that produces a power law spectrum over a limited frequency range can also be designed. Such a filter would represent an autoregressive process. The coefficients of the AR model for the filter are estimated as an intermediate step in the Burg-MEM algorithm.

Since the random number generator used to create the Gaussian white noise input can produce a virtually unlimited supply of independent random numbers, many different realizations of the colored noise (power law) process can be produced easily.

3. Computer Experiments

Using the program of McClellan et al. [1979], we designed filters whose squared magnitude had a power law response in the frequency range 0.01 to 0.5, with indices of 0.5 to 5 in steps of 0.5 (10 filters in all). The impulse response of the filter with index 2.0 is shown in Fig. 1(a). The impulse response is symmetric about the t = 0 point, and thus the frequency response is zero phase; that is, it is purely real. The frequency response is the Fourier transform of the impulse response, and it may be approximated as closely as desired by using an FFT of the impulse response augmented by a sufficiently large number of zeroes. The process of augmenting a discrete function with a large number of zeroes before performing a finite Fourier transform is called zero padding and results in a closer spacing of the transformed

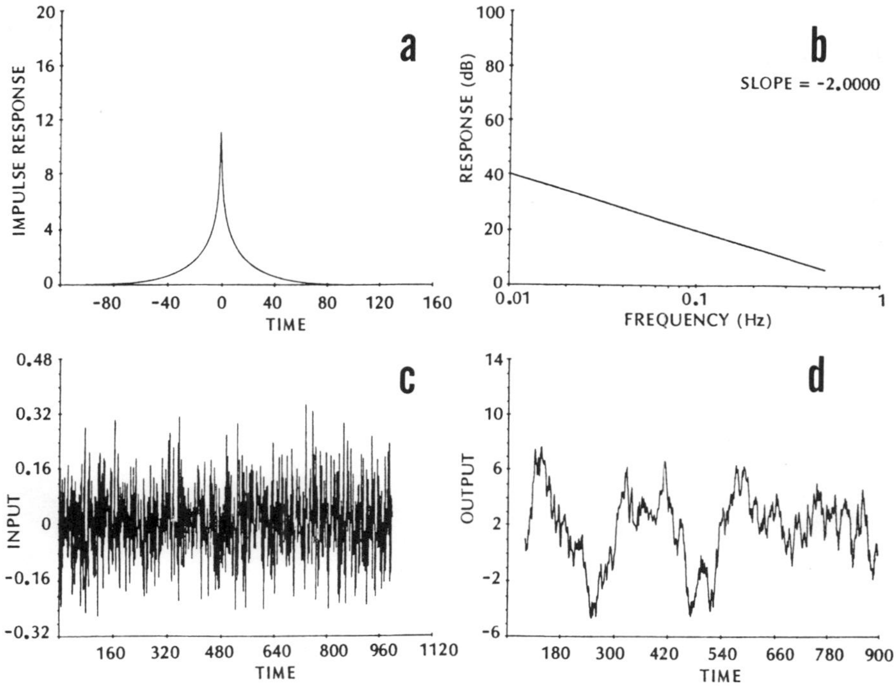

Figure 1. (a) Impulse response of a filter designed to have a squared power law frequency response with a slope of -2. A total of 201 weights are used. (b) Squared frequency response of the filter. The slope is -2.0000. (c) The 1000-point sample of Gaussian white noise used as input to the filter. (d) The 800-point output; at each end, 100 points are lost because the symmetrical filter is 201 weights long.

values. Effectively, zero padding in the input produces interpolation in the output. Since the frequency response of a digital filter is the discrete Fourier transform of the (essentially discrete) impulse response, with zero padding the discrete Fourier transform can be approximated as closely as desired by the finite Fourier transform. Figure 1(b) shows the squared frequency response of the filter with index 2, approximated using a 2048-point FFT. The actual slope, obtained by fitting a straight line to the computed points using least squares, is -2.0000.

Using the filter described in Figs. 1(a) and (b), the sample of Gaussian white noise, shown in Fig. 1(c), produced the realization of a power law process with index 2 (the FIR filter output) shown in Fig. 1(d). The maximum entropy spectrum using five prediction error filter weights is given in Fig. 2(a), and the periodogram (unaveraged, point-by-point square of magnitude of FFT) is shown in Fig. 2(b). The MEM spectrum is much smoother than the FFT spectrum: there is much less variance from one frequency estimate to the next, and the shape itself is nearly linear, reflecting the linearity of the true spectrum. For each spectrum the slope is found by fitting a straight line to the power spectral density (PSD) estimates using least squares. In this procedure, all of the computed points shown in Fig. 2(a) or (b) are used to fit a straight line in the form

$$\log y = -m \log x + A , \tag{2}$$

where m is the desired spectral index, using the method of least squares. The MEM slope is -1.9578, and the periodogram slope is -1.9397. Thus the smoothed behavior of the periodogram in this case is acceptable.

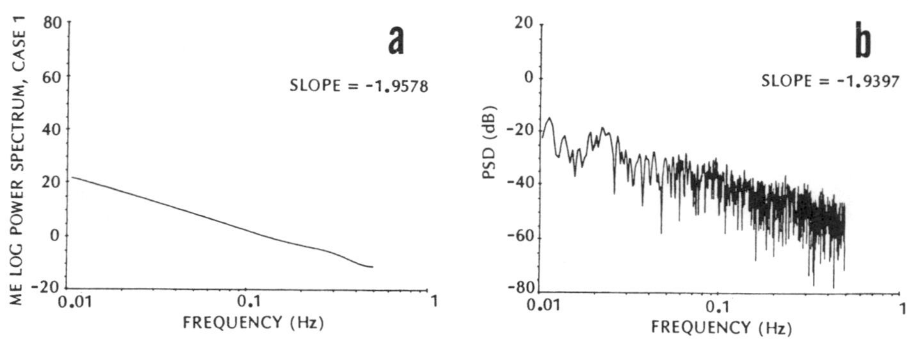

Figure 2. (a) Maximum entropy spectrum of the signal from Fig. 1(d). The ordinate is 10 $\log_{10}$ (PSD). (b) Periodogram of the same signal.

When the spectral index is increased to 4, the resulting impulse response, squared frequency response, white noise input, and red noise realization are those given in Fig. 3. The MEM spectrum, this time using 10 weights, and the periodogram are given in Fig. 4.

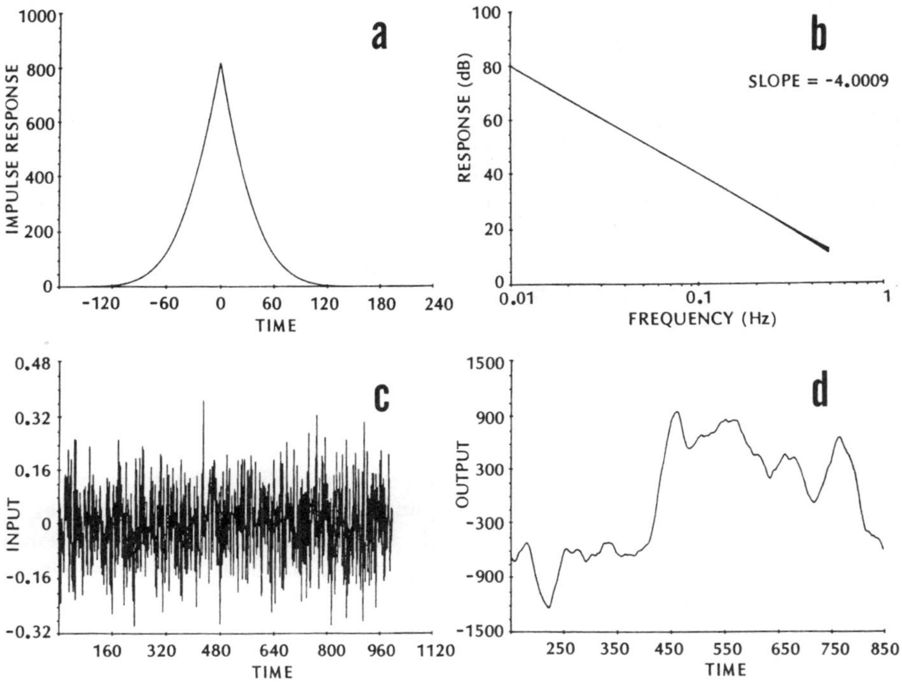

Figure 3. (a) Impulse response of a filter designed to have a squared power law frequency response with a slope of -4. A total of 301 weights are used. (b) Frequency response of the filter. The slope is -4.0009. (c) The 1000-point sample of Gaussian white noise used as input to the filter. (d) The 700-point output. At each end, 150 points are lost because the symmetrical filter is 301 weights long.

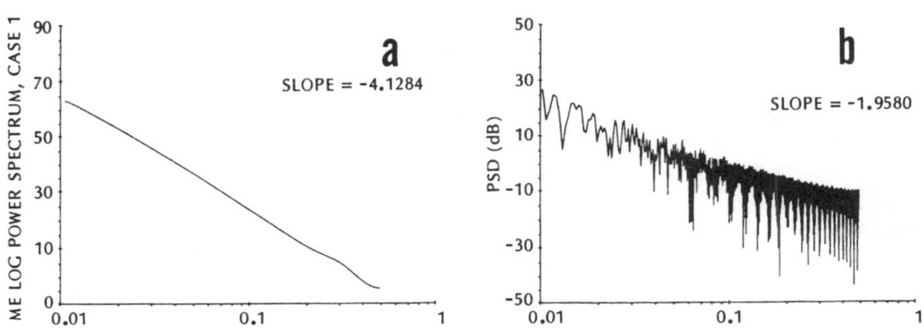

Figure 4. (a) Maximum entropy spectrum of the signal from Fig. 3(d). The ordinate is 10 $\log_{10}$ (PSD). (b) Periodogram of the same signal.

In this case the MEM spectrum is still quite acceptable, but the periodogram is not! There is essentially no relationship between the periodogram and the true spectrum. The true slope is -4, and the periodogram slope is only -1.9580.

An explanation of the periodogram difficulty derives from the fact that a finite data set is equivalent to an infinite data set multiplied by a rectangular window that is unity inside the measurement interval and zero outside. Multiplication of the actual time series by a window function implies that the overall transform is a convolution of the desired transform with the transform of the window function. The Fourier transform of the rectangular window has the form $\sin(\pi f)/\pi f$; this has a relatively narrow central lobe and very high side lobes. If the window were infinitely wide, the side lobes would disappear, and the central lobe would become a Dirac delta function; convolution with such a function would simply reproduce the correct spectrum. But for a finite-sized window the convolution produces some deleterious effects. The first is that the true spectrum is smeared out or defocused; spectral resolution is limited to the width of the main lobe $(N\Delta T)^{-1}$ Hz. The second, and more important for our discussion, is that the high side lobes increase the apparent power at points away from the central lobe. In effect, the decrease of power spectral density with increasing frequency, instead of characterizing the physical process, depends upon the window alone. Thus spectral indices greater than about 2 cannot be obtained using a rectangular window. Many other windows have been designed to offset this high side lobe behavior. The rectangular window gives the narrowest main lobe; other windows designed to reduce side lobes do so at the expense of increasing main lobe width. Nevertheless, some such window is essential: spectral estimation of red noise processes using periodograms requires the use of some nonrectangular window.

For systematic comparison of the two methods, each filter was used to produce 100 independent red noise realizations for each power law index. Four sets of experiments were run, characterized by differing treatments of the red noise before the spectral estimates. In all cases the order of the prediction error filter in MEM was set to 6.

Experiment A: raw data. The 1000 experiments of case A are summarized in Fig. 5(a). Here we see the spectral indices for periodograms versus those for MEM. The MEM indices are always reasonable, with relatively little scatter over the entire range of index 0.5 to 5.0. The periodogram indices always show greater scatter but are reasonable if the true index lies between 0.5 and 2.0, because of the side lobe behavior discussed above. As the true index becomes greater than 2, the periodogram results become worse; that is, an increase in the true index results in a decrease in the index approximated by the periodogram. Note that, for each point plotted, the same red noise realization was used for both MEM and the periodogram.

Another explanation of the difficulty with the periodogram results is that the periodogram technique produces a Fourier analysis of the data

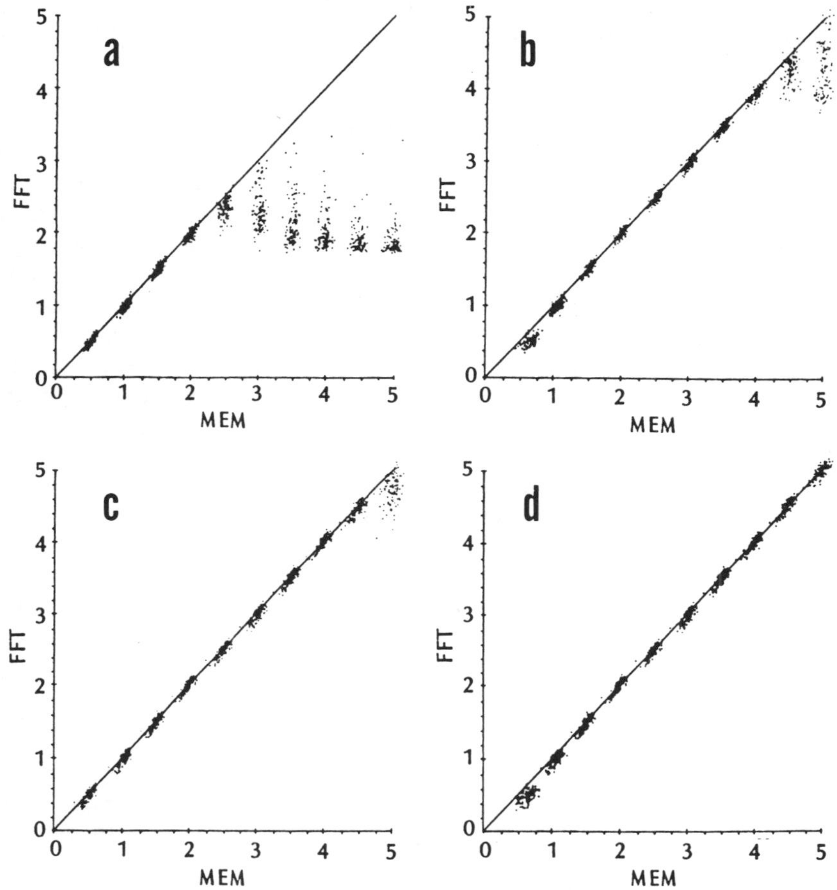

Figure 5. Observed FFT (periodogram) index versus the MEM index (index is the negative of the slope). There are 100 independent realizations of the power law process for each of the 10 indices 0.5, 1.0, ..., 5.0. In every case the same time series was used as input to MEM and to FFT (periodogram). (a) Raw data. (b) End-matched data. (c) Windowed data. (d) End-matched and windowed data.

sample. An inverse Fourier analysis, back into the time domain, produces a periodic function of time, whose period is equal to the duration of the original data. If the original data set does not look periodic, that is, if the first and last points are not equal, then the resulting discontinuity in value of the periodic function produces a distorted power spectrum estimate. If the discontinuity is large, the resulting spectrum distortion can be fatal. For an illuminating discussion of this problem, which is called "spectral leakage," see the paper by Harris [1978] on the use of windows.

Note that this phenomenon does not have any effect on the MEM spectrum, which is specifically designed to operate on the given data and only the given data. No periodic extension of the data is required for MEM as it is for the periodogram technique.

It is in the treatment of missing data that the sharpest and most obvious difference between the periodogram technique and MEM occurs. Jaynes' [1982] maximum entropy principle says that in inductive reasoning our general result should make use of all available information and be maximally noncommittal about missing information by maximizing the entropy of an underlying probability distribution. The MEM, developed by Burg in 1967, begins by assuming that the first few lags of an autocorrelation function are given exactly. Writing the power spectral density as the Fourier transform of the entire infinite autocorrelation function (ACF) and using an expression for the entropy of a Gaussian process in terms of its PSD, Burg solved the constrained maximization problem: Find the PSD whose entropy is maximum and whose inverse Fourier transform yields the given ACF. Maximization is with respect to the missing (infinite in number) ACF values. The result of all this is an extrapolation formula for the ACF. The MEM power spectrum is the exact Fourier transform of the infinitely extended ACF.

The discontinuity in value of the periodogram can be removed by "end-matching," in which a straight line is fit to the first and last data points and then this straight line is subtracted from the data. End-matching may be used to remove the low-frequency components, whose side lobes cause spectral distortion.

Experiment B: end-matching. Now the same 10 filters as in experiment A are used, but each red noise realization (time series) is end-matched before spectral estimation. Once again, the same data set is used as input to both MEM and the periodograms. Figure 5(b) shows the results, with a dramatic improvement in the periodogram indices up to a true index of 4.0. The results at 4.5 and 5.0 are again unacceptable. Note also that the MEM results are slightly but definitely worse. There is more scatter, and at an index of 0.5 the MEM results are biased upward.

Clearly then, end-matching is required for spectral estimation using periodograms: without it most of the spectrum estimates are unreliable. Except in the low index range, for true indices betweeen 0.5 and 2.0, the periodograms are distorted. A spectral index of 2.0 could have arisen from a true index anywhere between 2.0 and 5.0.

Just as clearly, end-matching should be avoided in the MEM calculations, where it does not help but degrades the spectral estimates.

Experiment C: windowing. The difficulty with the periodogram results at true indices of 4.5 and 5.0 may perhaps be explained by discontinuities in slope of the original data set at the beginning and end. The suggested cure is "tapering," or looking at the data through a window that deemphasizes the data at both ends by multiplying the data by a function that is near zero at the end points and higher in the center. This tapering does indeed reduce the tendency to discontinuity in slope.

The actual window used was recommended by Welch [1967] and is

$$W_j = 1 - \left[\frac{2j}{L+1} - 1\right]^2, \qquad j = 1,2,\cdots,L . \tag{3}$$

For those interested in "window carpentry," no fewer than 44 distinct windows are discussed in great detail in the review paper by Harris [1978]. The same 10 filters used in experiments A and B were used in experiment C. Here the above window was applied routinely to all red noise realizations before the spectra were estimated, but end-matching was not used. The results are shown in Fig. 5(c). The periodogram results are now usable for indices between 0.5 and 4.5, but the results at 5.0 are still biased. Once again, as with the use of end-matching, windowing degrades the MEM spectral estimate. The MEM scatter is even larger here than it was with end-matching.

Experiment D: both end-matching and windowing. Here the original red noise realizations are first end-matched and then windowed before the spectral estimation. Figure 5(d) shows the results. Now the periodogram results are usable for the entire range of spectral indices from 0.5 to 5.0. The MEM results show that end-matching and windowing, taken either singly or together, degrade the MEM spectral estimates.

Summary. Note that, in all cases, for all experiments, the MEM scatter is smaller than the periodogram scatter. The quantitative results for all four experiments are collected in Table 1, where the mean, standard deviation, and maximum and minimum values are given.

Table 2 summarizes the results of the simulation. The several experiments have shown that the Burg-MEM technique should be used on the raw data and that some form of windowing and detrending is required before the periodogram method can be employed for the estimation of the spectral index of a power law power spectrum. In Table 2 the standard deviation values were used to calculate the expected 90% confidence limits for the spectral index estimates. It is seen that the FFT or single-periodogram method produces an uncertainty that is almost twice that for the Burg-MEM technique (±0.015 versus ±0.009). The 90% confidence intervals (CI in the table) are very small, and either method must be judged acceptable.

A difficulty in the use of the Burg-MEM technique can also be seen in Table 2. The MEM estimates are biased. That is, the confidence interval does not include the true value. This small bias value is of the order of the uncertainty in the periodogram method. The bias can be changed by using a different order for the AR process. Recall that the order is set at 6; no attempt was made to optimize it.

But notice also that for MEM the error is negative in six cases and positive in four cases. For FFT the error is always positive, indicating a systematic bias. The periodogram systematically underestimates the spectral index. Most of the errors are quite close to the 10% confidence interval.

Table 1. Statistical Summary of the 4000 Experiments

True Index	Mean Index		Standard Deviation		Minimum		Maximum	
	MEM	FFT	MEM	FFT	MEM	FFT	MEM	FFT
Experiment A: Raw Data								
0.5	0.5155	0.4991	0.0541	0.0678	0.3488	0.3547	0.6160	0.6546
1.0	1.0308	0.9975	0.0534	0.0705	0.8905	0.8263	1.1400	1.1786
1.5	1.5108	1.4956	0.0530	0.0698	1.3712	1.3380	1.6316	1.6646
2.0	1.9938	1.9749	0.0546	0.0711	1.8833	1.8246	2.1248	2.1588
2.5	2.5063	2.3468	0.0550	0.1280	2.3802	1.9569	2.6248	2.6553
3.0	3.0241	2.3374	0.0548	0.2784	2.8773	1.7457	3.1507	2.9979
3.5	3.5119	2.1354	0.0579	0.3060	3.3660	1.7264	3.6810	3.2385
4.0	3.9855	2.0058	0.0612	0.2506	3.8402	1.7499	4.1375	3.3308
4.5	4.4798	1.9134	0.0627	0.1897	4.3133	1.6931	4.6044	3.0970
5.0	4.9813	1.8694	0.0637	0.1855	4.8104	1.7218	5.1130	3.2826
Experiment B: End-Matched Data								
0.5	0.6607	0.5096	0.0767	0.0699	0.4456	0.3480	0.8378	0.6743
1.0	1.0693	1.0003	0.0572	0.0661	0.9463	0.8597	1.2004	1.1852
1.5	1.5136	1.4964	0.0547	0.0695	1.3855	1.3518	1.6295	1.6906
2.0	1.9899	1.9928	0.0545	0.0708	1.8771	1.8062	2.1206	2.2038
2.5	2.5030	2.4940	0.0538	0.0697	2.3741	2.3082	2.6250	2.6812
3.0	3.0229	2.9923	0.0553	0.0770	2.8756	2.7881	3.1229	3.1696
3.5	3.5112	3.4860	0.0581	0.0751	3.3601	3.2890	3.6501	3.6359
4.0	3.9850	3.9550	0.0623	0.0775	3.8309	3.7450	4.1316	4.1355
4.5	4.4950	4.2794	0.0627	0.1629	4.3497	3.8593	4.6417	4.7137
5.0	4.9869	4.3143	0.0635	0.3632	4.8330	3.6926	5.1223	5.1244
Experiment C: Windowed Data								
0.5	0.5102	0.4875	0.0622	0.0794	0.3019	0.2930	0.6249	0.6516
1.0	1.0243	0.9861	0.0625	0.0809	0.8033	0.7873	1.1311	1.1298
1.5	1.5041	1.4865	0.0650	0.0837	1.2805	1.2798	1.6271	1.6369
2.0	1.9871	1.9865	0.0685	0.0869	1.7483	1.7680	2.1192	2.1505
2.5	2.5007	2.4871	0.0698	0.0861	2.2533	2.2679	2.6351	2.6547
3.0	3.0217	2.9878	0.0697	0.0890	2.7594	2.7572	3.1627	3.1685
3.5	3.5104	3.4891	0.0696	0.0915	3.2532	3.2527	3.6536	3.6713
4.0	3.9824	3.9872	0.0732	0.0905	3.7381	3.7532	4.1281	4.1648
4.5	4.4811	4.4415	0.0793	0.0983	4.2217	4.1652	4.6250	4.6105
5.0	4.9830	4.7165	0.0819	0.1989	4.7177	4.0513	5.1505	5.1166
Experiment D: End-Matched and Windowed Data								
0.5	0.6383	0.4877	0.0850	0.0792	0.4739	0.2928	0.8190	0.6519
1.0	1.0593	0.9858	0.0664	0.0811	0.8462	0.7873	1.2291	1.1301
1.5	1.5074	1.4864	0.0641	0.0837	1.2938	1.2796	1.6394	1.6370
2.0	1.9841	1.9865	0.0688	0.0867	1.7376	1.7695	2.1122	2.1509
2.5	2.4983	2.4871	0.0700	0.0861	2.2422	2.2622	2.6288	2.6556
3.0	3.0195	2.9878	0.0694	0.0894	2.7542	2.7558	3.1576	3.1696
3.5	3.5086	3.4886	0.0695	0.0920	3.2496	3.2525	3.6539	3.6630
4.0	3.9823	3.9861	0.0748	0.0923	3.7293	3.7641	4.1169	4.1504
4.5	4.4863	4.4926	0.0792	0.0955	4.2226	4.2289	4.6336	4.6677
5.0	4.9890	4.9884	0.0811	0.0952	4.7283	4.7529	5.1608	5.1671

Each entry is based upon a set of 100 realizations of the red noise process. FFT stands for FFT-based periodogram.

Table 2. Summary of Simulation Experiments

True Index	MEM (Experiment A) Mean	Error[1]	CI	FFT (Experiment D) Mean	Error[1]	CI
0.500	0.516	−0.016	0.009	0.488	0.012	0.013
1.000	1.031	−0.031	0.009	0.986	0.014	0.013
1.500	1.511	−0.011	0.009	1.486	0.014	0.014
2.000	1.994	0.006	0.009	1.987	0.013	0.014
2.500	2.506	−0.006	0.009	2.487	0.013	0.014
3.000	3.024	−0.024	0.009	2.988	0.012	0.015
3.500	3.512	−0.012	0.010	3.489	0.011	0.015
4.000	3.986	0.014	0.010	3.986	0.014	0.015
4.500	4.480	0.020	0.010	4.493	0.007	0.017
5.000	4.981	0.019	0.011	4.988	0.012	0.016

[1]Error = true index − mean index.

CI is the expected 90% confidence interval $t\sigma/\sqrt{n}$, where t is the 5% critical value of Student's distribution using 100 degrees of freedom, σ is the standard deviation from Table 1, and $n = 100$. For a single observation, CI = $t\sigma$, i.e., 10 times as large. Note that if $|\text{error}| > \text{CI}$, a bias exists.

For both MEM and FFT, however, the bias is quite small and should not prove troublesome.

It is noted that by averaging the spectral index estimates obtained from four periodograms the confidence bounds can be reduced to less than the bounds for a single Burg-MEM estimate. Using overlapping spectra as recommended [Nuttall and Carter, 1982], equivalent results could be obtained from averaged periodograms by using a data set 3 times the length of that required for the Burg-MEM analysis. The tradeoff between the use of the two techniques is evident. The nonparametric method employing windowed and averaged periodograms requires more data to produce the same result as can be obtained from the Burg-MEM algorithm. The parametric Burg-MEM algorithm, however, requires the selection of the correct order to produce unbiased results, but the order is not known a priori. When faced with a time series from an unknown process, both techniques should be applied, and the parameters of the models (such as order of the process) should be adjusted to provide consistent results [Jenkins and Watts, 1968].

Discussion. The 4000 runs, collected in four experiments of 100 runs each on 10 distinct spectral indices between 0.5 and 5.0, show that, correctly applied, the maximum entropy method and periodogram techniques yield results that may be thought of as complementary.

1. The MEM spectral shape is always smooth and nearly linear. The unaveraged periodogram shape is highly variable and noisy.

2. When straight lines are fit to the spectra, the resulting spectral indices are more variable with periodograms than with MEM.

3. Slight biases can result from the use of MEM unless care is taken in determining the order of the process to be analyzed.

Explanation. Since the difficulties with the periodogram-based techniques have been explained briefly, it seems in order to present an intuitive explanation of the success of the MEM technique. The MEM spectrum is based on the determination, from the data sample, of a prediction error filter, which finds the error in a one-step-ahead prediction as a linear combination of m previous sample values. The same filter is used to make predictions in both time directions (it is merely reversed to make predictions from the future into the past). The mean square prediction error in both time directions is minimized by varying the prediction error coefficients. Because the filter makes predictions of the time series based on previous values of the time series itself, it is also called an autoregressive filter.

Now the red noise process whose spectrum we are trying to estimate is also an autoregressive process, and thus MEM is ideally suited to the estimation of the autoregressive parameters and indeed produces spectra that are close to ideal.

4. Application to Geophysical Data

Ionospheric scintillation occurs when a radio wave, transmitted by a satellite toward a receiver on the earth, passes through a disturbed ionosphere. Both the phase and the amplitude of the wave suffer low-frequency (≃0.001 to 100 Hz) perturbations known as phase and amplitude scintillation, respectively. If the wave is detected in a suitable receiver, the scintillation can be separated from the carrier and can then yield important information on the nature of the ionospheric irregularities that are the source of the scintillation [Yeh and Liu, 1982].

The data to be analyzed here are amplitude scintillation data sampled at 36 Hz for 5 min from the MARISAT satellite in January 1981. Figure 6(a) shows the data set, which was chosen especially because it contains a quiet segment (from 0 to about 2-1/4 min), a moderately noisy segment (from 2-1/4 to 3-1/4 min), and a highly scintillating segment (from 3-1/4 to 5 min). The changes in character of the noise record are quite abrupt and easy to see in this time record.

It may legitimately be asked whether power spectrum analysis could be employed to monitor the development of such a process. A dynamic spectrum was constructed from the approximately 5-min data sample as follows. There were 10,981 observations in this data set, which was divided into 60 batches of 361 points (10 s each), overlapped by 181 points (5 s).

Some elementary statistics for each batch are shown in Fig. 6(b), which gives the maximum, standard deviation, and minimum. The input power to each power spectrum is simply the square of the standard deviation. Figures 6(a) and (b) show that the signal is approximately stationary, that is, the mean and standard deviation are approximately constant during the three separate time intervals 0 to 2-1/4 min, 2-1/4 to 3-1/4 min, and 3-1/4 to

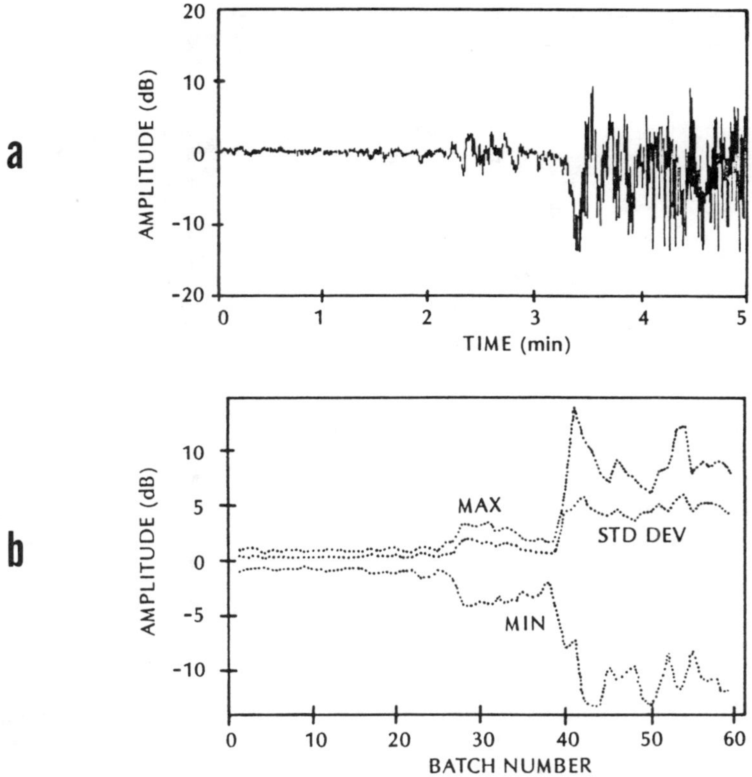

Figure 6. (a) Amplitude scintillation data taken from the MARISAT satellite in January 1981. Sampling rate is 36 Hz. There is background noise from 0 to 2-1/4 min, moderate scintillation from 2-1/4 to 3-1/4 min, and fully saturated scintillation from 3-1/4 to 5 min. (b) Maximum (top curve), standard deviation (middle), and minimum (lower) of each batch of 361 observations of data from (a). Batches are overlapped by 181 points.

5 min. These times correspond to batches numbered 1—27, 28—39, and 40—60. A power spectrum was then obtained for each batch. The 60 overlapped data sections are shown in Fig. 7(a), and the MEM and unaveraged periodogram spectra are shown in an identical format in Fig. 7(b). The smooth curve is the MEM result based on five prediction error filter weights applied directly to the raw data. The periodograms, which appear as the noisy curves in Fig. 7(b), were obtained on the same data sets after first end-matching and windowing using a three-term Blackman-Harris window. This is one of the "best" windows described by Harris [1978]. It can be seen that in every case the Burg-MEM result is an "ideally" smoothed version of the periodogram result or, more important, that the periodogram results show the statistical fluctuations to be expected when no form of

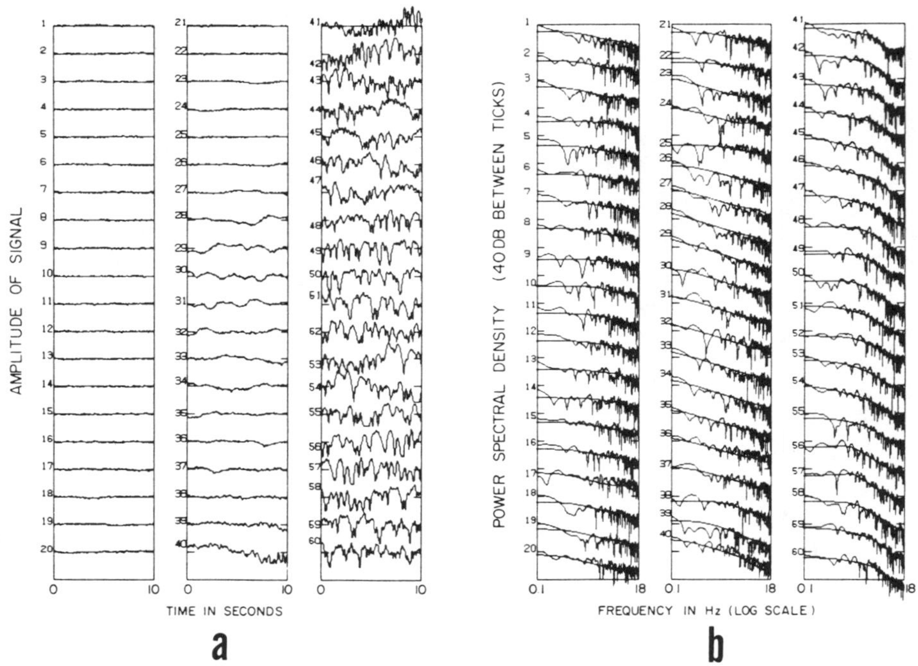

Figure 7. (a) The 60 overlapped data batches. (b) Smooth curves: Burg-MEM spectra of order 5 of the time series shown in (a). Noisy curves: periodograms of the same data after end-matching and windowing.

averaging is used. If one had only the periodogram results, one might be tempted to "see" significant spectral peaks, and these peaks in any case would tend to obscure real changes in slope in the underlying process. Such important changes in slope (and indeed changes in the character of the spectrum) are easy to identify in the MEM spectra and can be seen to correspond to changes in the character of the time series as seen in Fig. 6. The smoothness of the MEM spectrum is affected by the choice of the order of the spectrum. Since the process is nearly stationary over a number of spectra, additional information can be obtained by averaging the spectral estimates obtained from the periodograms. Notice that the spectra numbered 27—37 are all extremely close to pure power law spectra with a single spectral index over the entire frequency band of interest 0.1 to 18 Hz. This fact should serve to validate and motivate the simulation results presented earlier. Real geophysical data sets sometimes can be accurately represented as realizations of pure power law processes.

Notice that the character of the individual data sets changes quite sharply from batch 25 to batch 29. The Burg-MEM spectra similarly show changes of character from batch 25 to batch 29. Once again the character of the signal and of the associated spectra changes abruptly from pure power law to composite power law, approximated by two or more approximately linear sections at batch 38. Fully saturated scintillation is evident in batches 40 and 41 in the time record. The spectra here are composite power law, piecewise linear, with three different spectral indices, small, medium, and large, and a spectral minimum near the Nyquist frequency (18 Hz). These results, the spectra of batches 38—41, are expanded and shown in detail in Fig. 8(a). The sharp change of character from batches 38 and 39 to batches 40 and 41 is dramatic and easily visible. Figure 8(b) shows the periodogram results. To make the four curves distinguishable, a bias of 20, 40, and 60 dB was added to curves 39, 40, and 41, respectively. The change of character from curve 38 to curve 41 is obscured by the statistical fluctuation.

a

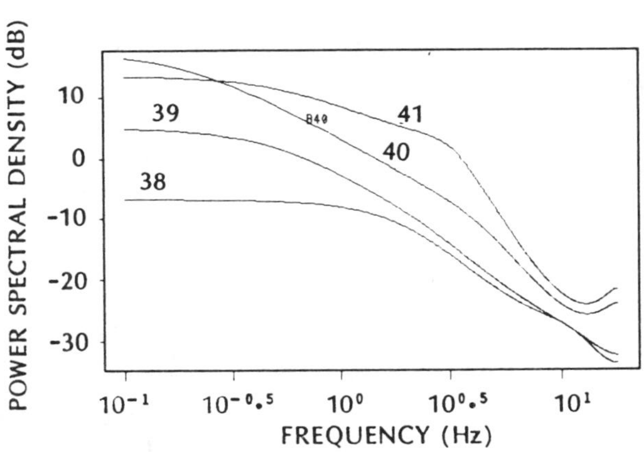

b

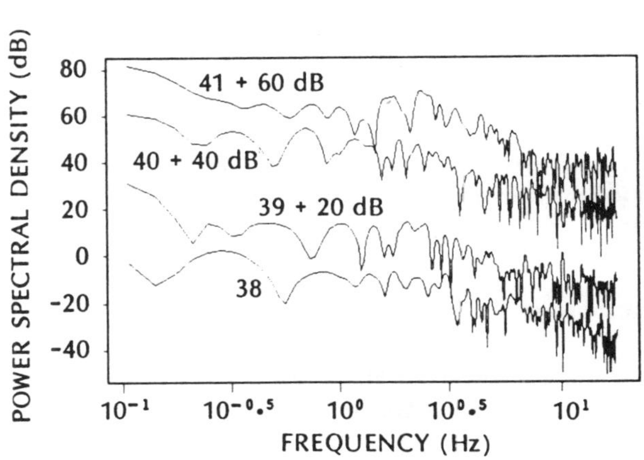

Figure 8. (a) Burg-MEM spectra of order 5 of data batches 38—41. (b) Periodograms of the same data after end-matching and windowing: 20, 40, and 60 dB have been added to batches 39, 40, and 41, respectively.

At the risk of belaboring the point, these qualitative changes in character of the spectrum are clearly and easily visible in the Burg-MEM results but are at least partially obscured in the fluctuations that are the constant companion of unaveraged periodogram results.

Note finally that certain features clearly visible in the Burg-MEM spectra could not even be imagined by examining the time series record. For example, the fully saturated three-component spectra evident in batches 41—47 change to predominantly two-component (white plus red) spectra in batches 48—54. Of course, it is the desire to see features not obvious in the time record that motivates the use of the power spectrum in the first place.

Another way of visualizing a dynamic power spectrum is a three-dimensional surface representation with hidden lines removed. Figure 9(a) shows such a representation of the 60 overlapped spectra using Burg-MEM. The qualitative picture is that of power spectral density changing smoothly with time with two regions of rapid but smooth increase. The companion picture from the overlapped periodograms, Fig. 9(b), likewise shows two regions of rapid increase, but the high-amplitude noise obscures all other features.

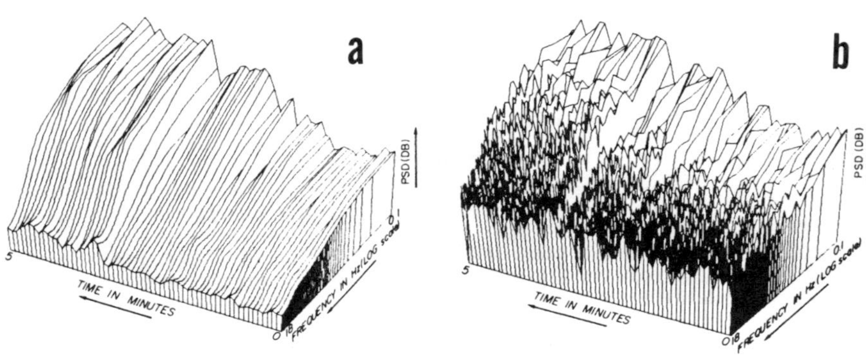

Figure 9. (a) Order 5, Burg-MEM dynamic spectra of all 60 data batches. (b) Periodogram dynamic spectra of the same data after end-matching and windowing.

5. Conclusions

1. The Burg-MEM applied to time series realizations of red noise processes produces consistently smooth power spectra.

2. Without averaging, the periodogram method applied to the same data sets produces power spectra with large statistical fluctuations that may obscure the true spectral variations. End-matching or windowing, or preferably both, is absolutely essential if meaningful periodogram results are to be obtained.

At this point we return briefly to the issue raised in the introduction, that of statistical consistency. It was mentioned that the periodogram produces a statistically inconsistent spectral estimate: the variability of the spectrum does not decrease as the data sample increases in size. This is a stochastic result that has nothing to do with the deterministic problems of "spectral leakage," which can be greatly reduced using end-matching or windowing or both.

This stochastic result has been well known for a long time. Subsequent statistical analysis has shown that there are two methods of reducing the variance in the periodogram spectral estimate. Both involve a smoothing procedure, one in the frequency domain, and the other in the time domain. In the frequency domain the variance can be reduced by applying various smoothing formulas to a set of adjacent spectral estimates in a periodogram. The simplest of the smoothing formulas is the running mean. The greater the number of ordinates that are smoothed, the greater the reduction in variance. Of course, at the same time the frequency resolution of the smoothed spectrum is similarly reduced. Thus a tradeoff must be made between decreased variance, which is desirable, and decreased resolution, which is not.

In the time domain the smoothing procedure that is useful for reducing spectral variance is that of averaging successive independent or overlapped periodograms. The price paid here is a corresponding reduction in time resolution. Thus if, as in the MARISAT data, the spectrum changes abruptly in time, because the underlying time series changes abruptly, the process of averaging n consecutive periodograms would blur any sharp changes in character of the spectrum.

If, however, the process under scrutiny were stationary and sufficient data were available, the method of averaging periodograms would produce a reasonable, smooth spectrum. To illustrate this point, we return briefly to our simulation results and show in Fig. 10(a) successively stacked and averaged periodograms using 1, 2, 3, 4, 5, 10, 25, 50, 75, and 100 independent periodograms and, on the bottom, a single maximum entropy spectrum for comparison. The original spectral index was 2.0 (slope of -2.0). The number given next to each spectrum is the rms deviation of the spectrum estimate from the power law spectrum determined by least squares. With 100 averaged periodograms the rms deviation is still a little larger than that of one maximum entropy spectrum. An adequate estimate of the spectral index, however, requires the use of only four periodograms, as indicated in the discussion of Table 2.

By way of a final resolution that may provide an intuitive understanding of the differences between maximum entropy and periodogram spectral analysis for red noise processes, Fig. 10(b) shows maximum entropy power spectra of a single realization of a simulated power law process with index 2.5. At the very bottom, for comparison, is the periodogram (after end-matching and windowing). As the number of filter weights increases (top to bottom), the MEM spectral appearance becomes more and more jagged until at 512 weights the MEM spectrum resembles quite closely the raw periodo-

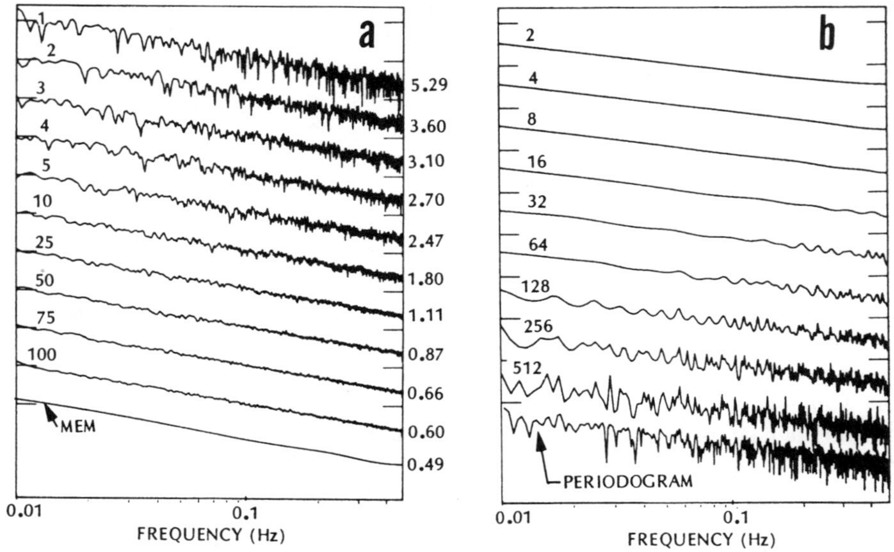

Figure 10. (a) Stacked and averaged periodograms (top 10 curves) of a simulated power law process with index 2. The number of independent spectra stacked is shown on the left (1, 2, 3, 4, ..., 100), and on the right are the rms deviations of the displayed spectrum from the power law spectrum, determined by least squares. The bottom spectrum is a single MEM spectrum. (b) Maximum entropy spectra (top 9 curves) of a simulated power law process with index 2.5. The number of prediction error filter weights is shown at the left (2, 4, 8, 16, ..., 512). The periodogram is shown at the bottom for comparison. In both graphs, the vertical scale is 30 dB between ticks.

gram. Note that very little change in the spectrum is shown in the top four curves, with 2, 4, 8, and 16 weights, but below that, for 32, 64, ... weights, more and more meaningless detail is displayed.

A complete FORTRAN package that finds Burg-MEM spectra has been prepared. Seriously interested scientists are invited to write to the author for a copy of the program and its documentation. Please indicate preferred tape density (800 or 1600 BPI) and code (EBCDIC or ASCII).

6. Acknowledgments

It is a pleasure to thank the reviewers, whose criticism and suggestions have helped me to produce a much better paper. I also want to thank Santimay and Sunanda Basu and Herbert C. Carlson for the use of the MARISAT data and for many very useful discussions. Finally, I am grateful to Celeste Gannon and Elizabeth Galligan for patiently and expertly typing the many, many draft versions as well as the final version of the manuscript.

7. References

Armstrong, J. W., and W. A. Coles (1972), "Analysis of three-station interplanetary scintillation," J. Geophys. Res. 77, pp. 4602-4610.

Burg, J. P. (1975), "Maximum Entropy Spectral Analysis," Ph.D. thesis, 123 pp., Stanford University, Stanford, Calif.

Burg, J. P. (1981a), "Maximum entropy spectral analysis," in D. G. Childers, ed., Modern Spectrum Analysis, IEEE Press, New York.

Burg, J. P. (1981b), "A new analysis technique for time series data," in D. G. Childers, ed., Modern Spectrum Analysis, IEEE Press, New York.

Childers, D. G., ed. (1981), Modern Spectrum Analysis, IEEE Press, New York.

Crane, R. K. (1976), "Spectra of ionospheric scintillation," J. Geophys. Res. 81, pp. 2041-2050.

Digital Signal Processing Committee, IEEE Acoustics, Speech, and Signal Processing Society (1979), Programs for Digital Signal Processing, IEEE Press, New York.

Harris, F. J. (1978), "On the use of windows for harmonic analysis with the discrete Fourier transform," Proc. IEEE 66, pp. 51-83.

Jaynes, E. T. (1982), "On the rationale of maximum-entropy methods," Proc. IEEE 70, pp. 939-952.

Jenkins, G. M., and D. G. Watts (1968), Spectral Analysis and Its Applications, Holden-Day, San Francisco.

Kay, S. M., and S. L. Marple, Jr. (1981), "Spectrum analysis—a modern perspective," Proc. IEEE 69, pp. 1380-1418.

Larsen, M. F., M. C. Kelley, and K. S. Gage (1982), "Turbulence spectra in the upper troposphere and lower stratosphere at periods between 2 hours and 40 days," J. Atmos. Sci. 39, pp. 1035-1041.

McClellan, J. H., T. W. Parks, and L. R. Rabiner (1979), "FIR linear phase filter design program," in Programs for Digital Signal Processing, IEEE Press, New York.

McKenzie, J. F. (1982), "Similarity solution for non-linear damping of Alfvén waves," J. Plasma Phys. 28, pp. 317-323.

Nuttall, A. H., and G. C. Carter (1982), "Spectral estimation using combined time and lag weighting," Proc. IEEE 20, pp. 1115-1125.

Rabiner, L. R., and B. Gold (1975), Theory and Application of Digital Signal Processing, Prentice-Hall, Englewood Cliffs, N.J.

Radoski, H. R., P. F. Fougere, and E. J. Zawalick (1975), "A comparison of power spectral estimates and applications of the maximum entropy method," J. Geophys. Res. 80, pp. 619-625.

Radoski, H. R., E. J. Zawalick, and P. F. Fougere (1976), "The superiority of maximum entropy power spectrum techniques applied to geomagnetic micropulsations," Phys. Earth Planet. Inter. 12, pp. 208-216.

Schuster, A. (1898), "On the investigation of hidden periodicities with application to a supposed 26 day period of meteorological phenomena," J. Geophys. Res. 3, pp. 13-41.

Shannon, C. E., and W. Weaver (1949), The Mathematical Theory of Communication, Universiity of Illinois Press, Urbana.

Welch, P. D. (1967), "The use of fast Fourier transform for the estimation of power spectra: a method based on time averaging over short, modified periodograms," IEEE Trans. Audio Electroacoust. AU-15, pp. 70-73.

Woo, R., and J. W. Armstrong (1979), "Spacecraft radio scattering observations of the power spectrum of electron density fluctuations in the solar wind," J. Geophys. Res. 84, pp. 7288-7296.

Yeh, K. C., and C. H. Liu (1982), "Radio wave scintillation in the ionosphere," Proc. IEEE 70, pp. 324-360.

RECENT DEVELOPMENTS AT CAMBRIDGE

Stephen F. Gull

Mullard Radio Astronomy Observatory, Cavendish Laboratory, University of Cambridge, Madingley Road, Cambridge CB3 0HE, England

John Skilling

Department of Applied Mathematics and Theoretical Physics, University of Cambridge, Silver Street, Cambridge CB3 9EW, England

C. R. Smith and G. J. Erickson (eds.),
Maximum-Entropy and Bayesian Spectral Analysis and Estimation Problems, 149–160.

1. Introduction

In recent years at Cambridge University we have had a small but vigorous team working on maximum entropy and related topics. Most of our work to date has concerned image processing in one form or another, and a progress report (to 1981) was given at the first of these meetings [Skilling and Gull, 1985]. Figure 1 depicts a selection of practical results, with examples of maximum entropy data processing taken from radio astronomy, forensic deblurring, medical tomography, Michelson interferometry, and "blind" deconvolution.

The purpose of this paper is to report on various other aspects of our group's work, with emphasis on our current thinking. First we consider problems involving multi-channel processing, where we wish to recover more than one reconstruction from data. "Blind" deconvolution is one such case, where both the point spread function and the underlying scene are unknown. We then turn in Sections 2 and 3 to the treatment of multifrequency (colored) image processing, and to pictures of polarized objects.

In the more theoretical sections (5, 6, and 7) we first remark that the maximum entropy principle seems applicable to any simple set of proportions, whether or not there is any direct probabilistic interpretation. The "traffic-jam" problem, invented by Ed Jaynes as a homework assignment for his students, has preoccupied us over the years, and we use it to illustrate why we have learned to be wary of the entropy concentration theorem [Jaynes, 1979]. The most satisfactory solution for the traffic jam (a uniform de Finetti generator [Jaynes, 1986]) has some practical consequences, and we give as a simple application of it a code-breaking problem—locating the byte-boundary in a string of bits.

Finally, in Section 8, we present an example taken from high-energy astrophysics, the acceleration of cosmic ray electrons. It now seems that, even in the depths of interstellar space, the maximum entropy principle reigns supreme!

2. Progress Toward "Blind" Deconvolution

Maximum entropy deconvolution with a known point spread function (PSF) is now a standard procedure in forensic analysis and elsewhere. We have a blurred picture $F_i = \Sigma_j f_j h_{i-j}$ of a scene $\{f_i\}$, convolved with PSF $\{h_i\}$. If $\{h_i\}$ is known, we can determine the most noncommittal scene by maximizing the configurational entropy:

$$S(f) = -\sum_i f_i \log f_i \qquad (1)$$

where f is normalized to $\Sigma f = 1$. However, in practice, $\{h_i\}$ is never known a priori; it has also to be estimated from the blurred picture. Usually, intelligent guesswork is used, and this has been quite successful, but ideally

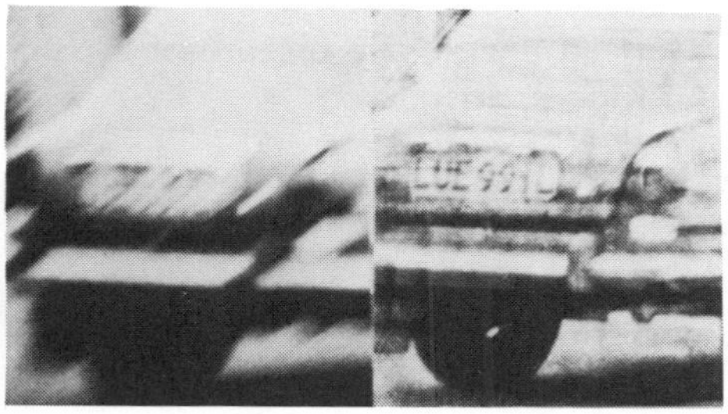

Maximum entropy deconvolution.
Left: before. Right: after. (UK Home Office)

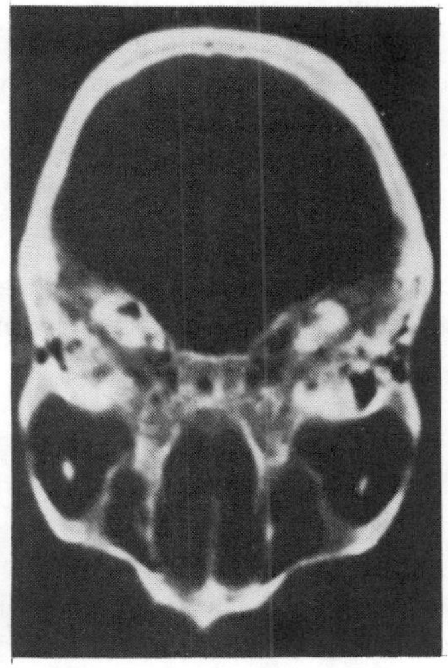

ME x-ray tomography
(skull in perspex, EMI Ltd.)

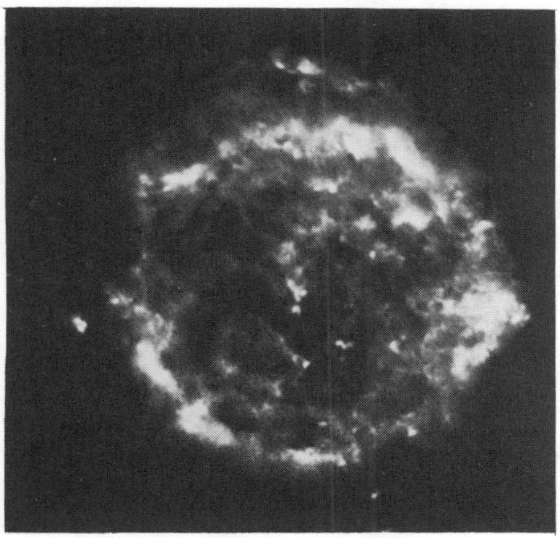

SNR Cas A at 5 GHz. 1024^2 maximum entropy image. (5-km telescope, MRAO, Cambridge)

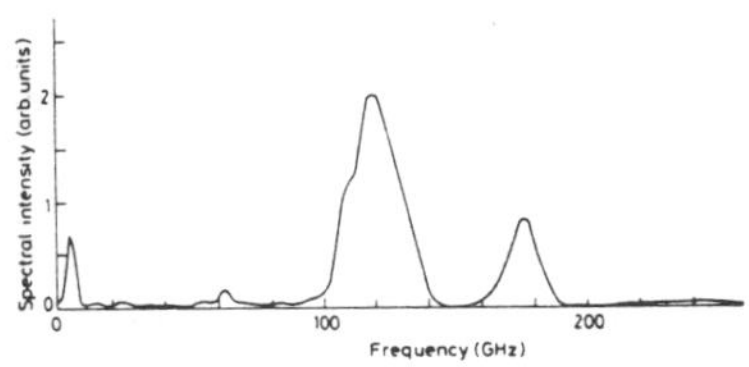

Millimeter-wave Michelson interferometer spectrum of cyclotron emission from DITE tokamak (Culham Laboratory)

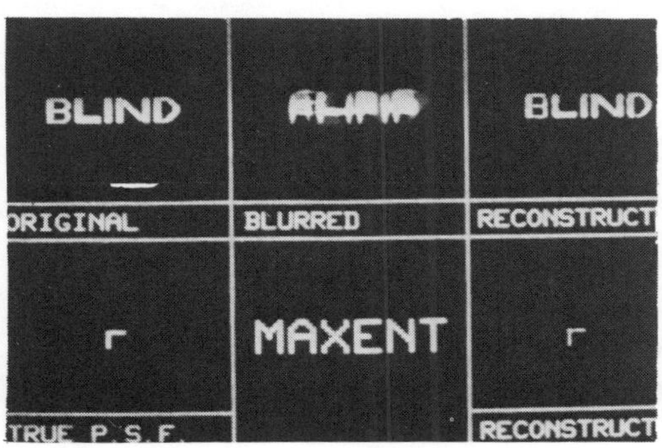

'Blind' deconvolution of unknown blurring.

Left: True image and blurring
Middle: Data as given to ME program
Right: Reconstructions
(T. J. Newton)

Figure 1. Selection of practical results in image processing.

one would wish to obtain a simultaneous reconstruction of the image and PSF: "blind" deconvolution.

In noncoherent optics, the PSF is itself an image, the picture of a point source. It therefore has its own configurational entropy:

$$S(h) = -\sum_i h_i \log h_i \tag{2}$$

where h is normalized to $\Sigma h = 1$. To perform blind deconvolution, we propose that the combined entropy be maximized:

$$S(\text{total}) = \theta\, S(f) + (1-\theta)\, S(h), \tag{3}$$

subject to the constraint of fitting the data adequately. The determination of θ is discussed below.

We can understand this formula in terms of a "monkey" model [Gull and Daniell, 1978] for the generation of images. If the picture and the PSF are separately generated by teams of monkeys throwing elements of luminance into the pixels, then $N(f)!/\Pi_i n_i(f)!$ and $N(h)!/\Pi_i n_i(h)!$ are the degeneracies of the images produced, where N(f) and N(h) are the total number of elements thrown by the two teams. The product of these degeneracies yields the joint degeneracy of a picture-PSF pair. Then, taking logarithms and using Stirling's approximation, we get

$$\log(\text{degeneracy}) = [N(f) + N(h)]\, S(\text{total})\,, \tag{4}$$

where $\theta = N(f)/[N(f) + N(h)]$. We have not yet determined θ, but there are further arguments that relate it to the relative amounts of structure appearing on the two images. We have, though, always fixed it as a controlling device.

An example of this [T. J. Newton, personal communication] is shown in the lower left corner of Fig. 1, in which a picture of the word BLIND was blurred with an L-shaped PSF. This type of asymmetric blur is notoriously difficult to estimate by conventional (cepstral) techniques. However, both the word and the blur are recovered by maximum entropy blind deconvolution using $\theta = 0.9$.

3. Images with Color

Suppose (as has often happened) that a radio astronomer has collected two data sets related to the same celestial object, but at different frequencies, e.g.:

(1) Very large array	Wavelength 21 cm	"Red data"
(2) Cambridge 5-km	Wavelength 6 cm	"Blue data"

How do we combine these disparate observations into a useful picture of the object, in particular to display the variations of color across it (the radio astronomers' spectral index map)? First, let us mention what we must not do: we must not make separate maximum entropy maps and display the ratio. The different antenna configurations of the telescopes give different residual sidelobes in different places, and the result will of course be disastrous. However, the "red" and "blue" images made in this way are useful, showing what can be concluded about the object on the basis of the red or blue data alone. But there is something else, which we can loosely call a "color" channel. If we wish to make reliable colored images, we must display only those images that are forced by the data to have colored features. We can do this by making the most nearly independent map.

Define:

$p(R,\mathbf{x})$ and $p(B,\mathbf{x})$	the red and blue images,
$p(R) = \Sigma\, p(R,\mathbf{x})$	total flux of the red map,
$p(\mathbf{x}) = p(R,\mathbf{x}) + p(B,\mathbf{x})$	total flux at pixel $\mathbf{x}$.

We now compare $p(C = R|B,\mathbf{x})$ with the independent map

$$m(C,\mathbf{x}) = p(C)\, p(\mathbf{x})\ , \tag{5}$$

maximizing the relative entropy

$$-\sum_{\mathbf{x}} \sum_{C} p(C,\mathbf{x}) \log\, [p(C,\mathbf{x})/m(c,\mathbf{x})]\ . \tag{6}$$

In this way we obtain the "least colored" image permitted by the data. Because of this, the image is able to take advantage of good data no matter in which channel it appears, and it therefore has the higher resolution available from either data set, but shows spectral features only where there is definite evidence for them in the data. The combination of "red," "blue," and "color" images must all be displayed, and all are relevant. The extension to multiple spectral channels is straightforward.

4. Polarization Images

A question posed immediately after we made the first radio-astronomical maps using maximum entropy was: "But my radio sources are polarized. How can I make a maximum entropy picture of the polarized flux?" The answer to this question has been a long time coming, but now seems very simple.

Consider first the case of unpolarized emission, where f_i is the flux density (in W Hz^{-1}) from pixel i. Define $p_i = f_i/\Sigma f$ to be the dimensionless pattern of proportions of intensity. We can then interpret $\{p_i\}$ in a probabilistic manner [Skilling and Gull, 1985] as the answer to the simple question: "Where would the next photon come from?" This provides a natural rationale for the application of the maximum entropy principle for determination of the pattern $\{p_i\}$, given incomplete information.

If the emission is polarized, we have to take into account the fact that this "next photon" can fall into any of the available eigenstates of polarization. The generalization of the pattern of proportions of flux density is the quantum-mechanical density matrix. For pixel i, in the circularly polarized representation, namely,

$$\left|\begin{matrix}1\\0\end{matrix}\right\rangle = \mathrm{LHC}\ ; \qquad \left|\begin{matrix}0\\1\end{matrix}\right\rangle = \mathrm{RHC}\ , \tag{7}$$

we have

$$\rho = \begin{pmatrix} I+V & Q-iU \\ Q+iU & I-V \end{pmatrix} \Big/ 2 \sum I\ , \tag{8}$$

where I, V, U, and Q are the usual Stokes parameters. This density matrix is the quantum generalization of the probability distribution of the next photon, satisfying Trace(ρ) = 1, and having entropy $S = -\mathrm{Trace}(\rho \log \rho)$. The maximum entropy principle is again appropriate for the determination of this matrix, given incomplete information. It simplifies nicely:

$$S(\mathrm{total}) = -\sum p_i \left[\log p_i + S(\mathrm{polarization})_i\right]\ , \tag{9}$$

where $p = I/\Sigma I$, and

$$S(\mathrm{polarization}) = -\frac{1-\alpha}{2} \log \frac{1-\alpha}{2} - \frac{1+\alpha}{2} \log \frac{1+\alpha}{2}\ . \tag{10}$$

The fractional polarization α is $(U^2 + Q^2 + V^2)^{1/2}/I$. The polarization part of the entropy can vary between 0 (fully polarized, independently of the specific state) and log 2 (unpolarized). There is a bonus of log 2 for being unpolarized!

5. Probabilities or Proportions?

In these next sections we turn to more general theoretical questions. We start with an issue that has caused a great deal of debate in the astronomical community.

The simple identification given in the last section of an image with a probability distribution—"Where would the next photon come from?"—seems to us to provide an intuitive and compelling justification for the use of the maximum-entropy principle in image processing. Other authors have objected to this argument [Cornwell, 1984; Nityananda and Narayan, 1983]. Some of these objections certainly stem from misunderstandings or from differences in psychology, but one of these [Cornwell, personal communication] deserves further discussion. All probability distributions are defined axiomatically to be proportions, but the set of proportions of intensity in a given image does not by itself make a particularly convincing probability distribution. At first sight, one is being asked to select "at random" a "small element of luminance," a procedure that does indeed seem rather artificial, even if it could be done! However, the probability connection and the selection idea are unnecessary. They appear only because of previous indoctrination that probability distributions must be concerned with unpredictably changing physical variables. It seems to us that the idea of "entropy" applies equally to any set of proportions, and so too does the principle of maximum entropy, without having to have unnecessary interpretations foisted upon them. Thus, the relative proportions $\{p_i\}$ of letters used in a book have entropy; in fact that was almost the first example used to illustrate the information-theoretic method! We do not have to imagine selecting a letter at random from the book. The proportions simply have entropy $S = -\Sigma_i\, p_i \log p_i$, and exp(S) has the interpretation of being the average number of letters used. Even (pro)portions of a cake have entropy in this sense. From this more general point of view the application of the maximum-entropy principle to the proportions of an image seems to us very natural.

6. The Trouble with N: The Traffic-Jam Problem

For many years one subject has caused more discussion in our group than any other: the meaning of N, that is, the infamous N in degeneracy = exp(NS). A related question we are often asked in image reconstruction is: "How much more likely is your maximum entropy image than any other possible one?" We know of (and have at various times given) two distinct answers:

(1) Use the entropy concentration theorem (ECT): $\mathrm{pr}(\{p\}) \propto \exp(NS)$.
(2) "It is no more likely than any other."

For image reconstruction and in most other circumstances, we believe that this second response is more nearly correct, but despite this, the maximum entropy image is still the best one to choose [Gull and Skilling, 1984].

To illustrate the difficulties we have with N, it is convenient to revisit a problem that was posed by Ed Jaynes many years ago: the "traffic-jam problem." A traffic jam of length 1 mile is composed of 388 cars, of three different varieties:

Fords	Length 4 yards
Mercedes	Length 5 yards
Lincolns	Length 7 yards

How many cars of each type are there?

Here are two different answers to the problem:

(1) Apply Bayes' theorem in the space $\{n_i\}$ = number of cars (i = 1,2,3). We have two constraints: $n_F + n_M + n_L = 388$, and $4n_F + 5n_M + 7n_L = 1760$. This restricts the solution to lie on a line in $\{n_i\}$ space, connecting the points (180, 208, 0) and (318.7, 0, 69.3), and we allow a small tolerance in its width to avoid Diophantine distractions. With an initially uniform prior probability in this space, we obtain by Bayes' theorem a uniform posterior $pr(\{n_i\})$ along the line. This answer appears to us to be perfectly reasonable and to accord with common sense.

(2) Apply the maximum entropy principle to the space $\{p_i\}$ of one car. In other words, what is the probability that any given car is of type 1, 2, or 3? To obtain suitable information relating to $\{p_i\}$, we have now to interpret the given data as implying that the average length of a car (ensemble average) is $1760/388 = \langle \ell \rangle = \Sigma\, p_i\, \ell_i$. Maximizing $S = -\Sigma\, p_i \log p_i$ under this and the normalization constraint, we obtain: $p_F = 8/13$, $p_M = 4/13$, $p_L = 1/13$. These are perfectly reasonable proportions of cars and, as Jaynes emphasizes [1978, Section B], there is no possible conflict between these solutions.

In one case we have asked for and obtained the probability distribution $pr(\{n_i\})$, given suitable conditioning information in $\{n_i\}$ space; in the other we chose to interpret our information as testable information [see Jaynes, 1978] about the probability distribution in the much smaller space of one car. Conflict arises only if we try to apply the second answer to the larger space of N cars and attempt to invoke the ECT to say that exp(NS) gives the uncertainty in our prediction in that space. With N = 388 the $\{n_i\}$ would be almost certain to lie very close to the entropy maximum (to within about $N^{1/2}$). This conclusion is indeed very strange: What if the jam were 10 times longer? Would we be even more certain as to its composition? Of course not!

But what has gone wrong with the ECT? If we knew in advance that the jam was in fact composed of cars taken individually and randomly from a reservoir with equal proportions of all the types of car, then the ECT would indeed be valid. But here we certainly aren't told that. We are very uncertain about the proportions of cars in the reservoir, though we might guess that the reservoir contains fewer cars of the more expensive types. However, if we were to put a uniform prior probability distribution on the composition of the reservoir (technically, a uniform de Finetti generator) and repeat the analysis, we would recover the first result.

The moral is that you can't get information along the constraint line just by thinking about it! In this case the degeneracies corresponding to rearrangements of the order of cars in the jam, though real enough, are simply not relevant for our inferences about the relative numbers of cars in the jam. We shall have more to say about this traffic jam later [Gull and Skilling, 1984; Jaynes, 1986], but in the meantime: Beware of the entropy concentration theorem!

7. Positive Aspects of the Traffic Jam

7.1. The Minimum Entropy Principle

Consider another simple illustration of our difficulty: An event has two outcomes. Call them Heads and Tails. It is repeated N times. What is a suitable prior for any given set of results $\{r_i\}$ (r_i = H,T; i = 1,...,N)? In the absence of any information at all, we must consider all results to be equally likely, so any given result $\{r_i\}$ has a probability of $1/2^N$. Expressing this another way, we find that results containing a total of n Heads have a probability that reflects the degeneracy factor:

$$\mathrm{pr}(n) = 2^{-N} \frac{N!}{n!(N-n)!} \,. \tag{11}$$

Suppose now that we are told there is an underlying similarity between the repetitions of the event: it's the same coin. There is then a well defined ratio q for the proportion of Heads, which is the probability of "Heads" for any one toss. This information dramatically changes our state of knowledge, and we can no longer tolerate the heaping up of probability near n = N/2 that is predicted by the degeneracy (monkey) factors. Since we don't know the ratio q (except that it lies between 0 and 1), we must take something like

$$\mathrm{pr}(n) = \frac{1}{N+1}\,, \qquad n = 0,\ldots,N \,. \tag{12}$$

This implies that the probabilities of sequences of results are no longer all equal:

$$\mathrm{pr}(\{r_i\}) = \frac{n!(N-n)!}{(N+1)!} \,. \tag{13}$$

We have generated, therefore, an "anti-entropy concentration" prior. A suitable approximation for the case where N is fairly large is

$$\mathrm{pr}(\{r_i\}) = \exp[-N\,S(q)] \,, \tag{14}$$

where $S(q) = -q \log q - (1-q)\log(1-q)$, the entropy of q. Of course, we could have arrived at this answer in other ways: by "probability of a probability" arguments, or by dignifying the procedure with the names of de Finetti or Dirichlet. The answer generalizes easily to the case of multiple outcomes, but when the number of possible outcomes M exceeds the number of repetitions it is wise to consider more carefully the "prior" for $\{q_i\}$ (i = 1,...,M) [Jaynes, 1968].

7.2. The Byte-Boundary Problem

Suppose a computer contains a string of 32,768 zeroes and ones that are known to be organized in patterns of eight (bytes), but unfortunately we have lost track of the byte-boundary marker. We wish to recover it. This is a simple task for the "anti-entropy concentration" prior. Suppose the boundary lies at position i (i = 0,...,7). Then, using Bayes' theorem:

$$\mathrm{pr}(i|\mathrm{string}) \propto \mathrm{pr}(i)\,\mathrm{pr}(\mathrm{string}|i)\,. \tag{15}$$

The prior is presumably constant (0.125), and the likelihood should, as above, be taken as $\exp[-N\ S(\{q\})]$, where N = 4096 (the number of bytes in the string) and $S(\{q\}) = -\Sigma\ q_i \log q_i$, the "entropy" of the string ($q_i = n_i/N$).

In other words, the byte boundary is at the position that puts the most structure into the string: the minimum entropy. Like many Bayesian solutions, this answer is so staggeringly obvious when you have seen it that there seems little point in testing it—but we did, and found that for 32k bits of ASCII the byte boundary is determined to a quite astronomical significance level. (The reader may like to contemplate why the position is still ambiguous to one of two positions.)

Other applications of this procedure abound—for example, tests of pattern length in strings and other tests of randomness.

8. Acceleration of Relativistic Electrons in a Plasma

We conclude with an example that is a good deal closer to physics, taken from our field of radio astronomy.

Most powerful radio sources emit their radiation by the synchrotron mechanism: Nonthermal radiation produced as relativistic (Lorentz factors $\sim$10,000) electrons travel around magnetic field lines. This radiation has a characteristic spectrum that allows us to calculate the energy spectrum of the emitting electrons: it is a power law. The number n(E) of electrons of energy E is $n(E) \propto E^{-\gamma}$. The energy index γ is, quite remarkably, almost constant. In our galaxy, for example, all sources (mainly exploded stars) have $\gamma \simeq 2.6$.

Why is there such a universal spectrum? At last we are beginning to understand this. It is interesting to discuss it in the light of the maximum-entropy principle.

Many of us believe that most, if not all, probability distributions that arise in physics are maximum entropy distributions, and that "all" we have to do is identify the appropriate constraints. In this case we need a constraint on $\langle \log E \rangle$ to give a power law spectrum rather than a Maxwell-Boltzmann one. Multiplicative (first-order Fermi) processes are needed: the relativistic electrons must receive a constant fractional acceleration irrespective of their energy. We must contrast this behavior with the approach to equilibrium in a gas. In a gas the molecules interact with each other, exchanging little packets of energy. These relativistic electrons, on the other hand, cannot be interacting directly with each other, but rather with something much bigger, which can well afford to lose (or acquire) any amount of energy that the individual particles possess.

The reason there is a constant spectrum (actually the Lagrange multiplier in the maximum-entropy principle) is now clearer: we think we can identify these "bigger partners" [Bell, 1978]. They are interstellar shock fronts. These shock fronts originate usually as the blast waves of supernovae, and propagate at anything up to several thousand kilometers per second. They are discontinuities in velocity maintained by plasma processes in the thermal electron plasma. Relativistic electrons, having greater momentum, are likely to pass through these shocks undisturbed, though they are eventually scattered and their directions become randomized. Because they move so fast compared to the thermal plasma, they must on average make many crossings of the shock before they are lost downstream. However, the velocity discontinuity at the shock means that the electrons receive a jolt every time they make a double crossing, playing "ping-pong" with the shock. This acceleration is indeed a multiplicative Fermi process, and it depends only on the compression ratio of the shock. But the probability of being lost downstream also depends only on the same compression ratio. This combination of constraints leads to a power-law spectrum [Heavens, 1983]. Further, the compression ratios for most shocks are the same; they are "strong" shocks, for which the compression in density is a factor of 4.

The nice thing is that the predicted spectrum agrees with what is observed.

9. Acknowledgments

We have benefited from many enjoyable discussions with our colleagues and wish in particular to thank Geoff Daniell, Tim Cornwell, and Ed Jaynes.

10. References

Bell, A. R. (1978), "The acceleration of cosmic rays in shock fronts," Mon. Not. R. Astron. Soc. 182, pp. 147-156, and 182, pp. 443-455.

Cornwell, T. J. (1984), "Is Jaynes' maximum entropy principle applicable to image construction?" in Indirect Imaging, J. A. Roberts, ed., Cambridge University Press.

Gull, S. F., and G. J. Daniell (1978), "Image reconstruction from incomplete and noisy data," Nature 272, pp. 686-690.

Gull, S. F., and J. Skilling (1984), "The maximum entropy method," in Indirect Imaging, J. A. Roberts, ed., Cambridge University Press.

Heavens, A. F. (1983), "Particle acceleration in shock waves," Ph.D. Thesis, University of Cambridge.

Jaynes, E. T. (1968), "Prior probabilities," IEEE Trans. SSC-4, pp. 227-241. Reprinted in Jaynes [1983].

Jaynes, E. T. (1978), "Where do we stand on maximum entropy?" in The Maximum Entropy Formalism, R. D. Levine and M. Tribus, eds., MIT Press, Cambridge, Mass. Reprinted in Jaynes [1983].

Jaynes, E. T. (1979), "Concentrations of distributions at entropy maxima," presented at the 19th NBER-NSF Seminar on Bayesian Statistics, Montreal. Reprinted in Jaynes [1983].

Jaynes, E. T. (1983), Papers on Probability, Statistics, and Statistical Physics, Synthese Library, vol. 158, R. D. Rosenkrantz, ed., D. Reidel.

Jaynes, E. T. (1986), "Monkeys, kangaroos, and N," in Maximum Entropy and Bayesian Methods in Applied Statistics, J. H. Justice, ed., Cambridge University Press, pp. 26-58.

Nityananda, R., and R. Narayan (1983), "Maximum entropy image reconstruction—a practical noninformation theoretic approach," Astron. Astrophys. 118, p. 194.

Skilling, J., and S. F. Gull (1985), "Algorithms and applications," Maximum-Entropy and Bayesian Methods in Inverse Problems, C. Ray Smith and W. T. Grandy, Jr., eds., D. Reidel, pp. 83-132.

PRIOR KNOWLEDGE MUST BE USED

John Skilling

Department of Applied Mathematics and Theoretical Physics, University of Cambridge, Silver Street, Cambridge CB3 9EW, England

Stephen F. Gull

Mullard Radio Astronomy Observatory, Cavendish Laboratory, University of Cambridge, Madingley Road, Cambridge CB3 0HE, England

C. R. Smith and G. J. Erickson (eds.),
Maximum-Entropy and Bayesian Spectral Analysis and Estimation Problems, 161–172.

1. Introduction: The Need for an Initial Model

Entropy is defined [Jaynes, 1962] as a relative quantity

$$S(p;m) = -\sum_{i=1}^{n} p_i \log(p_i/m_i), \qquad \sum p_i = \sum m_i = 1, \qquad (1)$$

which measures the (minus) configurational information of a set of n proportions p_i, relative to an initial model m_i. Any positive, additive scalar quantity can be identified with a set of proportions, so the entropy formula can be applied equally to probability distributions and to a wide variety of physical phenomena.

We use the maximum entropy method (MEM) in data analysis whenever we wish to estimate a set of proportions, but the relevant data are seriously incomplete—so incomplete that the set of "feasible" proportions p that could agree with the data is too large for easy comprehension. Writing the constraint obtained from the data as

$$C(p) \lesssim 1, \qquad (2)$$

we may be faced with the difficulty that very many proportions obey the condition. Practical necessity then forces us to make a selection from the feasible set which we must present as the "result" of the experiment that gave us the data.

Of course, this selection is not a matter of pure deduction, for we are logically free to make any selection we like. Nevertheless, we are led to prefer those particular proportions p that have maximum entropy. Several arguments lead to this selection. Perhaps the most appealing is the abstract argument of Shore and Johnson [1980; see also Johnson and Shore, 1983], based on simple and general consistency requirements.

However, we ought to supply some initial model m before we use MEM. For many practical cases, it is sufficient to ignore the problem and just let m_i be a constant, independent of i, in which case the model cancels out of the analysis and can be quietly forgotten. Although this is certainly the simplest way of dealing with the model, it is equally certainly not the best. On the other hand, we must be careful when we introduce initial models, precisely because they influence the final result. Indeed, any feasible proportions p can be obtained by suitable choice of initial model m. This gives MEM enormous flexibility but also introduces some danger. We are reminded of the moral that Jaynes [1978] draws from carpentry, that the introduction of more powerful tools brings with it the obligation to exercise a higher level of understanding and judgment in using them. If you give a carpenter a fancy new power tool, he may use it to turn out more precise work in greater quantity; or he may just cut off his thumb with it. It depends on the carpenter.

Unfortunately, our prior state of knowledge about what we are observing is often vague and amorphous, difficult to code into a precise initial model. Nevertheless, it is clear from studying some specific practical examples that an attempt must be made. Otherwise the full power of MEM will be wasted. In this paper, we explore possible ways of proceeding in a simple case. Generalization from this approach leads to the conclusion that MEM is an even more general and powerful tool than it appears at first sight.

2. Astronomical Photographs

In astronomy, we often see photographs that contain both point stars and extended objects such as nebulae and galaxies. Accordingly, we are not at all surprised to see bright points in the picture. Nevertheless, we often wish to remove some of the imperfections induced by the telescope and clarify the picture by the maximum entropy method. If we do not know beforehand where the stars will be, we seem forced to assign a uniform initial model m_i = constant. Our state of knowledge is translation invariant, so the model must also be translation invariant, hence constant.

Indeed, maximum entropy relative to a uniform model can demonstrably produce good astronomical images, including stars. There are, however, subtle imperfections in these images. Clearly, the point stars are the objects in the picture that have greatest surface brightness. Dividing the image into n cells, those cells that contain a star have a much greater proportion p of the total intensity than the others. The entropy gradient

$$\partial S/\partial p_i \;=\; -1 - \log(p_i/m_i) \tag{3}$$

is unusually strong there.

Now we usually set up our data constraint C(p) by comparing the noisy actual data $D_k \pm \sigma_k$ $(k = 1,2, \ldots, M)$ with the mock data P_k that would have been obtained by observing the proportions p with the telescope [Gull and Daniell, 1978]. The natural consistency test is the (normalized) chi-squared statistic

$$C(p) \;=\; \sum_{k=1}^{M} (P_k - D_k)^2/M\sigma_k^2 \qquad (M = \text{number of data})\,, \tag{4}$$

and we reject all images p for which C(p) significantly exceeds 1. The maximum entropy image is obtained by maximizing S(p;m) over C(p) and obeys the variational equation

$$\partial S/\partial p_i \;=\; \lambda\, \partial C/\partial p_i \tag{5}$$

for the appropriate Lagrange multiplier λ. If the entropy gradient $\partial S/\partial p_i$ is unusually large in some cells, the constraint gradient $\partial C/\partial p_i$ must also be large there, so the actual misfit will be unusually large.

Thus most of the misfit statistic C(p) may well become assigned to those small patches of the photograph that contain stars, so the stars are allowed to relax relatively far from the data and are reconstructed systematically too faint. Conversely, extended objects of lower surface brightness will be reconstructed too bright (to compensate for the stars' being too faint) and too noisy (because too much of the misfit, which allows for noise, has been assigned to the stars). The effect can be very significant, and although it can sometimes be masked by tinkering with the form of C(p) [Bryan and Skilling, 1980], the theoretical difficulty remains.

Somehow, the astronomer must tell the entropy not to discriminate against the stars, even though his state of knowledge is translation invariant.

A practical astronomer, uninhibited by theoretical dogma, would doubtless proceed as follows, or similarly. He would look at his reconstructed (maximum entropy) image, and see that it contained a sprinkling of bright points. He would be prepared to accept them as stars. He would then know how many bright stars there were, and roughly how bright they were, and where they were. He could remove them from his data, perhaps avoiding bias by using Bayesian parameter-fitting for the positions and brightnesses. Lastly, he could use his revised data to give him an uncorrupted maximum entropy image of the fainter objects such as nebulae and galaxies. (At Cambridge, we adopt this procedure routinely in radio astronomy.)

The crucial step occurs when the practical astronomer uses his first reconstruction to learn about the locations of the stars, and revises his data accordingly. Exactly the same effect can be accomplished entirely within the maximum entropy formalism by allowing the model m to be adjusted in the light of the reconstruction p. This change also lets us deal with nonlinear data, from which the stars could not be directly subtracted. We conclude that the astronomer's prior knowledge about stars must be encoded into a result-dependent model m(p).

3. Generalities

The same prescription will be needed in applications other than astronomical photography. Thus, geologists are not surprised to find rocks in relatively uniform layers, often roughly horizontal. Spectroscopists are not surprised to find sharp lines, or even doublets or multiplets of specified intensity ratios, in their spectra. Crystallographers are not surprised to find molecules built from connected atoms within their crystals. And so on. Each of these applications needs a result-dependent model m(p) if prior knowledge is to be used constructively.

Two difficulties spring to mind. There is the practical difficulty that it may be hard to encapsulate one's professional expertise in a particular discipline into a pre-assigned function m(p). There is also the philosophical difficulty that MEM has been developed and justified in terms of finding a result p conditional on a specific pre-assigned inital model m. By asking m to depend on p, we appear to be in serious danger of confusing our initial and final images, which runs counter to the normal philosophy of MEM. It is, however, quite clear that we must allow m to depend on p, and the philosophy will just have to accommodate the demand.

4. How to Set Up m(p)

We have argued that to obtain the best reconstructions, which use prior knowledge to the full, we must let m depend on p, and maximize

$$S(p) = -\sum_{i=1}^{n} p_i \log[p_i/m_i(p)] . \tag{6}$$

Clearly this procedure must be treated with great care. For example, the uniqueness theorem for linear data no longer holds. We now develop some preliminary thoughts on how to encode m(p) for certain particularly simple types of prior knowledge. To expound these ideas, we use the "monkey" approach, in which p is identified with a probability distribution. Specifically, we think of p_i as the probability that a sample chosen at random from the distribution would come from cell i.

The simplest case to consider is that of point stars, for which the prior knowledge was translation invariant. A single sample from the image could come from anywhere, but subsequent samples are more likely to come from exactly the same place. The space S of single samples is too small and impoverished to contain our prior knowledge, so we move to the larger space S^N of multiple samples, from which we can always recover distributions on S by marginalization [Jaynes, 1978].

Suppose we have prior knowledge that it is reasonable to assume that a fraction q of the image comes from a single one of the n cells, leaving the remaining probability to be distributed equally as $r = (1-q)/(n-1)$ among the other n-1 cells (Fig. 1). From this prior knowledge we can develop the

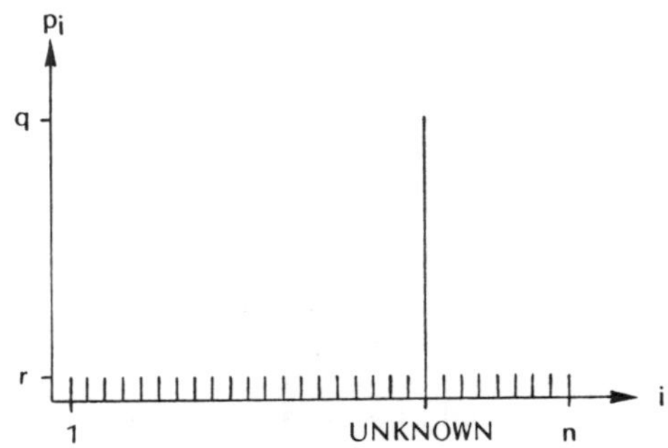

Figure 1. A fraction q of the initial model is assigned to a single (unknown) cell.

following probabilities for multiple samples:

$$\Pr(\text{star in cell k}) = \text{constant} = 1/n \tag{7}$$

$$\Pr(\text{1st sample in i} \mid \text{star in k}) = r + (q-r)\delta_{ik} \tag{8}$$

$$\Pr(\text{2nd sample in j} \mid \text{star in k}) = r + (q-r)\delta_{jk} \tag{9}$$

$$\Pr(\text{1st in i and 2nd in j} \mid \text{star in k}) = [r + (q-r)\delta_{ik}]\,[r + (q-r)\delta_{jk}] \tag{10}$$

$$\begin{aligned}\Pr(\text{1st in i and 2nd in j}) &= \sum_k \Pr(\text{1st in i and 2nd in j} \mid \text{star in k}) \Pr(\text{star in k}) \\ &= r^2 + \frac{2(q-r)r}{n} + \frac{(q-r)^2\delta_{ij}}{n} \;.\end{aligned} \tag{11}$$

This shows that prior knowledge of the desired type does show up as a non-uniform initial model

$$m_{ij} = r^2 + \frac{2(q-r)r}{n} + \frac{(q-r)^2\delta_{ij}}{n} \tag{12}$$

if we are prepared to work in the product-space of two (or more) samples from the image.

Clearly m_{ij} must always be symmetric because we may suppose successive samples from the image to be independent and exchangeable, and the general translation-independent initial measure on the double space is

$$m_{ij} = m_{|i-j|} \;. \tag{13}$$

Given a double-space model m_{ij}, any other probability distribution

$$p_{ij} = \Pr(\text{1st in i and 2nd in j} \mid \text{model m}) \tag{14}$$

will have entropy

$$S^{(2)} = -\sum_{ij} p_{ij} \log(p_{ij}/m_{ij}) \tag{15}$$

relative to the model. It would be catastrophically expensive if we had to compute directly in the very large double space to do maximum entropy in practice, but fortunately there are simplifications. We are supposing that successive samples are independent, so p_{ij} is merely the product

$$p_{ij} = p_i p_j, \; p_i = \Pr(\text{given sample in cell } i \mid \text{model } m) \; . \tag{16}$$

Hence the entropy becomes

$$S^{(2)}(p) = -2 \sum_i p_i \log p_i + \sum_{ij} p_i p_j \log m_{ij} \tag{17}$$

and the entropy gradient, which defines the reconstruction p through the variational Eq. (5), is

$$\partial S^{(2)}/\partial p_i = 2\,(-1 - \log p_i + \sum_j p_j \log m_{ij}) \; . \tag{18}$$

Comparing this with the ordinary entropy gradient, Eq. (3), we see that this gradient mimics the effect of an ordinary single-space entropy, Eq. (1), relative to a fixed initial model defined by the values

$$u_i = \exp(\sum_j p_j \log m_{ij}) \; . \tag{19}$$

This is the property we seek. Although the actual prior knowledge can be coded only in terms of a nonuniform model on a product space, it behaves like a result-dependent model in the single space. Furthermore, if the knowledge is translation-invariant, the evaluation of $\Sigma_j\, p_j \log m_{ij}$ is merely an n-cell convolution, which is entirely practical computationally.

It is interesting to compare the behavior of the ordinary entropy, Eq. (1), with the double-space entropy, Eq. (17), for a three-cell image (p_1,p_2,p_3). Taking $q = 0.95$ (95% of the proportion may come from one of the three cells), the two entropies are plotted in Figs. 2 and 3 respectively. The ordinary single-space entropy has the usual single maximum at (1/3, 1/3, 1/3). The double-space entropy, on the other hand, has four local maxima, and shows a pronounced tendency to favor images with one of the proportions larger than the other two.

We can extend the development to higher product spaces. The single-space model

$$m_i = \Pr(i) \tag{20}$$

codes prior knowledge of absolute positions. The double-space model

$$m_{ij} = \Pr(j \mid i)\,\Pr(i) \tag{21}$$

codes additional knowledge of relative positions in the form of two-point correlations such as bond lengths in a molecule. The triple-space model

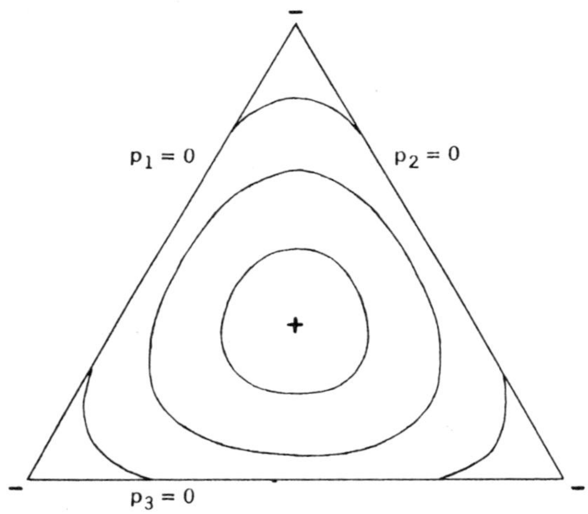

Figure 2. Contours of ordinary entropy $S = -\Sigma p \log p$ for a three-cell image. The domain $\Sigma p = 1$, $p \geq 0$, is plotted.

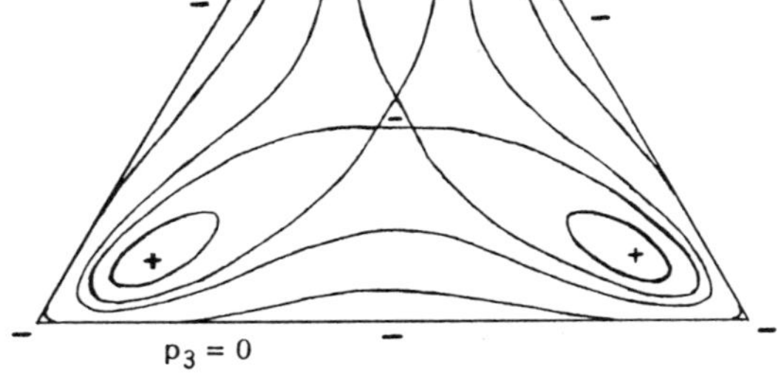

Figure 3. Contours of double-space entropy $S^{(2)}$ from Eq. (17) for a three-cell image with $q = 0.95$.

$$m_{ijk} = \Pr(k \mid ij)\Pr(j \mid i)\Pr(i) \tag{22}$$

codes additional knowledge of relative shapes such as bond angles in a molecule. It gives an entropy

$$S^{(3)} = -\sum_{ijk} p_i p_j p_k \log(p_i p_j p_k / m_{ijk}) \tag{23}$$

that mimics an ordinary entropy with model

$$u_i = \exp\left(\sum_{jk} p_j p_k \log m_{ijk}\right) . \tag{24}$$

Quadruple and higher models can allow for yet more detailed prior knowledge, and an N-space treatment mimics an ordinary entropy with model

$$u_i = \exp(\text{polynomial of degree } N-1 \text{ in } p) . \tag{25}$$

There is no natural limit to this extension, and as N becomes arbitrarily large, we reach a model that mimics

$$\begin{aligned} u &= \exp(\text{arbitrary function of } p) \\ &= \text{arbitrary positive function of } p . \end{aligned} \tag{26}$$

We have now succeeded in developing a formal maximum entropy approach that has the effect of allowing arbitrary result-dependent "initial" models m(p). The trick is to generalize from the space S of single samples to the richer space S^N of many samples. MEM, in principle carried out in the product space but in practice computed in the single space, will then produce an optimal reconstruction p on the basis of the knowledge coded into the model.

5. Can We Use an Arbitrarily Large Number of Samples?

Let us return to the simple case of knowing that a proportion q of the total may come from one unknown cell (Fig. 1). The probability formulation becomes more and more intricate as the number N of samples increases, but we can bypass this by proceeding directly to the entropy. In the "monkey" model, a set of occupation numbers

$$(g_1, g_2, \ldots, g_n) , \quad \sum g_i = N , \quad g_i = N p_i \tag{27}$$

will have an associated degeneracy

$$\Omega = N! \frac{m_1^{g_1} m_2^{g_2} \cdots m_n^{g_n}}{g_1! \; g_2! \cdots g_n!} \tag{28}$$

relative to an initial measure or model m, and the entropy is defined as

$$S = \lim_{N\to\infty} N^{-1} \log \Omega = -\sum p_i \log(p_i/m_i) \,. \tag{29}$$

Clearly this is a very natural way of working in the space S^N for large N.

Suppose the star is in cell k, so that

$$m_i^{(k)} = \begin{cases} q/h, & i = k \\ r/h, & i \neq k \end{cases} \tag{30}$$

where h is the quantum of measure. The degeneracy is now

$$\Omega^{(k)} = \frac{N! \, (r/h)^N \, (q/r)^{g_k}}{\Pi \, g_i!} \,. \tag{31}$$

The total degeneracy corresponding to our state of knowledge that the star could be in any of the cells may be obtained by <u>adding the individual degeneracies</u> to reach

$$\Omega = \sum_k \Omega^{(k)} = \frac{N!}{\Pi g_i!} (r/h)^N \sum_k (q/r)^{g_k} \,. \tag{32}$$

As usual, it is convenient to use the associated entropy

$$\begin{aligned} S^{(\infty)} &= \lim_{N\to\infty} (N^{-1} \log \Omega) + \log h \\ &= -p_m \log(p_m/q) - \sum_{i \neq m} p_i \log(p_i/r) \end{aligned} \tag{33}$$

where m is the cell containing the greatest proportion

$$p_m = \max\,(p_1, p_2, \ldots, p_n) \,. \tag{34}$$

This is precisely what we want since the best estimate of the position of the star is the position of the brightest cell. Figure 4 shows contours of $S^{(\infty)}$

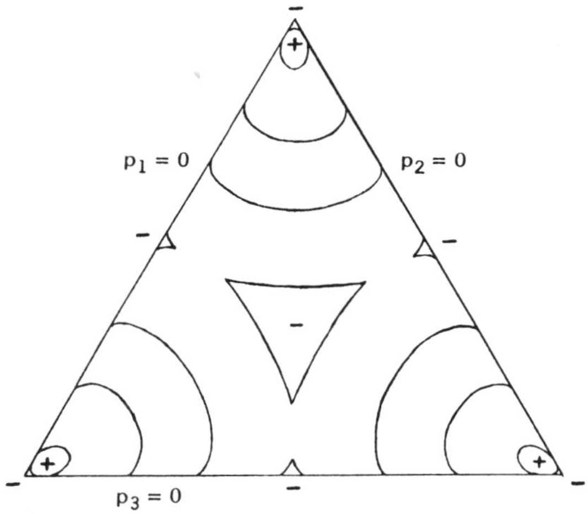

Figure 4. Contours of $S^{(\infty)}$ from Eq. (33) for a three-cell image with $q = 0.95$.

for a three-cell image with $q = 0.95$, for comparison with Figs. 2 and 3. The uniform image has become a local minimum, so there are now only three local maxima at positions $p_m = q = 0.95$, which is entirely reasonable.

The price to be paid for this neat adherence to the given state of prior knowledge is a loss of uniqueness. There may be several local maxima of entropy over the given constraint region $C(p) < 1$. If the brightest cell is only weakly indicated by the data, then there will be serious danger of error if we select just the highest overall maximum of entropy. A wiser course might be to display as many of the local maxima as are reasonably accessible to our computer programs, writing our preference for particular images in terms of their numerical values of entropy.

6. Conclusion: Anything Goes!

The maximum entropy method is sufficiently flexible to allow us to use result-dependent models m(p) without destroying the underlying philosophy. The method does not constrain the assumptions or the physics we may wish to use, and it will always give optimal reconstructions of any set of proportions p.

We have seen how to use degeneracy arguments to code a particularly simple type of prior knowledge into a specific model m(p), and this can be generalized immediately to somewhat more complicated problems. We may hope to generate and use more such models in the future. There is, however, a price to be paid in terms of safety. Users of result-dependent models should be prepared to be both careful and intelligent.

7. References

Bryan, R. K., and J. Skilling (1980), "Deconvolution by maximum entropy as illustrated by application to the jet of M87," Mon. Not. R. Astron. Soc. 191, pp. 69-79.

Gull, S. F., and G. J. Daniell (1978), "Image reconstruction from incomplete and noisy data," Nature 272, pp. 686-690.

Jaynes, E. T. (1963), Brandeis lectures, in Papers on Probability, Statistics, and Statistical Physics, R. Rosenkrantz, ed., Reidel (1983).

Jaynes, E. T. (1978), "Where do we stand on maximum entropy?" in Papers on Probability, Statistics, and Statistical Physics, R. Rosenkrantz, ed., Reidel (1983).

Johnson, R. W., and J. E. Shore (1983), "Comments and corrections to 'Axiomatic derivation of the principle of maximum entropy and the principle of minimum cross-entropy,'" IEEE Trans. Inf. Theory IT-29, pp. 942-943.

Shore, J. E., and R. W. Johnson (1980), "Axiomatic derivation of the principle of maximum entropy and the principle of minimum cross-entropy," IEEE Trans. Inf. Theory IT-26, pp. 26-37.

HOW THE BRAIN WORKS: THE NEXT GREAT SCIENTIFIC REVOLUTION

David Hestenes

Arizona State University, Tempe, AZ 85287

In spite of the enormous complexity of the human brain, there are good reasons to believe that only a few basic principles will be needed to understand how it processes sensory input and controls motor output. In fact, the most important principles may be known already! These principles provide the basis for a definite mathematical theory of learning, memory, and behavior.

C. R. Smith and G. J. Erickson (eds.),
Maximum-Entropy and Bayesian Spectral Analysis and Estimation Problems, 173–205.

1. Introduction

I am here to tell you that another major scientific revolution is well under way, though few scientists are aware of it even in fields where the revolution is taking place. In the past decade, a mathematical theory has emerged that bridges the gap between neurophysiology and psychology, providing penetrating insights into brain mechanisms for learning, memory, motivation, and the organization of behavior. It promises a formulation of fundamental principles of psychology in terms of mathematical laws as precise and potent as Newton's laws in physics. If the current theory is on the right track, then we can expect it to develop at an accelerating pace, and the revolution may be over by the turn of the century. We will then have a coherent mathematical theory of brain mechanisms that can explain a great range of phenomena in psychology, psychophysics, and psychophysiology.

To say that this conceptual revolution in psychology will be over is not to imply that all problems in psychology will be solved. It is merely to assert that the fundamental laws and principles of explanation in psychology will be established. To work out all their implications will be an endless task. So has it been in physics, where the laws of classical and quantum mechanics have been well established for some time, but even the classical theory continues to produce surprises. So has it been with the recent revolution in biology brought about by breaking the genetic code; though some principles of genetic coding are undoubtedly still unknown, the available principles are sufficient to provide the field with a unified theoretical perspective and determine the modes of acceptable explanation. Biology is now regarded as a more mature science than psychology, but we shall see that it may be easier to give psychology a mathematical formulation.

If indeed a conceptual revolution is under way in psychology and the brain sciences, you may wonder why you haven't heard about it before. Why hasn't it been bannered by Psychology Today or proclaimed by some expert on the Johnny Carson show? Why is it announced here for the first time to an audience of mathematicians, physicists, and engineers? Before these questions can be answered, we need to consider the status and interrelations of the relevant scientific disciplines.

2. The Science of Mind and Brain

Let us adopt the term neuroscience for the science of mind and brain. Neuroscience is the most interdisciplinary of all the sciences, and it suffers accordingly. The whole field has been carved into a motley assortment of subdisciplines that rarely communicate with one another. Consequently, most experts in one branch of neuroscience are profoundly ignorant about even closely related branches. Very few have a well grounded perspective on the field as a whole.

The neurosciences have accumulated overwhelming evidence that the characteristics of behavior observed, manipulated, and analyzed by psychologists are derived from the functioning of animal nervous systems or brains. This is the basis for the scientific conception of mind as a function of the brain, and it justifies regarding psychology as the science of mind. Despite this modern insight, the traditional academic division between psychology and biology perpetuates an artificial separation between mind and brain in research as well as knowledge.

Research on the structural properties of brains is carried out in the well established fields of neurophysiology and neuroanatomy as well as in a variety of related specialties such as electroencephalography. These fields subserve the medical profession, which, for the most part, pays scant attention to research in psychology and psychophysics. Psychophysiology attempts to bridge the mind-brain gap in research and knowledge, but the connections are academically tenuous.

The established disciplines in neuroscience from neurophysiology to psychology are predominantly empirical in content. They have accumulated a vast store of isolated facts about the structure and function of brains, but little in the way of coherent theory. They comprise the empirical component of neuroscience. The theoretical component of neuroscience is developing in the fledgling field of neural modeling. This field has yet to become a recognized academic discipline, so the academic respectibility of anyone who works in it is at risk.

One might expect the established neurosciences to encourage and support neural modeling. But the empiricist suspicion of theory in general and mathematical theory in particular is pervasive in these fields. I recently conversed at length with two capable young assistant professors at prominent universities. One was a neurophysiologist, the other a psychologist. Both wished to pursue research in neural modeling, but neither would dare to mention this to his colleagues or even consider beginning such research until he achieves tenure. As a consequence of this pervasive antitheoretical bias in the neurosciences, only a few tenured mavericks, like the physiologist Walter Freemann at Berkeley, have developed the mathematical skills needed for serious neural modeling. Most of the neural modeling is done by mathematically trained outsiders from engineering, mathematics, and physics.

As the name suggests, neural modeling is concerned with developing mathematical models of neurons and their interactions. The modeling proceeds at two levels, the single neuron level and the neural network level. These levels are concerned with different experimental and theoretical techniques, facts, and issues. The single neuron level is the physiological level of neural modeling, for it involves the complex details of cell and membrane physiology, chemistry, and physics. Neural modeling at this level has a measure of respectability in neurophysiological circles owing to the impressive success of Hodgkin and Huxley, who won a Nobel Prize for modeling and measuring the propagation of electrical signals along the axon of a neuron. The famous Hodgkin-Huxley equations are recognized as a paradigm for neural modeling at the neuron level.

The aim of modeling at the network level is to explain the information processing capabilities of macroscopic brain components as collective properties of a system of interacting neurons. This is the psychophysical level of neural modeling. It correlates the electrical activity of neural networks with their "mental" processing capabilities. At this level the fine physiological details of single neuron dynamics become unimportant.

Network modeling is the theoretical bridge between the microscopic and the macroscopic levels of brain activity, between neurophysiology and psychology. Consequently it is open to objections by empiricists at both ends who care little about connecting the antipodes. Little wonder that network modeling is mostly ignored by the neuroscience establishment! Little wonder that the establishment is unaware of the theoretical revolution taking place in its own fields!

I am especially pleased to tell this audience about the exciting developments in network modeling, because the field is wide open for theoretical exploration, and I doubt that I could find an audience more qualified to contribute. Neural modelers have yet to employ statistical concepts such as entropy with the skill and sophistication of the scientists gathered here. Indeed, I believe that maximum-entropy theory will play its greatest role yet in the neural network theory of the future.

3. Introduction to Grossberg

While neural modeling is ignored by the neuroscience establishment, the neural modelers tend to ignore one another. A more fragmented field would be hard to find. Almost everyone in the field seems to be pushing his or her own model, although a few small groups of interacting modelers have formed. One could read extensively in the literature without discovering that a coherent, general network theory already exists. The theory has been developed over the past two decades by Stephen Grossberg.

Grossberg is by far the most versatile and prolific of the neural modelers. He has written extensively on nearly every aspect of neural modeling. He has elevated the subject from a collection of isolated models to a genuine mathematical theory with a small set of general principles to guide the modeling of any brain component, and he has worked out many specific applications. He has thus produced the first truly <u>coherent</u> theory of learning, perception, and behavior. His theory provides coherent explanations for a wide range of empirical results from psychology, psychophysics, psychophysiology, and even neuropharmacology. And it makes a number of striking new predictions that are yet to be verified. Right or wrong, Grossberg has produced the first mathematical approach to psychology that deserves to be called a theory in the sense that the term "theory" is used in the physical sciences.

In spite of all this, Grossberg has been overlooked or ignored by most of the neural modeling community as well as the neuroscience establishment. Grossberg is seldom referenced by other neural modelers except for an occasional criticism, and he returns the favor. No doubt my remarks have strained your credulity to the limit, so let me try to explain why Grossberg is not more widely appreciated.

Let us first consider why Grossberg's impact on the neuroscience establishment has been so slight. In recent years, Grossberg has attempted to reach psychologists with several long reviews of his work. Unfortunately, few psychologists have the background needed to understand the mathematical core of his theory. Aware of this fact, Grossberg has pushed the mathematics into the background and presented a detailed qualitative account of this theory. But verbal arguments lack the logical force of mathematics at the most crucial points. Moreover, few psychologists are convinced that neural mechanisms are needed in psychological theory. So Grossberg's attempt to give a coherent account of the diverse phenomena in psychology looks to many psychologists like a series of extravagant claims of credit for every significant result in the field.

For quite different reasons, Grossberg is likely to be quickly dismissed by experts in neurophysiology. These experts are aware of many uncertainties and complexities in neuron dynamics, so they know that Grossberg's equations cannot be empirically justified by current physiological evidence, even though the questions are not inconsistent with the evidence. They look for "bottom-up" justification from physiology, whereas the main justification for Grossberg's equations is "top-down," from psychology. Grossberg's theory suggests constraints on physiological theory to make it consistent with psychological evidence. It therefore provides a guide for physiological research. But the physiologists are not looking for external guidance. And experts often point to the vast mass of partially digested data about brains as evidence that brains are much too complex for simple explanations. So they are skeptical of Grossberg's claim to explain brain functions with a small number of mathematical laws and organizing principles—too skeptical, probably, to give Grossberg the attention he needs in order to be understood.

It is harder to explain why Grossberg has been ignored by other neural modelers, since they share certain basic ideas about neural modeling and its significance. I believe the main reason is that very few have expended the substantial effort required to understand and evaluate Grossberg's work. Grossberg is not an easy read. To be understood, he must be studied—studied for weeks or months, not merely days. Let me cite my own experience by way of example.

In 1976 I made my initial foray into the neural modeling literature and came away with the disappointing impression that the field was hopelessly far from explaining anything important. I was unmoved by the only article of Grossberg's that I came across at the time. A couple of years ago I encountered a former student of mine, Bob Hecht-Nielsen, who spoke enthusiastically about Grossberg's work and told me how he was using it to design practical devices that learn to classify patterns. Since I had great respect for Bob's judgment, I decided to give Grossberg a closer look. Fortunately, Bob gave me sufficient insight into Grossberg's ideas to overcome the difficulties I again found in understanding his writings. Being a theoretical physicist, I had no difficulties at all with the mathematics in Grossberg's articles. But at first I had trouble in orienting myself to the thrust of his work and even in interpreting the variables in his equations. Interpretation and evaluation of Grossberg's equations require some familiarity with empirical data spanning the entire range from neurophysiology to psychology. I had a much stronger background in psychology and psychophysics than most physicists. Nevertheless, I initially had difficulty in Grossberg's papers distinguishing established empirical fact from tentative conjecture or even wild speculation.

I have since reviewed the experimental results relevant to Grossberg's work sufficiently to be confident that his citations of such results are apt and reliable. Indeed, I am impressed by his judgment in selecting results to explain with his theory. I believe his grasp of the entire empirical domain from neurophysiology to psychology is unsurpassed. But I am most impressed by the way he has gone about developing his theory. His theoretical style has all the elements of the best theoretical work in physics—studies of specific mathematical models in search of general principles, emphasis on functional relations without premature commitment to specific functional forms, and a high level of idealization to isolate major relations among variables, followed by successive elaborations to capture more details. Grossberg is especially clever at developing Gedanken experiments to motivate his theoretical constructions.

I don't expect you to accept my assessment of Grossberg at face value. My purpose here is to introduce you to Grossberg and supply some of the background you will need to make your own evaluation, to help you over the initial credibility barrier as Hecht-Nielsen helped me. The recent publication of Grossberg's collected papers [1982] makes it much easier to approach his work now. But I hope to give you a good idea of what you can expect to find there. I will introduce you to basic ideas generally accepted by neural modelers and discuss the distinctive contributions of Grossberg.

4. Empirical Background

Although the human brain is the most complex of all known systems, we need only a few facts about neurons and neuroanatomy to establish an empirical base for neural network theory. Only a brief summary of those facts can be given here. For more extensive background, see Kandel and Schwartz [1981].

The signal processing unit of the nervous system is the nerve cell or neuron. There are, perhaps, a thousand types of neurons, but most of them have in common certain general signal processing characteristics, which we represent as properties of an ideal neuron model. To have a specific example in mind, let us consider the characteristics of an important type of long range signaling neuron called a pyramid cell (schematized in Fig. 1).

(1) The internal state of a neuron is characterized by an electrical potential difference across the cell membrane at the axon hillock (Fig. 1). This potential difference is called the generating potential. External inputs produce deviations in this potential from a baseline resting potential (typically between 70 and 100 mV). When the generating potential exceeds a certain threshold potential, a spike (or action) potential is generated at the hillock and propagates away from the hillock along the axon.

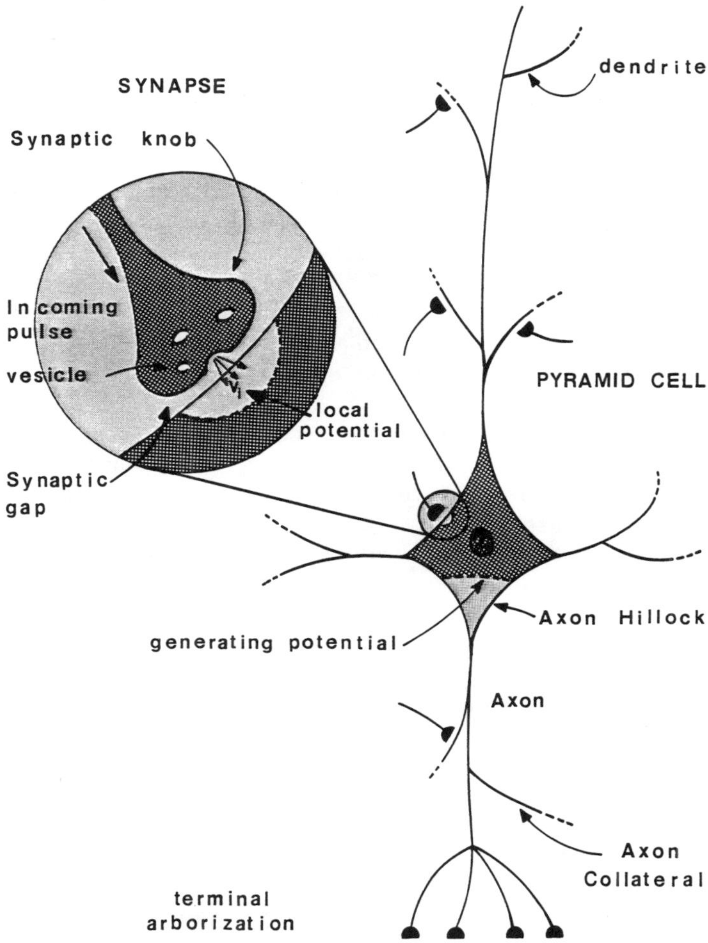

Figure 1. Anatomy of a pyramid cell.

(2) Axonal signals: The action potential is a large depolarizing signal (with amplitude up to 110 mV) of brief duration (1 to 10 ms). In a given neuron, every action potential travels with the same constant velocity (typically between 10 and 100 m/s) and undiminished amplitude along all axon collaterals (branches) to their terminal synaptic knobs.

Axonal signals are emitted in bursts of action potentials with pulse frequencies typically in the range between 2 and 400 Hz for cortical pyramid cells or between 2 and 100 Hz for retinal ganglion cells (see below). Single spike potentials are also spontaneously emitted, evidently at random. A single spike is not believed to carry information. It appears that all the information in an axonal signal resides in the pulse frequency of the burst. Thus, the signal can be represented by a positive real number in a limited interval.

(3) Synaptic inputs and outputs: The flow of signals in and out of a neuron is unidirectional. A neuron receives signals from other neurons at points of contact on its dendrites or cell body known as synapses. A typical pyramid cell in the cerebral cortex (see below) receives inputs from about 10^5 different synapses. When an incoming axonal signal reaches the synaptic knob it induces the release of a substance called a neurotransmitter from small storage vesicles. The released transmitter diffuses across the small synaptic gap to the postsynaptic cell, where it alters the local receptor potential across the cell membrane. The synaptic inputs have several important properties:

(a) Quantized transmission: Each spike potential releases (approximately) the same amount of transmitter when it arrives at the synaptic knob.

(b) Temporal summation: Changes in receptor potential induced by successive spike potentials in a burst are additive. Consequently, deviations of the receptor potential from the resting potential depend on the pulse frequency of incoming bursts.

(c) Synaptic inputs are either excitatory or inhibitory, depending on the type of interaction between neurotransmitter and receptor cell membrane. An input is excitatory if it increases the receptor potential or inhibitory if it decreases the receptor potential.

(d) Weighted spatial summation: Input-induced changes in the various receptor potentials of a neuron combine additively to drive a change in the generating potential.

Now let us identify some general anatomical characteristics of the brain to guide our constructions of networks composed of ideal neurons. We can do that by examining a major visual pathway called the geniculostriate system (Fig. 2). Light detected by photoreceptors in the retina drives the production of signals that are transmitted by retinal ganglion cells from the retina to the lateral geniculate nucleus (LGN), which (after some process-

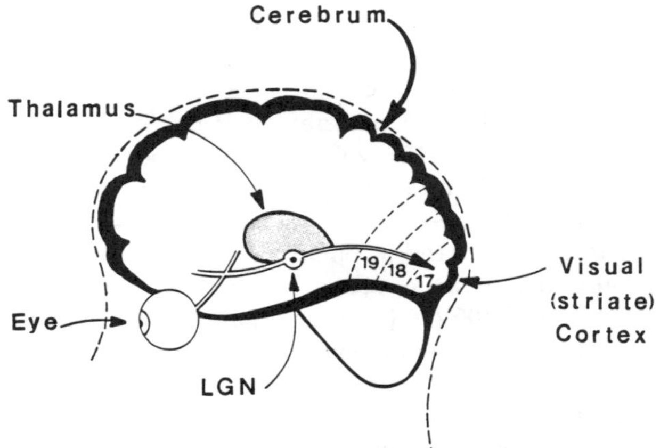

Figure 2. The geniculostriate system.

ing) relays the signal to the visual cortex (also known as the striate cortex or Area 17). From Area 17, signals are transmitted to Areas 18, 19, and other parts of the brain for additional processing.

In Fig. 3 the geniculostriate system is represented as a sequence of three layers of slabs connected by neurons with inputs in one slab and outputs in another slab. There are about 10^6 ganglion cells connecting the retina to the LGN, so we may picture the retina as a screen with 10^6 pixels. Let $I_k(t)$ be the light intensity input detected at the kth pixel at time t. Then the input intensity pattern or image displayed on the retina can be represented as an image vector with $n = 10^6$ components:

$$\mathbf{I}(t) = \{I_1(t), I_2(t), \ldots, I_n(t)\} . \quad (1)$$

This vector is filtered (transformed) into a new vector input to the LGN by several mechanisms: (a) the receptive fields (pixel inputs of the ganglion cells) overlap; (b) nearby ganglion cells interact, and (c) the output from each ganglion cell is distributed over the LGN slab by the arborization (branching) of its axon. Later we will see how to model and understand these mechanisms.

Actually, each of the three main slabs in the geniculostriate system is itself composed of several identifiable layers containing a matrix of interneurons (neurons with relatively short range interactions). We will take these complexities into account to some extent by generalizing our models

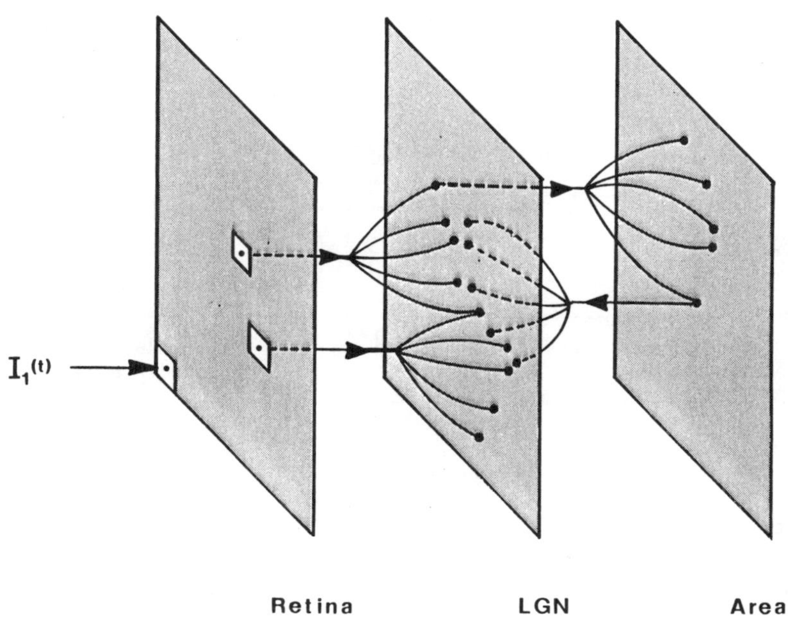

Figure 3. Layered structure of the brain (arrows indicate signal directions).

of slab-connecting neurons. There is little reason to believe that the details we omit from our models will detract from our general conclusions.

The layered structure in Fig. 3 is typical of the organization of major portions of the brain, so we can hope to learn general principles of brain design by understanding its functional significance.

Notice the reciprocal axonal connection from Area 17 back to the LGN. We shall see that the significance of that connection may be especially profound.

5. Neural Variables

We are now prepared to consider Grossberg's general principles for modeling a system of interacting neurons. The ith neuron is represented by a node v_i connected to node v_j by a directed pathway e_{ij} terminating in a synaptic knob N_{ij} as indicated in Table 1. The physical state of each of these three components of a neuron is characterized by a single real-valued state variable, which has a psychological as well as a physiological interpretation. These dual interpretations (mind-brain duality) provide the link between psychology and the brain sciences. The physiological interpretations are fairly evident from what we have said already. We shall see that the psychological interpretations convert the neural network theory into a genuine psychological theory.

Table 1. Neuron Components and Variables

	Node	Directed pathway	Synaptic knob	
Components:	v_i	e_{ij}	N_{ij}	v_j
Variables:	x_i	S_{ij}	z_{ij}	x_j

Name/variable	Physiological interpretation	Psychological interpretation
Activity x_i	Average generating potential	Stimulus trace or STM trace
Signal S_{ij}	Average firing frequency	Sampling or performance signal
Synaptic strength z_{ij}	Transmitter release rate	LTM trace

The "internal state" of node v_i is described by a variable x_i called the activity of the node. Physiologically, x_i can be interpreted as the deviation of a neuron generating potential from its equilibrium (resting) value. Psychologically, it can be interpreted as a stimulus trace or short term memory (STM) trace. The latter interpretation is especially interesting because it ascribes a definite physical referent for the STM concept. In cognitive psychology, the STM is regarded as some unspecified brain mechanism for the temporary storage of information. For example, a telephone number you have just heard is said to be stored in your STM for a time on the order of 10 seconds, after which it will soon be lost unless you re-store it in STM by rehearsing it or using it in some other way. Note that this way of speaking suggests that the STM is a special component of the brain to which a limited amount of information can be transferred for temporary storage. However, Grossberg claims that STM storage is simply an enhanced activity or activation of neurons somewhere in the brain, different neurons in different places for different concepts stored. He applies the term "STM storage" to any neuron activity x_i that is temporarily maintained at positive values by local feedback loops. Undoubtedly, the cognitive psychologists are unable to probe more than a limited subset of such activated brain states in their studies of STM storage. So Grossberg's STM concept is broader as well as more specific than the conventional STM concept.

A signal propagated along the directed pathway e_{ij} is represented by a nonnegative real variable S_{ij}. In Grossberg's theory S_{ij} is not an independent variable; it is some definite function of the node activity x_i. For the value of the signal when it reaches the synaptic knob we can write

$$S_{ij}(t) = f[x_i(t - \tau_{ij})]\, b_{ij} , \qquad (2)$$

where τ_{ij} is a "time delay constant" and b_{ij} is a "path strength constant" determined by physical properties of the pathway. In general, $f(x_i)$ is a sigmoid function, but for many purposes it can be approximated by the "threshold-linear function"

$$f(x_i) = [x_i - \Gamma_i]^+ , \qquad (3)$$

where Γ_i is a positive threshold parameter, and $[u]^+ = u$ for $u \geq 0$, $[u]^+ = 0$ for $u < 0$. Thus, the node v_i emits a signal only when its activity x_i exceeds the threshold Γ_i.

The variable S_{ij} [or $f(x_i)$, rather] is to be interpreted physiologically as the average firing frequency of a neuron. Therefore, it does not describe the sudden signal fluctuations in the bursts that are observed experimentally. Because those fluctuations are not believed to carry information, it is reasonable to suppress them in a theory that aims to characterize the information content of neuronal processes.

Psychologically, the variable S_{ij} may be interpreted as an information sampling signal when it is concerned with information input, or as a performance signal when it is concerned with output. We shall see that either of these interpretations might apply to a signal in a given pathway, depending on the state of the rest of the network.

The coupling of a synaptic knob N_{ij} to a postsynaptic node v_j is characterized by a positive real variable z_{ij} called the synaptic strength. Physiologically, it can be interpreted as the average rate of neurotransmitter release (per unit signal input) at the knob N_{ij}. This interpretation is tentative, however, because the biochemical processes at a synapse are complex and incompletely understood. In any case, a single variable should be sufficient to characterize the signal transmission rate across a synapse, whatever the underlying processes.

We shall see that the variable z_{ij} can be interpreted psychologically as a long term memory (LTM) trace. This is to assert that the long term storage of information in a brain takes place at the synapses, and that learning is a biochemical change of synaptic states. Although one cannot claim yet that this assertion is an established fact, there is considerable evidence to support it, much more than can be mustered to support any alternative hypothesis about the physiological basis for learning and memory. As a working hypothesis, our dual interpretations of the synaptic strength variable has immense implications for psychology. To begin with, we shall see that it has much to tell us about the way brains encode information.

6. Network Field Equations

Having identified the significant components of a neural network and appropriate variables to represent their properties, to complete the formulation of a network theory we need to postulate laws of interaction and equations of motion for the variables. The main facts and hypotheses about neurons that we have already mentioned are accounted for by Grossberg's field equations for a neural network with n nodes:

$$\dot{x}_i = -A_i x_i + \sum_{k=1}^{n} S_{ki} z_{ki} - \sum_{k=1}^{n} C_{ki} + I_i(t) \tag{4}$$

$$\dot{z}_{ij} = -B_{ij} z_{ij} + S'_{ij}[x_j]^+ , \tag{5}$$

where the overdot denotes a time derivative and $i,j = 1,2,\ldots,n$.

Grossberg's equations are generic laws of neural network theory in the same sense that $F = ma$ is a generic law of Newtonian mechanics. To construct a mathematical model in mechanics from $F = ma$, one must introduce a force law that specifies the function form of F. Similarly, to construct a definite network model from Grossberg's equations, one must introduce laws of interaction that specify the functional dependence of the quantities A_i, S_{ki}, C_{ki}, B_{ij}, and S_i on the network variables x_i and z_{ij}. To investigate these laws of interaction is a major research program within the context of

Grossberg's theory, just as a theoretical and experimental investigation of the force laws "realized" in nature has been a major research program in physics since it was initiated by Newton.

Before examining specific interactions, we should be clear about the general import of Grossberg's equations. Let us refer to Eq. (4) as the activity equation of node v_i. The right-hand side of this equation describes interactions of the node or, if you prefer, inputs to the node. The first thing to note about the equation is the additivity of inputs from different sources, which represents the basic experimental fact of the spatial summation of synaptic inputs to a neuron. The term $I_i(t)$ represents input from sources outside the network, usually some other neurons but sometimes sensory transducers such as photoreceptors in the retina. The other terms represent internal interactions within the network.

The term $-A_i x_i$ characterizes self-interactions of the node v_i. In the simplest case when the node represents a single neuron, A_i is a positive constant, so the term represents the passive decay that is inevitable in a dissipative system. More generally, it will often be convenient to use a single node to represent a lumped subsystem or pool of interneurons coupled to a single output neuron such as a cortical pyramid cell. The pool can be designed with feedback to make the node capable of STM and more complex responses to inputs. The net result can be described simply by making the decay coefficient A_i into some function $A_i = A_i(x_i)$ of the activity x_i. It is reasonable to suppose that, on account of the additivity of interactions, a neuron pool will have an activity equation of the same general form as that of a single neuron, if the time scale for integrating inputs to the pool is sufficiently short. As the theory develops, it should become possible to replace such reasonable assumptions by rigorous "lumping theorems."

For a neuron pool, the activity x_i of the output neuron may be proportional to the number of excited interneurons in a subpopulation of the pool. In that case, it may be more useful to interpret x_i as the number of excited states in the pool rather than as the potential of a single neuron, especially when characterizing the self-interaction properties of the pool.

The term $S_{ki}z_{ki}$ describes an excitatory node-node interaction as indicated by the plus sign preceding it in the activity equation (4). It expresses the effect of node v_k on node v_i mediated by the signal S_{ki} as given by Eq. (2). The synaptic strength z_{ki} plays the role of a variable coupling constant in the activity equation. Typically the time variation of z_{ki} is slow compared to that of x_i. In many cases z_{ki} is essentially constant, and we say that the connection from v_k to v_i is hardwired. This includes the common case when there is no direct connection from v_k to v_i if we regard it as a case with $z_{ki} = 0$. The multiplicative form of the interaction $S_{ki}z_{ki}$ expresses the temporal summation of synaptic inputs. Grossberg describes the role of the synapse by saying that "z_{ki} gates the signal S_{ki}."

The term C_{ki} in the activity equation (4) describes an inhibitory node-node interaction as indicated by the minus sign preceding it, supplemented by the assumption that $C_{ki} \geq 0$. We can interpret C_{ki} as a signal function similar to S_{ki}. The symmetry as well as the generality of the activity equation could be increased by allowing for a variable inhibitory coupling constant. However, the available evidence suggests that inhibitory connections are usually (if not invariably) hardwired. So we restrict our considerations to that case and incorporate the fixed coupling constant into the signal function C_{ki}.

From a general theoretical perspective it is crucial to realize that inhibitory interactions are essential for the stability of the network activity equations, just as attractive forces are essential for bound systems in physics. This can be proved as a mathematical theorem of great generality, and we shall see its significance later on in Grossberg's solution to the noise-saturation problem. In spite of the fact that this has been known to some neural modelers for a long time, papers that are flawed by a failure to take stability requirements into account are still being published.

Now let us turn to Eq. (5) and refer to it as the learning equation in anticipation of support for our interpretation of z_{ij} as an LTM trace. We can identify $S'_{ij}[x_j]^+$ as the learning term in the equation since it drives an increase in z_{ij}. The factor S'_{ij} is a nonnegative signal similar to the signal S_{ij} in the activity equation, but possibly differing in its dependence on the presynaptic activity x_i, owing to details of the underlying biochemical processes. The multiplicative form of the learning term implies that learning takes place only when the presynaptic learning signal S'_{ij} and the postsynaptic activity x_j are simultaneously positive. Thus, learning at a synapse is driven by correlations between presynaptic and postsynaptic activities.

The term $-B_{ij}z_{ij}$ can be regarded as a forgetting term, describing passive memory decay when B_{ij} is a positive constant. By allowing B_{ij} to be a more general function of the network parameters, Grossberg allows for the possibility of modulating memory loss.

For the case of constant $B_{ij} = \beta$, the learning equation (5) can be given the enlightening integral form

$$z_{ij}(t) = e^{-\beta t} z_{ij}(0) + \int_0^t e^{-\beta(t-\tau)} S'_{ij}(\tau)\, [x_j(\tau)]^+ \, d\tau \,. \tag{6}$$

This is an integral equation rather than a solution of Eq. (5) because $x_j(\tau)$ depends on $z_{ij}(\tau)$ in the activity equation (4). However, it shows that in the long run the initial STM trace $z_{ij}(0)$ is forgotten and $z_{ij}(t)$ is given by a time correlation function of the presynaptic signal S'_{ij} with the postsynaptic activity x_j.

Special cases and variants of Grossberg's equations have been formulated and employed independently by many neural modelers. But Grossberg has gone far beyond anyone else in systematically analyzing the implications of such equations. In doing so, he has transformed the study of isolated ad hoc neural models into a systematic theory of neural networks. But it will be helpful to know more about the empirical status of the learning equation before we survey the main results of Grossberg's theory.

7. Hebb's Law

From the viewpoint of neural network theory, Hebb's law is the fundamental psychophysical law of associative learning. It should be regarded, therefore, as a basic law of psychology. However, most psychologists have never heard of it because they do not concern themselves with the neural substrates of learning.

Hebb [1949] first formulated the law as follows: "If neuron A repeatedly contributes to the firing of neuron B, then A's efficiency in firing B increases." The psychological import of this law comes from interpreting it as the neural basis for Pavlovian (classical) learning. To see how, recall Pavlov's famous conditioning experiment. When a dog is presented with food, it salivates. When the dog hears a bell, it does not salivate initially. But after hearing the bell simultaneously with the presentation of food on several consecutive occasions, the dog is subsequently found to salivate when it hears the bell alone. To describe the experiment in more general terms, when a conditioned stimulus (CS) (such as a bell) is repeatedly paired with an unconditioned stimulus (UCS) (such as food) that evokes an unconditioned response (UCR) (such as salivation), the CS gradually acquires the ability to evoke the UCR.

To interpret this in the simplest possible neural terms, consider Fig. 4. Suppose the firing of neuron B produces the UCR output, and suppose the UCS input fires neuron C, which is coupled to B with sufficient strength to make B fire. Now if a CS stimulates neuron A to fire simultaneously with neuron B, then, in accordance with Hebb's law, the coupling strength z_{AB} between neurons A and B increases to the point where A has the capacity to fire B without the help of C. In actuality, of course, there must be many neurons of types A, B, and C involved in the learning and controlling of a molar behavioral response to a molar stimulus, but our reduction to the interaction of just three neurons assumes that the learning actually takes place at the synaptic level.

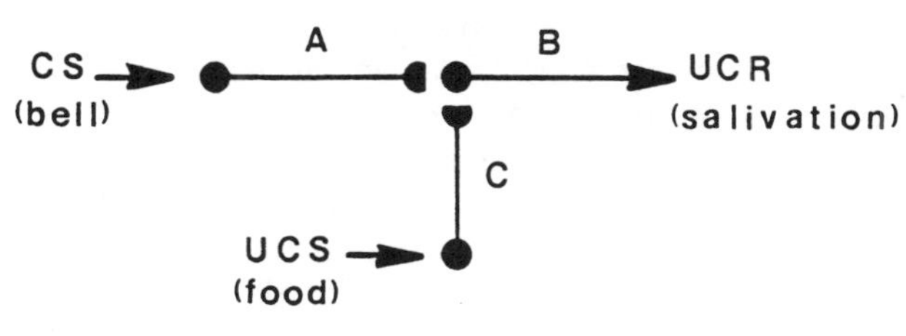

Figure 4. A neural interpretation of Pavlovian learning.

Thus, the molar association strength between stimulus and response that psychologists infer from their experiments is a crude measure of the synaptic coupling strength between neurons in the central nervous system (CNS). The same can be said about all associations among ideas and actions. Thus, the full import of Hebb's law is this: All associative (long term) memory resides in synaptic connections of the CNS, and all learning consists of changes in synaptic coupling strengths.

This is a strong statement indeed! Although it is far from a proven fact, it is certainly an exciting working hypothesis, and it provides a central theme for research in all the neurosciences. It tells us where to look for explanations of learning and memory. It invites us to do neural modeling.

Ironically, cognitive psychologists frequently dismiss Pavlovian learning as too trivial to be significant in human learning. But Hebb's law tells us that Pavlovian learning is simply an amplified form of the basic neural process underlying all learning. Neural modeling has already advanced far enough to give us good reasons for believing that the most complex cognitive processes will ultimately be explained in terms of the simple neural mechanisms operating in the Pavlovian case.

Direct physiological verification of the synaptic plasticity required by Hebb's law has been slow in coming because the experimental difficulties are extremely subtle and complex. Peripheral connections in the CNS are most easily studied, but they appear to be hardwired as one would expect because the delicate plastic synapses must be protected from destructive external fluctuations. Though some limited experimental evidence for synaptic plasticity exists, there are still considerable uncertainties about the underlying physiological mechanism. There are still doubts as to whether plasticity is due to a pre- or postsynaptic process, though a postsynaptic process is most likely [Stent, 1973].

Considering the experimental uncertainties, many neuroscientists are reluctant to take Hebb's law seriously. They fail to realize that the best evidence for Hebb's law is indirect and theory dependent. Hebb's law should be taken seriously because it is the only available theoretical construct that provides plausible, coherent explanations for psychological facts about learning and memory. Indeed, that is what led Hebb to the idea in the first place. Empiricists may regard such inverse arguments from evidence to theory as unconvincing or even inadmissible, but history shows that inverse arguments have produced the most profound advances in physics, beginning, perhaps, with Newton's law of gravitation. As an example with many parallels to the present case, recall that the modern concept of an atom was developed by a series of inverse arguments to explain observable macroscopic properties of matter. From macroscopic evidence alone, remarkably detailed models of atoms were constructed before they could be tested in experiments at the atomic level. Similarly, we should be able to infer a lot about neural networks from the rich and disorderly store of macroscopic data in psychology. Of course, we should not fail to take into account the available microscopic data about neural structures.

Hebb's original formulation of the associative learning law is too general to have detailed macroscopic implications until it is incorporated in a definite network theory. Grossberg has given Hebb's law a mathematical formulation in his learning equation (5). Hebb himself was not in position to do that, if only because the necessary information about axonal signals was not available. Of course, Hebb's formulation is vague enough to admit many different mathematical realizations. Grossberg has chosen the simplest realization compatible with the known physiological facts. No doubt Grossberg's law of synaptic plasticity is a crude description of real synaptic activity, but it may be sufficient for the purposes of network theory. In any case, it is good research strategy to study the simplest models first. Recall that Lorentz's classic theory of optical dispersion is based on an electric dipole oscillator model of an atom. The dipole is hardly more than a caricature of a real atom, but Lorentz's dispersion theory was so successful that it is still used today, and failures of the theory were important clues in the development of quantum theory. Grossberg's theory may not characterize neurons any better than a classic dipole characterizes an atom, but it may nevertheless have great success, and any clear failures will be important clues to a better theory. That's how theories progress.

8. The Outstar Learning Theorem

There is one obvious property of neurons that we have not yet incorporated into the network theory, and that is the treelike structure of an axon. What are its implications for information processing? Grossberg's "outstar theorem" shows that the implications are as profound as they are simple.

Consider a slab of noninteracting nodes $V = \{v_1, v_2, \ldots, v_n\}$ with a time varying input image $\mathbf{I}(t) = [I_1(t), I_2(t), \ldots, I_n(t)]$ that drives the nodes above signal threshold. The total intensity of the input is $I = \Sigma_k I_k$, so we can write $I_k = \theta_k I$ where $\Sigma_k \theta_k = 1$. Thus, the input has a reflectance pattern $\boldsymbol{\theta}(t) = [\theta_1(t), \theta_2(t), \ldots, \theta_n(t)]$.

Now consider a node v_0 with pathways to the slab as shown in Fig. 5. This configuration is called an outstar because it can be redrawn in the symmetrical form of Fig. 6. When an "event" $I_0(t)$ drives v_0 above threshold, a

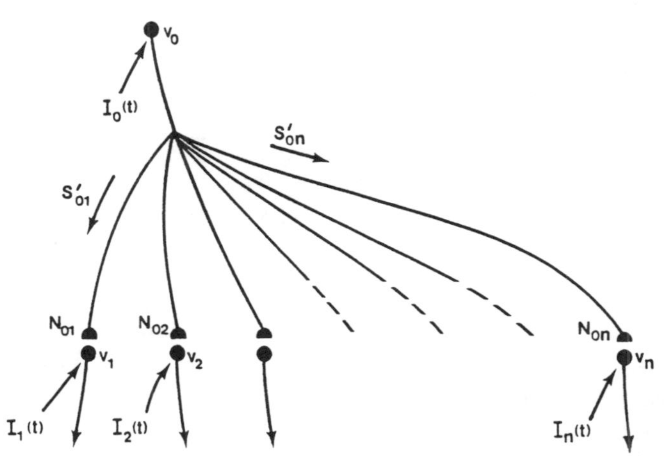

Figure 5. The outstar is the minimal network capable of associative learning.

learning signal S_{0k} is sent to each of the synaptic knobs to trigger a sampling of the activity pattern "displayed" on the slab by driving changes in the synaptic strengths z_{0k}.

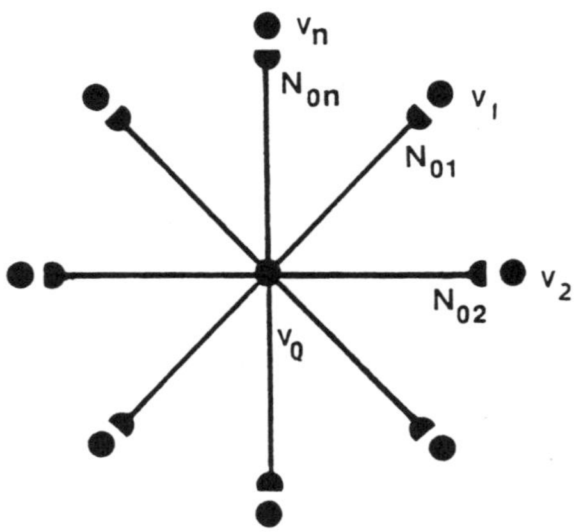

Figure 6. Symmetry of the outstar anatomy.

The outstar learning theorem says that an outstar learns a weighted average of reflectance patterns $\overline{\boldsymbol{\theta}} = (\overline{\theta}_1, \overline{\theta}_2, \ldots, \overline{\theta}_n)$ displayed on the slab in the sense that

$$Z_{0k}(t) \xrightarrow[t\to\infty]{} \overline{\theta}_k \tag{7}$$

where

$$Z_{0k} = z_{0k}\left(\sum_k z_{0k}\right)^{-1} . \tag{8}$$

Grossberg calls the Z_{0k} 'stimulus sampling probabilities' to emphasize the statistical aspect of the learning process.

The outstar has truly learned the pattern $\overline{\boldsymbol{\theta}}$ in the sense that it can recall the pattern exactly in the following way. Suppose that, at some time after learning, the external slab input $\mathbf{I}(t)$ vanishes but $I_0(t)$ is sufficient to stimulate signals S_{0k} from v_0. The signals S_0 read out an activity pattern on the slab that is proportional to the synaptic strengths z_{0k} and hence to the pattern θ_k. When the S_{0k} are sufficiently strong by themselves to drive the slab above threshold, they are called performance signals.

It will be recognized that outstar learning is an instance of Pavlovian learning, where v_0 corresponds to neuron A in Fig. 4 and instead of a single neuron B we have a whole slab of neurons. Accordingly, we can interpret the slab output as the UCR controlled by a UCS input $\mathbf{I}$. When the CS input I_0 is synchronized with the UCS input $\mathbf{I}$, the outstar gradually gains control over the UCR.

To see how the outstar theorem follows from Grossberg's equations, let us consider the simplest case, where the signals $S'_{0k} = S'_0$ are the same for all pathways, and all nodes have identical constant self-interaction coefficients. Then the outstar network equations are

$$\dot{x}_0(t) = -\alpha x_0(t) + I_0(t) \tag{9a}$$

$$\dot{x}_k(t) = -\alpha x_k(t) + S_0(t)\, z_{0k}(t) + I_k(t) \tag{9b}$$

$$\dot{z}_{0k}(t) = -\beta z_{0k} + S'_0(t)\, x_k(t)\ . \tag{9c}$$

The internal processes in a neuron are comparatively fast, so it is often a good approximation to suppose that the relaxation time α_k^{-1} is short compared to the time variations of the input $I_k(t)$. Then we can use the equilibrium solutions of the activity equations (9b). Therefore, for $S_0 = 0$, Eq. (9b) gives us

$$x_k(t) = \alpha^{-1} I_k(t) = \frac{I(t)}{\alpha}\, \theta_k(t)\ . \tag{10}$$

Thus, the activity pattern across the slab is proportional to the reflectance pattern.

Suppose now that the same reflectance pattern $\boldsymbol{\theta}$ is repeatedly presented to the slab, so the θ_k are constant but the intensity $I(t)$ may vary wildly in time. For the sake of simplicity, suppose also that the signal S_0 does not significantly perturb the activity pattern across the slab. Then, the integral of the learning equation (9c) has the form of Eq. (6), and gives the asymptotic result

$$z_{0k}(t) = N(t)\, \theta_k \tag{11a}$$

where

$$N(t) = \int_0^t d\tau\, e^{-\beta(t-\tau)}\, S'_0(\tau)\, \alpha^{-1}\, I(\tau)\ . \tag{11b}$$

According to Eq. (11a), the outstar learns the reflectance pattern exactly. The same result is obtained with more mathematical effort even when perturbations of the slab activity pattern are taken into account.

Note that, according to Eq. (11b), the magnitude of $N(t)$, and therefore the rate of learning, is controlled by the magnitudes of the sampling signal $S'_0(t)$ and the total input intensity $I(t)$. Stronger signals, faster learning!

If the θ_k are not constant, the learning equation will still give an asymptotic result of the form (11a) if θ_k is replaced by a suitably defined average θ_k. To see this in the simplest way, suppose that two different but constant patterns $\boldsymbol{\theta}^{(1)}$ and $\boldsymbol{\theta}^{(2)}$ are sampled by the outstar at different times. Then the additivity property of the integral in Eq. (6) allows us to write

$$z_{0k} = N\bar{\theta}_k = N_1\theta_k^{(1)} + N_2\theta_k^{(2)}, \tag{12}$$

where N_1 and N_2 have the same form as N in Eq. (11b), except that the integration is only over the time intervals when $\theta_k^{(1)}$ (or $\theta_k^{(2)}$) is displayed on the slab. Thus, the pattern $\bar{\boldsymbol{\theta}}$ stored in the outstar LTM is a weighted average of sampled patterns. Note that this is a consequence of formulating Hebb's law specifically as a correlation function.

Outstar learning has a number of familiar characteristics of human learning. If the outstar is supposed to learn pattern $\boldsymbol{\theta}^{(1)}$, and $\boldsymbol{\theta}^{(2)}$ is an error, then according to Eq. (12), repeated sampling of $\boldsymbol{\theta}^{(1)}$ will increase the weight (N_1/N) of $\boldsymbol{\theta}^{(1)}$ and $\bar{\boldsymbol{\theta}} \to \boldsymbol{\theta}^{(1)}$. Thus outstar learning is "error correcting" and "practice makes perfect." Or if $\boldsymbol{\theta}^{(1)}$ is presented and sampled with sufficient intensity, the outstar can learn $\boldsymbol{\theta}^{(1)}$ on a single trial. Thus, "memory without practice" is possible. On the other hand, if an outstar repeatedly samples new patterns, then the weight of the original pattern decreases. So we have an "interference theory of forgetting." Forgetting is not merely the result of passive decay of LTM traces.

Grossberg has proved the outstar theorem for much more general functional forms of the network equations than we have considered. For purposes of the general theory, it is of great importance to determine the variations in physical parameters and network design that are compatible with the pattern learning of the outstar theorem.

The outstar theorem is the most fundamental result of network theory because it tells us precisely what kind of information is encoded in LTM, namely, reflectance patterns. It goes well beyond Hebb's original insight in telling us that a single synaptic strength z_{0k} has no definite information content. All LTM information is in the pattern of relative synaptic strengths z_{0k} defined by Eq. (8).

To sum up, we have learned that the outstar network has the following fundamental properties:

- Coding: The functional unit of LTM is a spatial pattern.
- Learning: A pattern is encoded by a stimulus sampling signal.
- Recall: Probed read-out of LTM into STM by a performance signal.
- Factorization: A pattern is factored into a reflectance pattern, which is stored, and the total intensity I, which controls the rate of learning.

The outstar is a universal learning device. Grossberg has shown how to use this one device to model every kind of learning in the CNS. The kinds of learning are distinguished by the different interpretations given to the components of the outstar network, which, in turn, depend on how the components fit into the CNS as a whole. To appreciate the versatility of the outstar, let us briefly consider three examples of major importance:

Top-down expectancy learning: Suppose the slab $V = \{v_1, v_2, \ldots, v_n\}$ represents a system of sensory feature detectors in the cerebral cortex. A visual (or auditory) event is encoded as an activity pattern $\mathbf{x} = (x_1, x_2, \ldots, x_n)$ across the slab, where x_k represents the relative importance of the kth feature. The outstar node v_0 can learn this pattern, and when the LTM pattern is played back on V, it represents a prior expectancy of a sensory event.

Motor learning and control. Suppose the slab V represents a system of motor control cells, so each v_k excites a particular muscle group and the activity x_k determines its rate of contraction. Then the outstar command node v_0 can learn to control the synchronous performance of a particular motion with factored rate control modulated by the strength of the performance signal.

Temporal order encoded as spatial order. If the slab V represents a sequence of codes for items on a list (such as a phone number) and the relative magnitudes of the activities x_k reflect the temporal order of the items, then the outstar node v_0 can learn the temporal order.

9. The Network Modeling Game

Once the outstar is recognized as the fundamental device for learning and memory, it becomes evident that information is represented by spatial patterns in the CNS. Information is encoded in STM activity patterns and stored in LTM synaptic strength patterns. Information is processed by filtering, combining, and comparing patterns with patterns and more patterns. Thus, we come to formulate the Network Modeling Game as follows: Explain all learning and behavior with modular network models composed of outstars. The term "behavior" is to be understood here in the broadest sense of any output pattern. The emphasis on outstars is an important refinement of the game introduced by Grossberg.

The aim of the Network Modeling Game is to reduce psychology to a theory of neural mechanisms. Grossberg has been playing the game for a long time and has built up an impressive record of victories. He plays the game systematically by formulating a sequence of design problems for neural networks. For each design problem he finds a minimal solution that provides a design principle for constructing a network with some specific pattern processing capability. He has already developed too many design principles for us to review them all here. But we will look at a few of the most basic design problems and solutions to see how the game gets started.

We will consider the following network design problems:

- Pattern registration. Considering the limited dynamic range of a neuron, how can a slab be designed to register a well defined pattern if the input fluctuates wildly?
- STM storage. How can an evanescent input be stored temporarily as an activity pattern for further processing or transfer to permanent LTM storage?
- Code development. How can a network learn to identify common features and classify different patterns?
- Code protection and error correction. How can a network continue to learn new information without destroying what it has already learned?
- Pattern selection. How does a network decide what is worth learning?

9.1. Pattern Registration and STM Storage

A neuron has a limited dynamic range; its sensitivity is limited by noise at low activity and by saturation at high activity. How, then, can neurons maintain sensitivity to wide variations in input intensities? Grossberg calls this the noise-saturation dilemma, and he notes that it is a universal problem that must be solved by every biological system. Therefore, by studying the class of neural networks that solve this problem, we may expect to discover a universal principle of network design. Grossberg has attacked this problem systematically, beginning with a determination of the simplest possible solution.

Grossberg has proved that the minimal network solving the noise-saturation dilemma is a slab of nodes with activity equations

$$\dot{x}_i = -Ax_i + (B - x_i)\, I_i - (x_i + C) \sum_{k \neq i} I_k , \tag{13}$$

where $i = 1,2,\ldots,n$, and A,B are positive constants while C may be zero or a positive constant. It is easy to see that the form of Eq. (13) limits solutions to the finite range $B \geq x_i(t) \geq -C$, whatever the values of inputs $I_k(t)$.

Two things should be noted about the form of Eq. (13). First, the excitatory input I_i to each node v_i is also fed to the other nodes as an inhibitory input as indicated in Fig. 7. A slab with this kind of external interaction is called an on-center off-surround network. Second, both excitatory and inhibitory interactions include terms of two types, one of the form BI_i and the other of the form x_iI_k. Grossberg calls the first type an additive interaction and the second type a shunting interaction. Accordingly, he calls a network characterized by Eq. (13) a shunting on-center off-surround network.

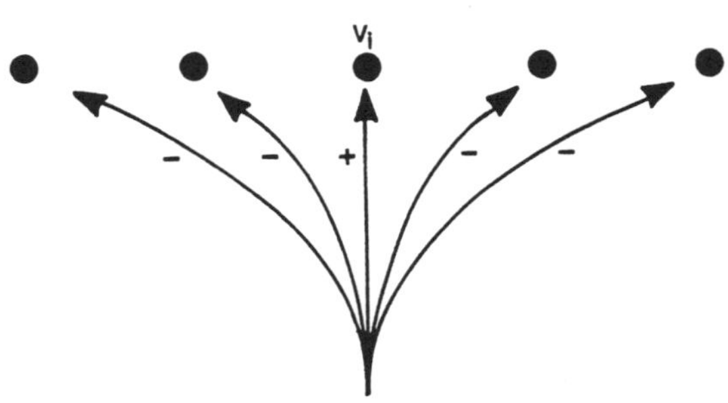

Figure 7. A nonrecurrent on-center off-surround anatomy.

This network can be regarded as a simple model of a retina. The existence of on-center off-surround interactions in the retina is experimentally well established in animals ranging from the mammals to the primitive horseshoe crab. The nodes in our model correspond roughly to the retinal ganglion cells with outputs in the LGN as we noted in Sec. 4. The distribution of inhibitory inputs among the nodes is biologically accomplished by a layer of interneurons in the retina called horizontal cells. Our model lumps these interneurons together with the ganglion cells in the nodes. It describes the signal processing function of these interneurons without including unnecessary details about how this function is biologically realized.

It is well known among neuroscientists that intensity boundaries in the image input to an on-center off-surround network are contrast-enhanced in the output. So it is often concluded that the biological function of such a network is contrast enhancement of images. But Grossberg has identified a more fundamental biological function, namely, to solve the noise-saturation dilemma. However, shunting interactions are also an essential ingredient of the solution. Many neural modelers still do not recognize this crucial role of shunting interactions and deal exclusively with additive interactions. It should be mentioned that the combination of shunting and additive interactions appears in cell membrane equations that have considerable empirical support.

To see how Eq. (13) solves the noise saturation dilemma, we look at the steady-state solution, which can be put into the form

$$x_i = \frac{(B + C)I}{A + I}\left(\theta_i - \frac{C}{B + C}\right). \tag{14}$$

Here, as before, the θ_i are the reflectances and I is the total intensity of the input. This solution shows that the slab has the following important image processing capabilities:

(1) Factorization of pattern and intensity. Hence, for any variations in the intensity I, the activities x_i remain proportional to the reflectances θ_i [displaced by the small constant $C(B + C)^{-1}$]. Thus, the slab possesses automatic gain control, and the x_i are never saturated by large inputs.

This factorization matches perfectly with the outstar factorization property, though they have completely different physical origins. Since an outstar learns a reflectance pattern only, it needs to sample from an image slab that displays reflectance patterns with fidelity. Thus, we have here a minimal solution to the pattern registration problem.

(2) Weber law modulation by the factor $I(A + I)^{-1}$. Weber's law is an empirical law of psychophysics that has been found to hold in a wide variety of sensory phenomena. It is usually given in the form $\Delta I/I$ = constant, where ΔI is the "just noticeable difference" in intensity observed against a background intensity I. This form of Weber's law can be derived from Eq. (14) with a reasonable approximation.

(3) Featural noise suppression. The constant $C(B + C)^{-1}$ in Eq. (14) describes an adaptation level, and $B \gg C$ in vivo. The adaptation level $C(B + C)^{-1} = n^{-1}$ is especially significant, for then if $\theta_i = n^{-1}$, one gets $x_i = 0$. In other words, the response to a perfectly uniform input is com-

pletely quenched. It has, in fact, been experimentally verified that when the human retina is exposed to a perfectly uniform illumination, the subject suddenly sees nothing but black. Evidently, the human visual system detects only deviations from a uniform background.

(4) Normalization. From Eq. (14), the total activity is given by $x = \Sigma_k x_k = [B - (n-1)C](A+1)^{-1}$, which is independent of n if C = 0 or $C(B+C)^{-1} = n^{-1}$. In that case, the total activity has an upper bound that is independent of n and I, and we say that the activity is normalized. For some purposes it is convenient to interpret a normalized activity pattern as a probability distribution.

A slab with the pattern processing capabilities that we have just mentioned can perform a variety of different functions as a module in a larger network. It can be used, for example, to compare patterns. Two distinct inputs I_i and J_i produce a composite input $I_i + J_i$. If the patterns mismatch, then the peaks of one pattern will tend to fall on the troughs of the other and the composite will tend to be uniform and so be quenched by the noise saturation property. But if the patterns match, they will be amplified. Indeed, if they match perfectly, then Eq. (14) gives

$$x_i = \frac{(B + C)(I + J)}{A + I + J}\left(\theta_i - \frac{C}{B + C}\right), \tag{15}$$

where I and J are the intensities of the two patterns. The slab output, quenched or amplified, amounts to a decision as to whether the patterns are the same or different. This raises questions about the criteria for pattern equivalence that can be answered by more elaborate network designs. Note that this competitive pattern matching device compares reflectance patterns rather than intensity patterns.

From his analysis of the minimal solution to the noise-saturation dilemma, Grossberg identifies the use of shunting on-center off-surround interactions as a general design principle solving the problem of accurate pattern registration. He then employs this principle to construct more general solutions with additional pattern processing capabilities. He introduces recurrent (feedback) interactions to give the slab an STM storage capability. By introducing distance-dependent interactions he gives the network edge-enhancement capabilities. Beyond this, he develops general theorems about the form of interactions that produce stable solutions of the network equations.

With the general network design principle for accurate pattern registration in hand, we are prepared to appreciate how Grossberg uses it to solve the STM storage problem. The main idea is to introduce interactions among the nodes in a slab in a way that is compatible with accurate pattern registration. Accordingly, we introduce an excitatory self-interaction (on-center) for each node and inhibitory lateral interactions (off-surround), as indicated in Fig. 8. Such a network is said to be recurrent because some of its output is fed back as input.

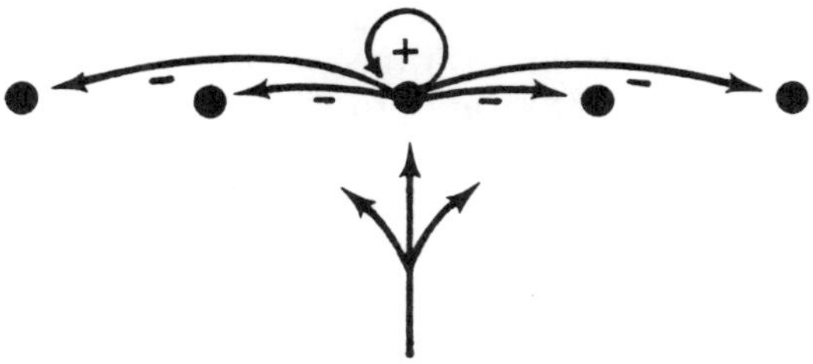

Figure 8. A recurrent on-center off-surround anatomy.

The minimal solution to the STM storage problem is given by the network equations

$$\dot{x}_i = -Ax_i + (B - x_i)[I_i + f(x_i)] - (x_i + C)[J_i + \sum_{k\neq i} f(x_k)] , \qquad (16)$$

where $I_i \geq 0$ and $J_i \geq 0$ are excitatory and inhibitory inputs respectively, and $f(x_k)$ is the signal from the kth node. Grossberg has systematically classified signal functions $f(x)$ according to the pattern processing characteristics they give to the network. His main result is that a sigmoid form for the signal function is essential for stability of the network. In spite of this definitive result and its important implications, many modelers continue to work exclusively with linear feedback models.

Besides solving the STM storage problem, the network equations (16) endow the slab with additional pattern processing capabilities. Grossberg has proved that the sigmoid signal function has a definite quenching threshold (QT). Activities below the QT are quenched while those above the QT are sustained. Moreover, the network can easily be given a variable QT controlled by external parameters acting on the interneurons that carry the feedback. This improves the pattern-matching capabilities of the slab, which we have already mentioned. The variable QT provides a tunable criterion for pattern equivalence that can be used as a partial pattern-matching mechanism.

9.2. Pattern Classification and Code Development

Now that we know how to hardwire a slab for accurate pattern registration and STM storage, we are ready to connect slabs into larger networks capable of global pattern processing. Here we shall see how to design a two-slab network that can learn to recognize similar patterns and distinguish between different patterns. This network is a self-organizing system capable of extracting common elements from the time-varying sequence of input patterns and developing its own pattern classification code. Thus, the network learns from experience. No programmer is necessary. We call such a network an adaptive pattern classifier.

The simplest version of an adaptive pattern classifier is a feed-forward network composed of two slabs with the specifications listed in Fig. 9. To emphasize their functions, let's refer to slabs S_1 and S_2 as the image slab and feature slab, respectively. As indicated in Fig. 9, each node in the image slab is connected to the nodes in the feature slab by pathways terminating in plastic (modifiable) synapses. The configuration is identical to that of the outstar discussed earlier, but we shall not refer to it as an outstar, because its function is not to learn patterns, as we shall see.

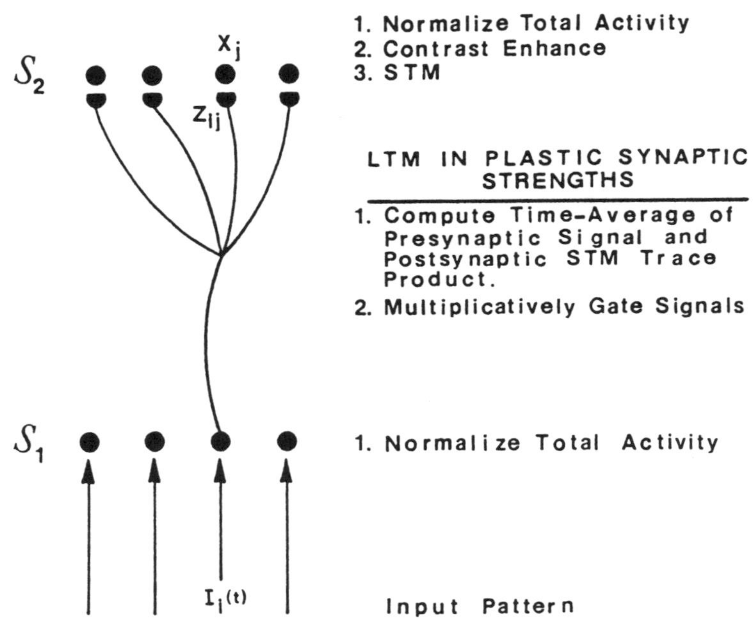

Figure 9. A feed-forward adaptive pattern classifier.

To have a definite example in mind, we can interpret slab S_1 as the LGN and slab S_2 as the visual cortex in Figs. 2 and 3. The input to S_1 can be regarded as the visual input to the eye after some "preprocessing" in the retina. Assuming that the relaxation time for nodes in S_1 is sufficiently short, the activity pattern across S_1 will be proportional to the reflectance pattern $\boldsymbol{\theta} = (\theta_1, \theta_2, \ldots, \theta_n)$ of the input. In other words, the input image $\boldsymbol{\theta}$ is accurately registered on S_1. That's why we call S_1 the image slab.

When an image is registered on S_1, each node v_k in S_2 receives a signal S_{ik} from every node u_i in S_1, as indicated in Fig. 10. The signals S_{ik} are gated by synaptic strengths z_{ik} when they arrive to produce a total gated input to v_k:

$$T_k = \sum_{i=1}^{n} S_{ik} z_{ik} = \mathbf{S}_k \cdot \mathbf{z}_k , \qquad (17)$$

where the sum is over all nodes in the image slab, and $\mathbf{S}_k = (S_{1k}, S_{2k}, \ldots, S_{nk})$, $\mathbf{z}_k = (z_{1k}, z_{2k}, \ldots, z_{nk})$.

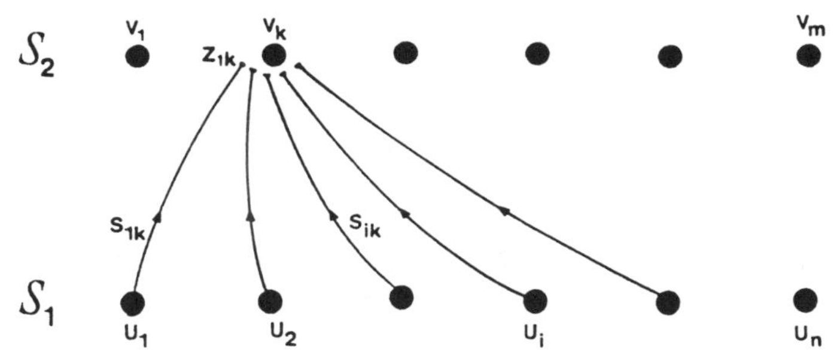

Figure 10. Gated image input to a feature detector.

The excitatory inputs T_k induce a pattern of activities x_k on the feature slab. However, the nodes v_k compete for the input by lateral inhibitory interactions as in Fig. 8. The greater the activity of one node, the more it inhibits the activities of its neighbors. Consequently, nodes with the largest inputs tend to increase their activities and suppress the activities of others until equilibrium is reached. The normalization property limits the total activity of the slab. So an increase in the activity of one node implies a decrease in the activities of others. The normalization property, therefore, implies that the activities of only a limited number of nodes can be driven above their signal thresholds. Thus, the input singles out or chooses these nodes. If the threshold is sufficiently high, then only the node with the greatest activity will be chosen. For the sake of simplicity, let's limit our considerations to this case.

A variety of different input patterns can maximally activate the same node v_k. The node v_k evidently responds to some common feature of these patterns, so we call it a feature detector, and we interpret its output as a signal that a pattern with this feature has been detected. To see how the feature is described mathematically, we look at the activity equation for v_k, which must have the general form of Eq. (4). After v_k has been chosen and the slab is in equilibrium, the lateral interactions C_{ik} in Eq. (4) vanish, and the activity x_k will be proportional to the gated input T_k given by Eq. (17). Also, the signal S_{ik} is a function of the reflectance θ_i, and it will simplify our job of interpretation if we assume that the function is linear, though this is by no means necessary. Thus, the activity of v_k is given by

$$x_k = \lambda \boldsymbol{\theta} \cdot \mathbf{z}_k , \tag{18}$$

where λ is a positive constant. This enables us to describe pattern classification in more specific terms.

The feature detector v_k <u>classifies</u> (recognizes as equivalent) all patterns in the set

$$P_k = \{\boldsymbol{\theta} : \boldsymbol{\theta} \cdot z_k > \max(\mu, \boldsymbol{\theta} \cdot z_j) \quad \text{for all } j \neq k\} , \tag{19}$$

where $\mu > 0$ defines a recognition threshold. Clearly, the pattern class is determined by the vectors $\mathbf{z}_j$, so we call them <u>classifying vectors</u>. But note that the pattern class P_k depends on the whole set of classifying vectors and not on $\mathbf{z}_k$ alone. The size of Σ determines how similar patterns must be to be classified by the same v_k.

Since the product $\boldsymbol{\theta} \cdot \mathbf{z}_k$ can never be negative, the set P_k defined by Eq. (19) is <u>convex</u> in the sense that the pattern $\alpha \boldsymbol{\theta}_1 + (1-\alpha)\boldsymbol{\theta}_2$ is in P_k if $\boldsymbol{\theta}_1$ and $\boldsymbol{\theta}_2$ are in P_k and $0 \leq \alpha \leq 1$. This convex set defines the <u>feature</u> identified by v_k.

The classification rule in Eq. (19) partitions the set of all input patterns into <u>mutually exclusive and exhaustive convex</u> subsets $P_0, P_1, P_2, \ldots, P_M$, where, for $1 \leq k \leq M$, v_k detects patterns in the set P_k, and P_0 is the set of undetectable patterns that cannot drive any v_k over threshold. Thus, the feed-forward pattern classifier is capable of <u>categorical perception</u>. The number of feature detectors m determines the maximum number of categories $M \leq m$. The choice of classifying vectors $\mathbf{z}_k$ determines how different the categories can be.

The boundaries between the P_j are categorical boundaries. If a pattern input is deformed across a boundary, then there will be a sudden shift in category perception, as occurs when we view those ambiguous figures exhibited in introductory psychology textbooks. The definition of a category P_k by Eq. (19) shows that the boundary of P_k depends on the whole set of classifying vectors and not on z_k alone. Thus, the representation of a single "word" depends on the entire network "vocabulary."

Now that we know how patterns are classified, we are ready to see how new classifications can be learned. When a particular feature detector v_k detects a pattern, it is driven by the gated image as indicated in Fig. 10. This network system is called an <u>instar</u> because it can be redrawn in the symmetrical form of Fig. 11. It differs from the outstar of Fig. 6 only in the signal direction. This is expressed by saying that outstar and instar are <u>dual</u> to one another. Their anatomical duality is matched by a functional duality, the <u>duality of recall and recognition</u>. An outstar can learn to recall a given pattern but cannot recognize it whereas an instar can learn to recognize a pattern but not recall it. The outstar is blind. The instar is dumb.

While the instar to the feature detector v_k is operating, the input pattern drives changes in the classification vector $\mathbf{z}_k$ determined by the learn-

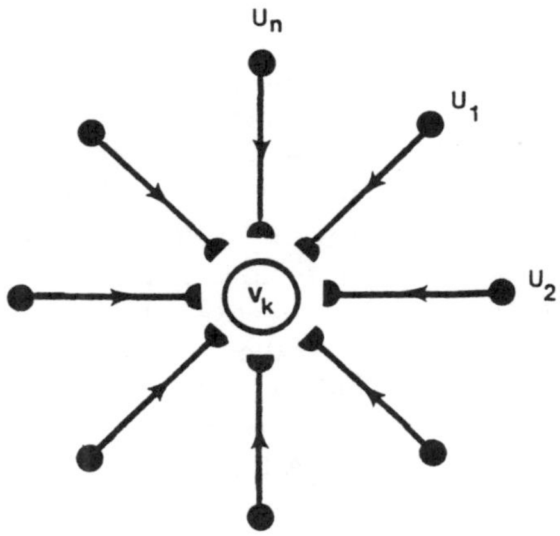

Figure 11. Symmetry of the instar anatomy.

ing equation (4). Noting that normalization will keep the feature detector activity at a fixed value $x_k = \mu^{-1}C_k$ and assuming, for simplicity, that the learning signal has the linear form $S_{ik} = \mu\theta_i$, we can put the instar learning equations into the vectorial form

$$\dot{\mathbf{z}}_k = -\beta \mathbf{z}_k + C_k \boldsymbol{\theta} , \tag{20}$$

where β is a positive constant.

For constant reflectance, Eq. (20) integrates to

$$\mathbf{z}_k(t) = \mathbf{z}_k(0)e^{-\beta t} + (C_k/\beta)(1 - e^{\beta t})\boldsymbol{\theta} , \tag{21}$$

which holds no matter how wildly the input intensity $I(t)$ fluctuates. For $t \gg \beta^{-1}$, this reduces to $\mathbf{z}_k = \beta^{-1}C_k\boldsymbol{\theta}$. Thus, the classification vector $\mathbf{z}_k$ aligns itself with the input reflectance vector $\boldsymbol{\theta}$. More generally, it can be shown that the classification vector $\mathbf{z}_k$ aligns itself asymptotically with a weighted average $\overline{\boldsymbol{\theta}}$ of reflectance vectors that activate the feature detector v_k. This is the instar code development theorem. It is dual to the outstar learning theorem. Like the outstar theorem, it can be proved rigorously under more general assumptions than we have considered here. Note that the instar factorizes pattern and intensity just like the outstar, but the physical mechanism producing the factorization is quite different in each case. For the instar, factorization is produced by lateral interactions in the image slab.

The code development theorem tells us that an adaptive classifier tunes itself to the patterns it "experiences" most often. When a single classifying vector is tuned by experience, it shifts the boundaries of all the categories P_k defined by Eq. (19). "Dominant features" will be excited most often, so they will eventually overwhelm less salient features. Thus, the adaptive classifier exhibits a progressive sharpening of memory. The degree of sharpening can be manipulated by varying the QT. The higher the QT, the more sharply tuned the evolving code.

A computer simulant of a feed-forward additive pattern classifier was constructed and tested by Christopher von der Malsburg [1973] with a striking result. He endowed the system with a random set of initial classifying vectors (synaptic strengths). Then he "trained" the system by presenting it with a sequence of nine different input patterns in a "training session." Each pattern displayed on the two-dimensional image slab was a straight line with a particular orientation. He followed the tuning of feature detectors with each training session. His main results after 100 sessions are displayed in Fig. 12. Each point on the figure represents one of the 169 feature detectors in Malsburg's feature slab. A line through a point indicates that the detector is optimally sensitive to a line image with that orientation. Points without lines through them do not respond significantly to any of the patterns. The figure shows many feature detectors for each pattern instead of a single one as in our discussion. That is because Malsburg's hardwired inhibitory interactions were limited in range to near neighbors, so widely separated nodes hardly affect one another. This is more realistic than the simpler case we considered.

Figure 12. Selective sensitivity of simple cells in the striate cortex.

The striking thing about Fig. 12 is its "swirling" short range order. The points tend to be organized in curves with gradual changes in the direction of the "tangent vector" as one moves from one point to the next. This is striking because it is qualitatively the same as experimental results for which Hubel and Wiesel received a Nobel prize in 1981. Into the striate cortexes (Fig. 2) of cats and monkeys, Hubel and Wiesel inserted microelectrodes that can detect the firing of single neurons called "simple cells." Simple cells are selectively responsive to bars of light with particular orientations that are moved across an animal's retina. Moreover, they found that the responses of neighboring simple cells are related as in Fig. 12. Malsburg's computer experiment suggests an explanation for their results, and so provides one among many pieces of evidence that brains actually process patterns in accordance with the principles we have discussed.

The Adaptive Classifier is a general-purpose pattern processing device. We have discussed the processing of spatial patterns by way of illustration, but Grossberg shows how the same device can be used to process temporal patterns. Moreover, by connecting classifiers in series, one can construct a device that can decompose patterns into features and assemble the features into a hierarchy of perceptual units [Fukushima, 1980]. Computer simulations like Malsburg's show that Adaptive Pattern Classifiers work the way the theory says they should. The development of practical artificial devices for pattern learning and recognition based on the same design principles is under way [Hecht-Nielsen, 1983]. Such devices will be quite different from devices based on the current theory of pattern recognition in the field of artificial intelligence. They will mark the beginning of a new species of computer designed with principles of neural network theory.

9.3. Code Stabilization and Pattern Selection

The feed-forward adaptive pattern classifier of Fig. 9 has a severe limitation. The number of different patterns it can learn cannot exceed the number of nodes in the feature slab. Consequently, when the number of different patterns it experiences approaches the number of feature detectors, the system begins a massive recoding that destroys things it has already learned. This presents us with a new design problem: How can we design a pattern classifier that is plastic enough to learn from experience but stable enough to retain what it has learned? Grossberg calls this the stability-plasticity dilemma. His solution to the dilemma is full of surprises and brings forth a new set of network design problems whose solutions result in network capabilities with increasing similarities to real biological systems. Indeed, Grossberg has developed the theory to the point where it has many implications for animal learning theory in psychology, perhaps only a step or two away from a viable theory of higher order cognitive processes in human beings. Unfortunately, we do not have sufficient space here to do more than indicate the direction of Grossberg's research program.

Grossberg solves the stability-plasticity dilemma by introducing feedback from the feature slab to the image slab. Each feature detector is then the command cell for an outstar that can read out a learned pattern or template on the image slab. An outstar of this sort from the visual cortex to the LGN is indicated in Fig. 3. The template can be interpreted as an expectation; it is the pattern that the feature detector "expects to see." The template is superimposed on an external input image. If the match between template and image is sufficiently close, then the instar signal to the feature detector is amplified and fed back by the outstar to amplify the template. Thus, a feedback loop of sustained resonant activity is set up, and it drives a recoding of the classification vector as well as the template in the direction of the input image. Grossberg calls this resonant state an adaptive resonance. He suggests that every human act of conscious recognition should be interpreted biologically as an adaptive resonance in the brain. This brilliant idea has a host of implications that will surely enable us to tell someday whether the idea is true or false. For example, adaptive resonances must be accompanied by distinctive electric fields, and the experimental study of such fields is beginning to show promising results. In the meantime, the adaptive resonance idea raises plenty of questions to keep theorists busy.

Adaptive resonances process expected events (that is, recognized patterns). An unexpected event is characterized by a mismatch between template and image. In that case, the mismatch feature detectors must be shut off immediately to avoid inappropriate recoding. Here we have a new design problem for which Grossberg has developed a brilliant solution involving two new neural mechanisms that play significant roles in many other network designs.

The first of these new mechanisms Grossberg calls a gated dipole. The gated dipole is a rapid on-off switch with some surprising properties. There is already substantial indirect evidence for the existence of gated dipoles, but Grossberg's most incisive predictions are yet to be verified. In my opinion, direct observations of gated dipole properties would be as profound and significant scientifically as the recent "direct" detection of electroweak intermediate bosons in physics.

The second new mechanism Grossberg calls nonspecific arousal. It is a special type of signal that Grossberg uses to modulate the quenching threshold of a slab. It is nonspecific in the sense that it acts on the slab as a whole; it carries no specific information about patterns. As Grossberg develops the theory further, he is led to distinguish several types of arousal signals, and their similarities to what we call emotions in people becomes increasingly apparent. Grossberg shows that arousal is essential for pattern processing. This has the implication that emotions play an essential role in the rational thinking of human beings. Reason and emotion are not so independent as commonly believed.

The processing of unexpected events requires more than refinements in the design of adaptive classifiers. It requires additional network components to decide which events are worth remembering and how they should be encoded. To see the rich possibilities opened up by Grossberg's attack on this problem, I refer you to his collected works.

10. Playing the Game

The Classical Mechanics Game is played by Newton's rules. The Relativity Game is played by Einstein's rules. The MAXENT Game is played by Jaynes' rules. If you want to play the neural Network Modeling Game, you had better learn Grossberg's rules or you're not likely to win any of the prizes.

11. References

Scott [1977] discusses single neuron models from the viewpoint of a physicist. The articles by Amari in Metzler [1977] and by Kohonen in Hinton and Anderson [1981] are good samples of work on network modeling by Grossberg's "competitors." The "bottom-up" approach in Freeman [1975] appears to be converging on the same idea of adaptive resonance that came from Grossberg's "top-down" approach, but much theoretical work needs to be done to relate the approaches.

Freeman, W. J. (1975), Mass Action in the Nervous System, Academic Press, New York.

Fukushima, K. (1980), "Neocognitron, a self-organizing neural network model for a mechanism of pattern recognition unaffected by shift in position," Biol. Cybernet. 36, pp. 193-202.

Grossberg, S. (1982), Studies of Mind and Brain, D. Reidel, Dordrecht.

Hebb, D. O. (1949), The Organization of Behavior, Wiley, New York, p. 62.

Hecht-Nielsen, R. (1983), "Neural analog processing," Proc. SPIE 360.

Hinton, G. E., and J. A. Anderson, eds. (1981), Parallel Models of Associative Memory, Lawrence Erlbaum, Hillsdale, N.J.

Kandel, B., and J. Schwartz, eds. (1981), Principles of Neural Science, Elsevier/North Holland, New York.

von der Malsburg, Ch. (1973), "Self-organization of orientation sensitive cells in the striate cortex," Kybernetic 14, pp. 85-100.

Metzler, J., ed. (1977), Systems Neuroscience, Academic Press, New York.

Scott, A. C. (1977), Neurophysics, Wiley, Interscience, New York.

Stent, G. S. (1973), "A physiological mechanism for Hebb's postulate of learning," Proc. Nat. Acad. Sci. 70, p. 997.

MAXIMUM ENTROPY IN STRUCTURAL MOLECULAR BIOLOGY: THE FIBER DIFFRACTION PHASE PROBLEM

Richard K. Bryan

European Molecular Biology Laboratory, Meyerhofstrasse 1, 6900 Heidelberg, West Germany

The maximum entropy method of image processing is applied to the problem of calculating a map of electron density from its x-ray fiber diffraction pattern. Native and heavy atom derivative data, although insufficient for the usual multiple isomorphous replacement method, are used in combination with a constraint on the maximum particle radius, and with a constraint on the maximum electron density enforced by a "Fermi-Dirac" form of entropy. The problem is illustrated by application to the filamentous bacterial virus Pf1.

C. R. Smith and G. J. Erickson (eds.),
Maximum-Entropy and Bayesian Spectral Analysis and Estimation Problems, 207–228.

1. Introduction

Structural molecular biology is a field rich in inverse problems. Molecules and assemblies thereof are bombarded with electromagnetic radiation of every wavelength from radio frequencies to x-rays, electrons, neutrons, and so on. These various experimental techniques provide data on a variety of different aspects of the structures, from direct images (in optical or electron microscopy) to subtle aspects of molecular interactions (for example, fluorescence decay rates). The problems associated with the interpretation of the experimental data can broadly be put into three categories: removing instrumental response or degradations, performing an idealized inverse problem, and interpreting the results in structural terms. The various techniques exhibit these problems to different degrees.

The usual method for the high-resolution determination of structures is x-ray crystallography. The data are usually extremely good, and there are well established methods for solving, or rather circumventing, the associated phase problem. Here we will consider a close relative, x-ray fiber diffraction, which is used in the study of particles with helical symmetry. In principle, the method is capable of yielding as good results as crystallography, but the experimental problems are greater: getting well aligned specimens is difficult, and the data are noisier since the 'amplification factor' in the data is due to one-dimensional periodicity, instead of three. On the other hand, the continuity in the data makes the phase problem more amenable to direct attack. Moreover, there are many objects of biological importance that have helical symmetry in their in vivo state (few are 3-D crystals!): DNA, the muscle proteins actin and myosin, collagen, keratin, many complete viruses of rod or filamentous forms, parts of other more complex viruses, and various appendages of cells. Here, maximum entropy is applied to the fiber diffraction problem in the context of the filamentous virus Pf1, a member of a class of bacterial viruses that has been studied for more than 20 years [reviews in Denhardt et al., 1978, and DuBow, 1981] and in particular at the European Molecular Biology Laboratory (EMBL) in the past few years. Considerable experimental advances have been made, which have enabled diffraction patterns to be obtained that yield data of sufficiently high quality for an attempt to be made to calculate a medium resolution structure. A large amount of information about the general properties of this virus has been accumulated, with which the details of any calculated structure should agree.

The layout of this paper is as follows. In Section 2 we establish the terminology, outline the theory of fiber diffraction, and discuss the experimental imperfections that must be taken into account. The maximum entropy method is set up in Section 3, and Section 4 gives the details of its application to the fiber diffraction problem. The results of the Pf1 structural calculation are discussed in Section 5.

2. Fiber Diffraction

2.1. Helical Diffraction Theory

The x-ray diffraction pattern of a particle is the intensity of the Fourier transform of its electron density, ρ. Here, the fiber diffraction problem is considered, that is, when diffraction experiments are performed on arrays of particles, each with helical symmetry, which are aligned in direction but not in azimuth, and are sufficiently separated that the diffraction from different particles adds incoherently. Thus the recorded diffraction pattern is the cylindrical average of that from a single particle. The notation used here, and development of the theory, follows Klug, Crick, and Wyckoff [1958].

A particle with helical symmetry is one in which the density ρ at cylindrical coordinates (r,ϕ,z) obeys the relation

$$\rho(r,\phi,z) = \rho(r, \phi + 2\pi u_t, z + u_r), \tag{1}$$

where u_r is termed the unit rise and u_t the unit twist. The unique line specified by the z axis is called the helix axis or fiber axis. Using capitals for coordinates in reciprocal space, the Fourier transform of ρ is

$$F(R,\Phi,Z) = \int_{-\infty}^{\infty}\int_{0}^{2\pi}\int_{0}^{\infty} \rho(r,\phi,z)\, e^{2\pi i[zZ+rR\cos(\phi-\Phi)]}\, r\,dr\,d\phi\,dz\,. \tag{2}$$

Using the generating function for Bessel functions

$$e^{iu\cos v} = \sum_{n=-\infty}^{\infty} e^{in(v+\frac{\pi}{2})} J_n(u) \tag{3}$$

and the helical symmetry, this becomes

$$F(R,\Phi,Z) = \sum_{n=-\infty}^{\infty} e^{in(\Phi+\frac{\pi}{2})} \sum_{j=-\infty}^{\infty} e^{2\pi ij(nu_t-Zu_r)} G_n(R,Z)\,, \tag{4}$$

where

$$G_n(R,Z) = \int_{0}^{u_r}\int_{0}^{2\pi}\int_{0}^{\infty} \rho(r,\phi,z)\, e^{i(2\pi zZ-n\phi)} J_n(2\pi Rr)\, r\,dr\,d\phi\,dz \tag{5}$$

(usually called Bessel function terms). The second sum can be evaluated as

$$\sum_{m=-\infty}^{\infty} \delta(nu_t - Zu_r + m) , \tag{6}$$

giving

$$F(R,\Phi,Z) = \sum_{n=-\infty}^{\infty} \sum_{m=-\infty}^{\infty} \delta(nu_t - Zu_r + m)\, G_n(R,Z)\, e^{in(\Phi+\frac{\pi}{2})} . \tag{7}$$

Thus the transform is nonzero only on layer planes perpendicular to the helix axis, with nth-order Bessel function terms falling at a Z position given by the selection rule

$$Z = \frac{nu_t + m}{u_r} , \qquad m \text{ integer} . \tag{8}$$

The helix parameters are experimentally determined quantities, deduced from the diffraction patterns [Franklin and Holmes, 1958], and will have a certain accuracy. There is no loss of precision, and it is usually more convenient, to assume a rational approximation to the selection rule, $N/K = u_r/u_t$, so that there is an exact repeat after N units in K turns over the repeat length $c = Nu_r$. The positions of the layer planes can now be given an integral index $\ell = Zc$, the selection rule becomes $\ell = nK + mN$, and let

$$G_{n\ell}(R) = G_n(R,\ell/c) . \tag{9}$$

Finally, averaging over the azimuthal angle Φ, the diffraction intensity becomes

$$I_\ell(R) = \langle |F(R,\Phi,\ell/c)|^2 \rangle_\Phi = \sum_n |G_{n\ell}(R)|^2 , \tag{10}$$

with the summation only over the values of n fitting the selection rule for the given ℓ. The notation Σ_n will from now on imply the use of the selection rule in determining the range of summation. As the Φ dependence has been removed, a single axial slice through reciprocal space yields all the possible diffraction information. This can be recorded on a film in a single diffraction experiment, whereas to collect data over all of reciprocal space requires rotation of the specimen relative to the incident beam. The geometry of the experiment usually means that a curved surface is actually recorded, and some data are missing from near the Z axis (the meridian). The lines of intersection with the layer planes are layer lines, and the $\ell = 0$ layer line is the equator. Since $J_n(x)$ is small for $x \lesssim n$, to a given resolution (that is, reciprocal space radius), only a few of the possible Bessel function terms are nonzero on each layer line.

However, real fibers are imperfect in that the individual particles may not be exactly oriented and in that there is a limited number of coherently diffracting units in each particle. These effects cause layer line spreading in the directions of the polar angle and meridian. Examples of diffraction patterns from magnetically oriented fibers of bacteriophage Pf1 [Nave et al., 1981] are shown in Fig. 1. These are extremely good fiber patterns, but they still show some disorientation, particularly visible at large radius. The form of the point-spread function has been calculated in various approximations by Deas [1952], Holmes and Barrington Leigh [1974], Stubbs [1974], and Fraser et al. [1976], and a fast algorithm, accurate over the required domain, has been programmed by Provencher and Gloeckner [1982]. The recorded diffraction pattern is also degraded by film background noise and air scatter of the x-ray beam. For the well oriented patterns used here, this background is easily estimated by fitting a smoothly varying function of reciprocal space to the film density in the parts not occupied by layer line intensity, and interpolating between.

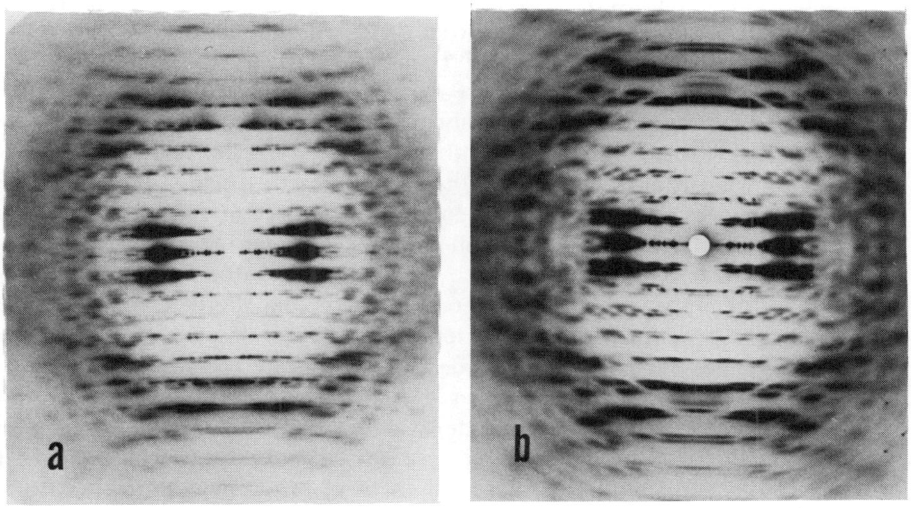

Figure 1. Diffraction patterns of (a) native Pf1 and (b) iodine derivative, from Nave et al. [1981]. The fiber axis is vertical but tilted slightly from the film plane so that the pattern is asymmetric. This enables meridional data to be collected in the 5 Å region.

The problem of deducing an electron density therefore falls into two parts: the deconvolution of the layer lines from the point-spread function, and then the calculation of the electron density from the layer line intensities. This is by no means the end of the story, though. Interpretation in terms of atomic positions is essential to understanding the function of the structure in a biological context. This can often be a very time-consuming activity, which can be aided by starting with the best possible electron density that can be deduced from the data. From our point of view, this means using maximum entropy to solve the inverse problem! First, though, we give a brief summary of the traditional method of phasing intensity data, the method of isomorphous replacement.

2.2. The Method of Multiple Isomorphous Replacement

We first outline this method in the context of protein crystallography, and then discuss the modifications needed for application to the fiber problem. Complete descriptions are given in most textbooks on protein crystallography, for example Blundell and Johnson [1976].

An isomorphous heavy atom derivative of a protein is one in which one or more atoms of large atomic number (usually metal atoms) are added to each native molecule, without significantly changing the structure. This is often a considerable achievement in protein chemistry! If diffraction data are collected for the native and one or more derivatives, they can be used to calculate the phases in the following way. An approximation to the autocorrelation map of the heavy atom density is calculated by taking the inverse transform of the square of the difference of the derivative and native amplitudes (the "difference Patterson"). This map is corrupted by protein-heavy atom cross vectors, and is usually noisy, as it is calculated from the differences of experimental quantities. If the heavy atom sites are few, their positions can often be deduced from this map. There is often some arbitrariness in the origin, even after the crystal symmetry is taken into account, and in the enantiomorph chosen, so if more than one derivative is used, the relative positions of the various heavy atoms must be established, by, for instance, cross-difference Pattersons between the various derivatives. The transform of the located heavy atoms is then calculated and used to phase the native amplitudes, illustrated by the Harker construction [Harker, 1956], Fig. 2. A single derivative clearly gives a twofold ambiguity in phase for each amplitude, which a second derivative is needed to resolve. In practice, owing to noisy data and uncertainties and errors in locating the heavy atoms, the phases are still in error, even with more than the theoretical minimum of two derivatives, and a probabilistic calculation is used to find the "best" phase [Blow and Crick, 1959]. The amplitudes are weighted according to the estimated reliability of the phase. The electron density is then calculated by a straightforward Fourier transform of the phased amplitudes. Although the resultant map is best in a "least squares" sense, one can criticize several aspects of this procedure: there is truncation error, owing to the (inevitable) finite resolution of the data, giving ripples on the map and areas of physically impossible negative density; the weighting scheme means that the amplitudes of the transform of the map are not measured amplitudes although the weights tend to decrease at high resolution, reducing the truncation ripple at the expense of resolution; and

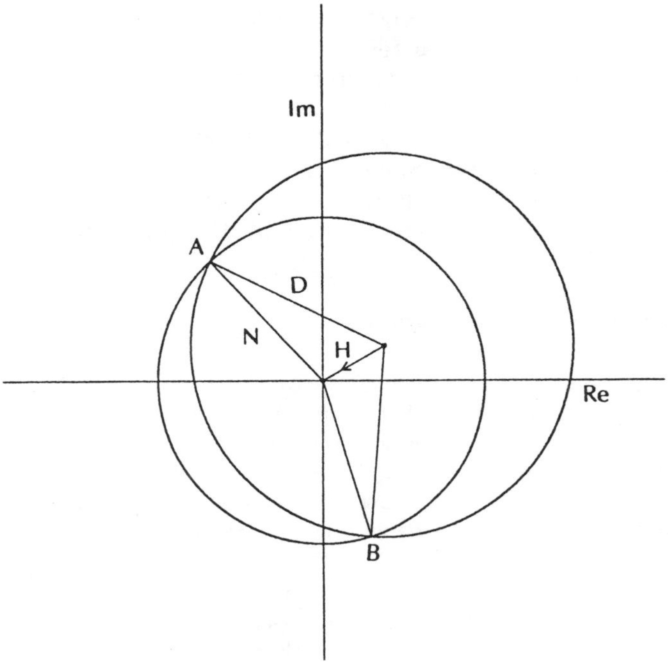

Figure 2. The Harker construction in the complex plane for a single reciprocal space point. H is the complex heavy atom amplitude. Circles of radius N, the native amplitude, and D, the derivative amplitude, are drawn centered at the origin and at -H respectively. The points of intersection A and B, where the derivative is the sum of the native and heavy atom vectors, give the possible phase solutions.

the native data often extend to higher resolution than the derivatives, so cannot be used until an atomic model has been built and is being refined against the entire native data set. Despite these criticisms, around 200 protein structures have been solved so far by this or closely related methods.

The application to fiber diffraction is very similar, but there are significant differences [Marvin and Nave, 1982]. Since a continuous distribution of intensities is measured, instead of the integrated intensities over distinct diffraction spots, the noise on fiber data is usually considerably higher than for crystalline diffraction, and the phases are consequently less well determined. The finite radius of the structure implies continuity in R of the Bessel function terms. Point-by-point construction of phases and weighting of amplitudes does not necessarily ensure this continuity of the complex amplitudes. Even seeking a Bessel function term with minimum curvature in an Argand diagram plot [Stubbs and Diamond, 1975] does not mean that it is a possible transform of a structure of finite radius. Performing a Fourier inverse will then give structure outside the known maximum radius, as will truncation error due to limited resolution data. Such results clearly contradict known properties of the structure. Finally, if more than one Bessel function term contributes on a layer line (to the required resolution), the Harker construction becomes a higher dimensional problem, and to separate each extra term requires two more derivatives [Holmes et al., 1972]. So far, only one complex biological structure, tobacco mosaic virus, has been solved to high resolution by application of this technique [reviewed in Holmes, 1982], after many years of effort.

3. The Maximum Entropy Algorithm

There has been so much discussion of the maximum entropy algorithm method in various applications [for example, Frieden, 1972; Gull and Daniell, 1978; Skilling et al., 1979; Bryan and Skilling, 1980; Skilling, 1981; Willingale, 1981; Burch et al., 1983; Minerbo, 1979; Kemp, 1980; Collins, 1982; Wilkins et al., 1983] that only a brief summary is given here, with an outline of a very successful numerical algorithm for solving the problem with a convex data constraint.

One wishes to reconstruct a set of positive numbers ρ, for which there is some experimental data $\mathbf{D}$, subject to noise ε of known statistical distribution, related to ρ by a transform Γ, so that

$$\mathbf{D} = \Gamma(\rho) + \varepsilon . \tag{11}$$

This relation allows a set of feasible reconstructions to be defined as those which predict the observed data to within the noise, according to some statistical test, typically a chi-squared test. For a selected confidence level, the feasible set can be represented as $C(\rho;\mathbf{D}) \le C_0$. From this set, the map having the greatest configurational entropy is chosen as the preferred reconstruction. This has been shown to be the only consistent way to select a single map by a variational method [Shore and Johnson, 1980; Johnson and Shore, 1983; Gull and Skilling, 1984]. If the feasible set is convex, there will be a unique maximum entropy map, which will either be the global entropy maximum in the unlikely event that it is within the set, implying extremely uninformative data, or lie on the boundary $C = C_0$. The usual problem is therefore to maximize the entropy S over $C = C_0$.

The numerical algorithm [Bryan, 1980; Skilling, 1981; Burch et al., 1983; Skilling and Bryan, 1984] attempts to find iterative increments $\delta\rho$ such that $\rho^{(n+1)}$ is a better approximation to the solution than $\rho^{(n)}$, as is usual with nonlinear optimization algorithms. The increment is taken as a linear combination of search directions $\mathbf{e}_\mu$

$$\delta\rho = x^\mu \, \mathbf{e}_\mu , \tag{12}$$

for some set of coefficients $\{x^\mu\}$. Quadratic models of S and C are constructed in the subspace spanned by the search directions. Because the quadratic approximation is only locally accurate, a limit is put on the step length at each iteration by imposing $|\delta\rho|^2 \leq \ell_0^2$. The length is evaluated using the second derivative of S as a metric, the 'entropy metric' [Bryan, 1980; Skilling and Bryan, 1984], now also interpreted as a second-order approximation to the relative entropy between successive iterates [Skilling and Livesey, 1984; Bricogne, 1984]. Within the region defined by the distance limit, the x^μ are selected to give an increment toward the local maximum of S over C, while C is reduced toward its target value C_0. Skilling and Bryan [1984] give a comprehensive account of this quadratic optimization step. For a problem with a convex constraint, the set of search directions is constructed at each iteration from the contravariant gradients of S and C, and $\nabla\nabla C$ acting on these. A total of three or four such directions is usually sufficent. The convergence of the algorithm is checked by calculating the angle between ∇S and ∇C at the solution. These vectors should be parallel if S is truly a maximum.

4. Maximum Entropy and the Fiber Diffraction Problem

In principle, one could calculate the electron density directly from one or more of the background-corrected diffraction patterns, using maximum entropy with the forward transform

$$\rho \xrightarrow{F \text{ trans}} G \xrightarrow{|\;|^2} \text{Layer lines} \xrightarrow{\text{Convolution}} \text{Diffraction pattern} . \qquad (13)$$

However, the computational requirements for this are vast! Experience shows that solving the phase problem (the first two steps in the above transform) requires many times the number of iterations as would a similar sized linear problem. Moreover, the transform from layer lines to diffraction pattern, although linear, has a space-variant point spread function and is computationally very slow. With realistic computing resources, the only feasible approach was to break the calculation in two: to the deconvolution of the layer lines from the diffraction pattern, and the calculation of the electron density from corrected layer line data. Since the layer lines must be positive, and indeed obey the additive axioms of a probability density, this is an ideal and straightforward application of maximum entropy deconvolution to which the algorithm of Section 3 can be applied directly. A disadvantage of splitting the problem in this way is that the layer line data are no longer calculated as the Fourier transform of a structure of limited radius. This means that where, say, layer lines corresponding to a high and a low order Bessel function are close together (of the order of the spread distance), the deconvolution may incorrectly show intensity on the near-meridional region of the higher order term. On the other hand, it is necessary to scale together different layer line data sets (for example, from native and derivative fibers) and to identify heavy atom positions, which is most conveniently done from the layer line data themselves.

The main part of this discussion will therefore be concerned with the more challenging problem of calculating the electron density from the layer line intensities.

The data constraint requires a comparison of the data that would be observed from the estimated structure ρ with the actual data. The data are sampled in R, so R will now be used as a subscript to denote a discrete point in reciprocal space. As described above, the native layer line intensities are $\Sigma_n |G_{n\ell R}|^2$. The derivative structure is simply the sum of the density of the heavy atoms at the appropriate coordinates—whose transform will be denoted by $H_{n\ell R}$—and the native density, giving predicted derivative data

$$\sum_n |G_{n\ell R} + H_{n\ell R}|^2 \ . \tag{14}$$

Any data that are already phased, such as the equator, are included in a separate term. These are compared with the measured data by means of the traditional χ^2 test [Ables, 1974; Gull and Daniell, 1978]:

$$\chi^2 = \chi_P^2 + \chi_N^2 + \chi_D^2, \tag{15}$$

where P, N, and D stand for phased, native, and derivative respectively,

$$\chi_P^2 = \sum w_{n\ell R}^P (G_{n\ell R} - E_{n\ell R})^2 ,$$

$$\chi_N^2 = \sum w_{\ell R}^N \Big(\sum_n |G_{n\ell R}|^2 - I_{\ell R}^N\Big)^2 , \tag{16}$$

$$\chi_D^2 = \sum w_{\ell R}^D \Big(\sum_n |G_{n\ell R} + H_{n\ell R}|^2 - I_{\ell R}^D\Big)^2 ,$$

the E's are the phased data amplitudes, the I's are the measured intensities, the w's are the weights to be given to the respective measurements, taken as the inverse variances, and the summations are over the data points only. A further term like χ_D^2 is included for each extra heavy atom derivative. A statistic comparing the predicted and measured <u>amplitudes</u> has also been suggested for the phase problem [Gull and Daniell, 1978; Wilkins et al., 1983]. A statistic defined on the actual measured quantities, in this case the intensities, is obviously to be preferred. The amplitude statistic also has the disadvantage of being nondifferentiable if any amplitude is zero, so it is impossible to start the iterative algorithm from a flat map, which is desirable if the phases are not to be biased to those of a particular model starting map.

The maximum entropy algorithm requires that S be maximized over the volume defined by $\chi^2 \leq \chi_0^2 = N$, the total number of data points. The derivatives of χ^2 are required in the numerical algorithm, and examination of them will give some insight into the modifications needed for this problem. Let F be the operator taking ρ to $G_{n\ell R}$. Then, using a rather informal notation and specializing to one Bessel function term per layer line to avoid a plethora of confusing subscripts, we write

$$\nabla \chi_D^2 = 4F^T w\{(|G + H|^2 - I^D)(G + H)\} , \tag{17}$$

and for any real-space vector **a**, with $A = F\mathbf{a}$,

$$\nabla\nabla \chi_D^2 \mathbf{a} = 4F^T w\{(|G + H|^2 - I^D)A + 2[(G + H)^* A](G + H)\} . \tag{18}$$

The expressions for χ_N^2 are obtained by setting H to zero. The second derivative is (complex) diagonal in data space. If there is more than one Bessel function term per layer line, it becomes block diagonal, the block size being the number of Bessel function terms contributing to the layer line. The second derivative term shows that it is possible to have negative curvature if $|G + H|^2$ is smaller than I, so this constraint is not convex. Consequently, there may be more than one local entropy maximum in the feasible set.

Brief consideration of the set of search directions suggested for the convex problem will show that this set is no longer adequate. Assume the starting map is uniform. ∇S is zero, $\nabla \chi^2$, in reciprocal (data) space, has its components lined up with the heavy atom vector, and $\nabla\nabla \chi^2$ is diagonal. Thus the first map increment has the same phase as the heavy atom vector. Further search directions derived from $\nabla \chi^2$ will continue to show this alignment, although the nonlinearity of the entropy will break it via a ∇S search direction, unless some additional condition holds, such as a symmetric map causing all search directions to be symmetric. For Pf1, the initial positions of the heavy atoms were symmetric, so this condition pertains, as it also would if no derivative data were used. Therefore, to break this phase alignment, additional search directions must be used that have a "phase" component and that will be related to eigenvectors of $\nabla\nabla \chi^2$ with a negative eigenvalue.

More generally, the algorithm works by following a path of local S maxima on surfaces of constant χ^2 for a decreasing series of χ^2 values. Negative χ^2 curvature may mean that the S maximum turns into a saddle point, giving a bifurcation in the solution trajectory (Fig. 3). The enantiomorphic ambiguity will give rise to one such point, but there may be others. It is important to detect these bifurcations by finding eigenvectors of $\nabla\nabla \chi^2$ with negative eigenvalues; otherwise a path of local S minima may be followed in error. The current algorithm will choose only one of the possible descent directions from a saddle point. Which one it chooses is a matter of chance or rounding error. Apart from the breaking of phase alignment, which just determines which enatiomorph is calculated, these various choices may lead to significantly different solutions. Starting the algorithm from

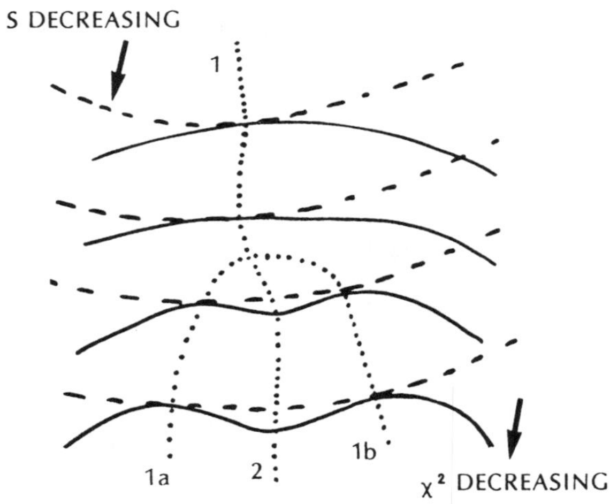

Figure 3. Bifurcation of the maximum entropy solution. The solid lines are contours of constant χ^2, the dashed lines are contours of S, and the dotted lines are trajectories of local stationary S over constant χ^2. Branch 1 is the line of local entropy maxima when χ^2 is convex. When this constraint becomes nonconvex, the line of maxima bifurcates into two branches, 1a and 1b, and a line of local S minima arises, branch 2.

various maps, other than the flat map, is one way of checking whether there are other possible solutions, and for the Pf1 study this was done to a limited extent. A program to investigate systematically all the branches, specifically for the crystallographic problem, has been proposed [Bricogne, 1984].

To attempt to find the directions of negative curvature, the search direction set was supplemented as follows:

(a) A direction that has a reciprocal space "phase" component. This can be constructed in various ways, for example, specific reciprocal space components, random search directions, and asymmetrizing the map both systematically and randomly. Bryan [1980] describes many attempts at such constructions. The efficacy varies slightly, but so far no "magic" direction has been found. Although simple constructions like these usually have only a small component in a negative curvature direction, they are essential as "seed" directions for (b).

(b) Contrary to the recommendations made for the convex case [Skilling and Bryan, 1984], it is here essential to carry forward from iteration to iteration some information on the constraint curvature, namely the directions associated with negative constraint curvature. This is done by calculating the k eigenvectors in the search subspace with the k lowest eigenvalues, and using these as additional search directions at the next iteration. The direction from (a) provides an initial perturbation, and, in practice, the negative curvature directions are built up over several iterations. There is, of course, no guarantee that all negative curvature directions will be found; some may be orthogonal to every search direction constructed. For the Pf1 calculation, $k = 2$ was used, but there is no difficulty in using larger k; $k = 1$ is empirically not so good, as there is then more chance that a negative curvature direction will be "missed."

One consequence of this method of detecting negative curvature directions is that the algorithm must be run much more slowly than for a convex problem; that is, the trust region, defined by $|\delta\rho|^2 \leq \ell_0^2$, must be smaller, or otherwise a saddle point may be crossed in a single step along the line of minima, instead of a sideways increment being made in the negative curvature direction. The solution of this problem typically takes 500 to 1000 iterations, instead of around 20 for a convex problem with similar signal-to-noise ratio. It is also much more important in the phase problem than in a convex problem to ensure parallelism of ∇S and $\Sigma\chi^2$ at the solution; there can still be considerable phase changes after $t = 1 - \cos\alpha$ (α is the angle between the gradients) is reduced below 10^{-4}!

5. Calculation of the Pf1 Structure

Pf1 is a filamentous bacterial virus of a class that has been studied extensively over the past 20 years. The structure of these viruses is of interest, as it provides information both on the mechanism of infection and on the assembly of the virus as it emerges through the bacterial membrane. The Pf1 virus consists of a cylindrical shell of coat protein encapsulating a single-stranded DNA molecule. The protein coat is composed of several thousand identical subunits, each of 46 amino-acid residues, in a helical array, together with minor proteins at each end of the virus. It is the structure of the major protein, which makes up about 95% by weight of the virus and so dominates the diffraction pattern, which is calculated here.

Fiber diffraction from magnetically aligned Pf1 [Nave et al., 1981] gave patterns with resolution to beyond 4 Å, a considerable improvement on earlier results. However, there were still too few data to use the multiple isomorphous replacement method directly at this resolution. Native data were available to 4 Å resolution (only one Bessel function term on each layer line is significant at this resolution) from two derivatives, a doubly substituted iodine and a partly substituted mercury. As had previously been determined [Nave et al., 1981], the mercury site was approximately one-half a helical symmetry operation from the mean iodine position and thus yielded no independent phasing information (the heavy atom vectors differ by 180°). Moreover, owing to a slight change in symmetry between the native and derivatives, the isomorphism cannot be expected to hold at high resolution, but will at low, so low-resolution derivative data can be used if the shift in Z of the Bessel function terms is taken into account [see Nave et al., 1979, for discussion of this point]. Since the iodine was the "better" derivative, it was decided to calculate the structure using the 4 Å native data and the iodine derivative data extending to 5 Å in the meridional direction and 7 Å in the equatorial direction.

At this stage it is appropriate to consider how the existing knowledge of the Pf1 structure can be used. The maximum radius, about 35 Å, was deduced from the positions of the Bessel function term maxima and from the positions of crystalline sampling points in diffraction from dried fibers [Marvin, 1978]. This gives an important constraint on the continuity of the layer line amplitudes. Analysis of the equator had fixed the radial positions of the iodine atoms, and enabled the (real) equator to be phased [Nave et al., 1981]. Consideration of the difference in intensities on the layer lines with ℓ values a multiple of 6 had unambiguously determined the virus symmetry, and together with the chemical knowldege that the two iodine atoms were both substituted on the Tyr-25 residue, and therefore about 6 Å apart, enabled their approximate relative positions to be established. General properties of the diffraction pattern [Marvin et al., 1974] (strong intensity in the 10 Å near-equatorial and 5 Å near-meridional regions) led to the conclusion that the structure was mostly α-helical, with these helices slewing around the virus helix axis. The amino acid sequence of the coat protein was known. Since the possible conformations of a protein, in terms of bond lengths and angles, are extremely well characterized, this should be a strong constraint on the solution. It is however, not yet apparent how descriptive information of this nature should be used [but see Skilling, 1983, for progress in a similar, but simpler, problem]. It can nevertheless be used to estimate the total integrated electron density in the asymmetric unit and an upper bound for the electron density. Since the viruses are immersed in solvent, the measured diffraction is relative to the solvent electron density [Marvin and Nave, 1982]. Therefore the solvent density (0.334e $Å^{-3}$) must be subtracted from the calculated density before transformation. This affects only the equatorial amplitudes.

For completeness, we include some more details on the numerical methods. An important aspect is to arrange the array storage to optimize the calculation of the Fourier transform and its transpose. The helical symmetry must of course be exploited. The selection rules typical for biological specimens (for example, 71/13 for Pf1) imply that only a few values of n need be included for each ℓ. For the current work only one is needed; perhaps up to three are needed in other foreseeable applications (higher resolution work on Pf1, TMV, ...). Although fast Hankel transform routines are available [Siegman, 1977], they require sampling at exponential intervals in radius in both real and reciprocal space. This is inconvenient, as the data are uniformly sampled, and would also give unnecessarily fine sampling at low radius in real space. The net computational saving would be small, for considerable additional complication. There is also the shape of the stored asymmetric unit to consider. In principle it can be any shape, but the only practical choice is a cylinder of one unit rise height, stored on a cylindrical polar grid, which allows the maximum radius constraint to be imposed easily. A fast Fourier transform (FFT) is used on the polar angle coordinate, a DFT is used on the z coordinate (using the selection rule, so that only the n-values associated with a given ℓ are transformed), and a final Hankel transform in radius is performed using a table of stored Bessel function values.

With regard to grid size, the aim is for a maximum spacing in the reconstruction of 0.25 times the resolution length. Since the resolution of the data is 4 Å, the maximum spacing is $\Delta = 1$ Å. The minimum number of angular samples N_ϕ, can then be calculated from $N_\phi \geq 2\pi r_{max}/\Delta$. Using $r_{max} = 35$ Å gives $N_\phi \geq 220$, conveniently rounded to 256 for the angular FFT. Bessel function terms up to about n = 40 are required to represent the structure. However, with the selection rule 71/13, a Bessel function term n is on the same layer line as one of order 71-n, so separation of terms with n in the range 30 to 40 is particularly difficult. In the Pf1 data, such layer lines happened to be very weak and were omitted from the data. This causes a slight drop in resolution at large radius. Also omitted were meridional parts of layer lines (for example, $\ell = 45$) that showed particularly strong crystal sampling. (The patterns in Fig. 1 were not the ones used for data collections. They were made from drier and hence more closely packed fibers, which show stronger intensities and are thus more convenient for the purposes of illustration, but also show more crystal sampling.) Three z-sections are required to represent the unit rise of 3.049 Å. In view of the polar grid used, the entropy has to be weighted by the pixel size. Instead of working directly with the density ρ_i, one can use $\eta_i = r_i\rho_i$, proportional to the integrated density in the pixel, enabling the entropy to be written as

$$S = -\sum_i \frac{\eta_i}{E} \log \frac{\eta_i}{r_i E}, \tag{19}$$

where

$$E = \sum_j \eta_j . \tag{20}$$

This also enables the "r" in the Hankel transform kernel to be absorbed into the stored density value.

As applied to the Pf1 data, this algorithm was spectacularly unsuccessful, and produced a solution with a single blob of density near the heavy atom coordinates. Clearly, the phases of the solution map were still lined up on the heavy atom phases. The peak density exceeded any possible protein density, and it seemed obvious to apply a maximum density limit to the reconstruction. Although such a limit could be imposed by individual constraints on each pixel, such an approach is computationally intractable, and it is more in keeping with the philosophy of the method to modify the entropy expression (which describes our real-space knowledge) to a "Fermi-Dirac" form [Frieden, 1973]

$$S_{FD} = -\sum_i \frac{\eta_i}{E} \log\frac{\eta_i}{r_i E} - \sum_i \frac{\eta'_i}{E'} \log \frac{\eta'_i}{r_i E'}, \tag{21}$$

where

$$\eta_i' = \rho_{max} r_i - \eta_i ,$$
$$E' = \sum_j \eta'_j , \tag{22}$$

thus putting the upper bound on the same status as the positivity constraint implicit in the ordinary entropy expression. The entropy metric in the algorithm is changed to the second derivative of S_{FD}, proportional to

$$\frac{1}{\eta(r\rho_{max} - \eta)} . \tag{23}$$

Unlike the usual (Shannon) entropy, the Fermi-Dirac entropy is invariant if the sign of ρ is reversed (and the origin suitably shifted). For the pure phase problem (native data only), the fit to the data will also be unchanged. Only chemical knowledge will distinguish the true solution. In the problem considered here, an absolute sign is allocated by having a positive heavy atom density. The heavy atoms also fix the origin in ϕ and z, and the only orientational ambiguity left is a twofold axis through the mean heavy atom position perpendicular to the helix axis.

The Fermi-Dirac entropy was used with a ρ_{max} of 1.7e Å^{-3}, a value typical of protein structures at this resolution. The initial map calculation was followed by refinement of the heavy atom positions, minimizing χ_D^2 over the heavy atom coordinates, keeping the map transform $G_{n\ell}(R)$ fixed. The map was then recalculated using the new heavy atom contribution. A few such cycles resulted in convergence, with the net change to the heavy atom positions mostly in the radial direction. The result is shown contoured in cross section in Fig. 4. The helical symmetry can be used to build a complete map of the structure, Fig. 5. It clearly shows rods of density in curved paths around the helix axis. At this resolution, such a core of density is typical of an α-helix structure. The gaps between are filled with closely interdigitated sidechains, invisible at this resolution. A preliminary atomic model (Fig. 6) has been built into this density [Bryan et al., 1983], showing that it can be fitted by a curved α-helical molecule. The resolution is sufficient to see the backbone twist of the α-helix, which is known to be right-handed, and thus enables an absolute hand to be assigned to the virus symmetry.

Previously, data to 8 Å resolution had been used to calculate a structure [Nave et al., 1981] and also to 7 Å for a form of Pf1 with a slightly different symmetry [Makowski et al., 1980] by assuming an initial α-helical model. There was, however, some controversy as to the exact shape of the protein segments, and as to the influence of the initial model on the solution. Here, using the maximum entropy method, no such initial model was required, yet the structure was easily recognizable. Figure 7 shows the fit to a few of the stronger layer lines achieved by the calculated electron density and the model. For a more complete discussion of the structural interpretation, see Bryan et al. [1983].

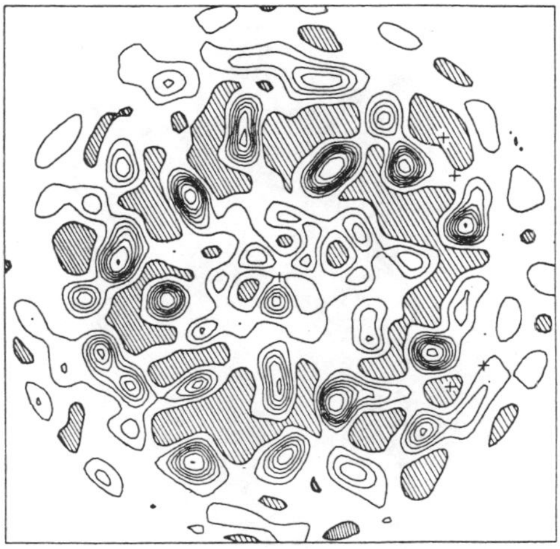

Figure 4. Contour map of solution of Pf1 electron density [from Bryan et al., 1983]. This is a section perpendicular to the z axis, contoured at intervals of 0.2e $Å^{-3}$. Density below 0.2e $Å^{-3}$ is shaded to show regions of low electron density, probably concentrations of hydrophobic sidechains [Marvin and Wachtel, 1975; Nave et al., 1981]. The positions of two symmetry-related pairs of iodine, which are within 3 Å of this section, are shown by crosses.

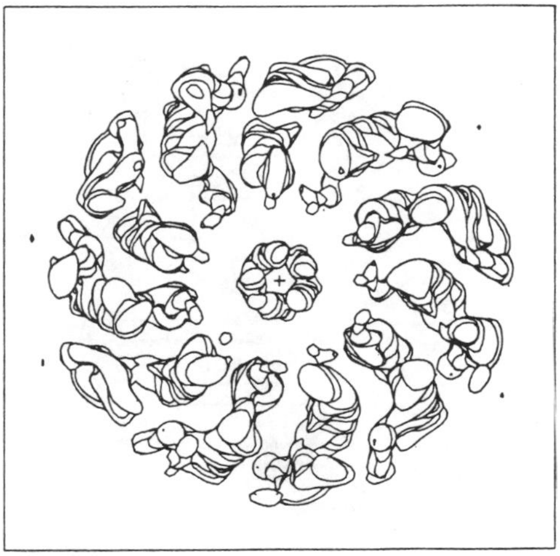

Figure 5. Three-dimensional plot of cores of density. This is a view down the virus axis of a set of 20 sections spaced 1 Å apart in z, contoured at 0.8e $Å^{-3}$, with hidden contours removed. The rods of density follow curved paths around the axis, and an approximate 5 Å helical pitch can be seen within the rods.

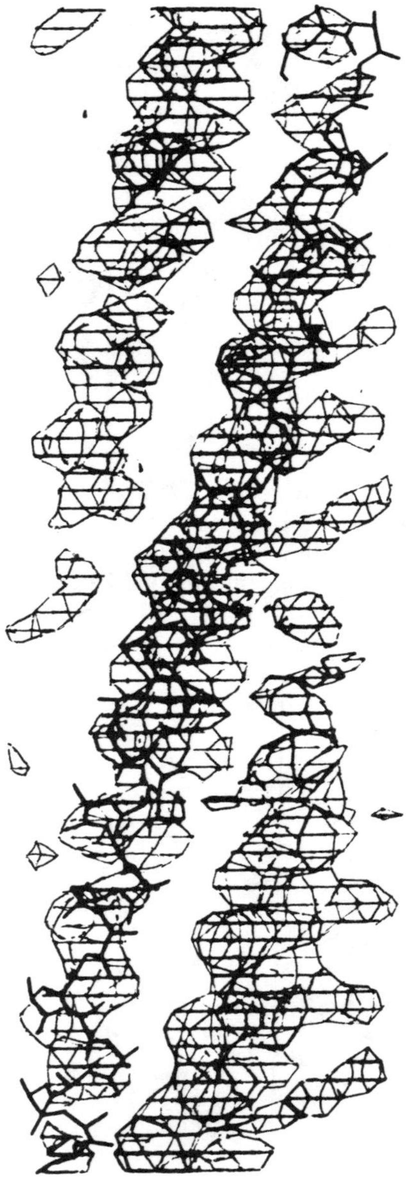

Figure 6. View of a 3-D contour map of the electron density, with the backbone and β-carbons of a mostly α-helical atomic model of the coat protein built in. Parts of three adjacent units are shown, looking from the outside of the virus, with the z axis vertical.

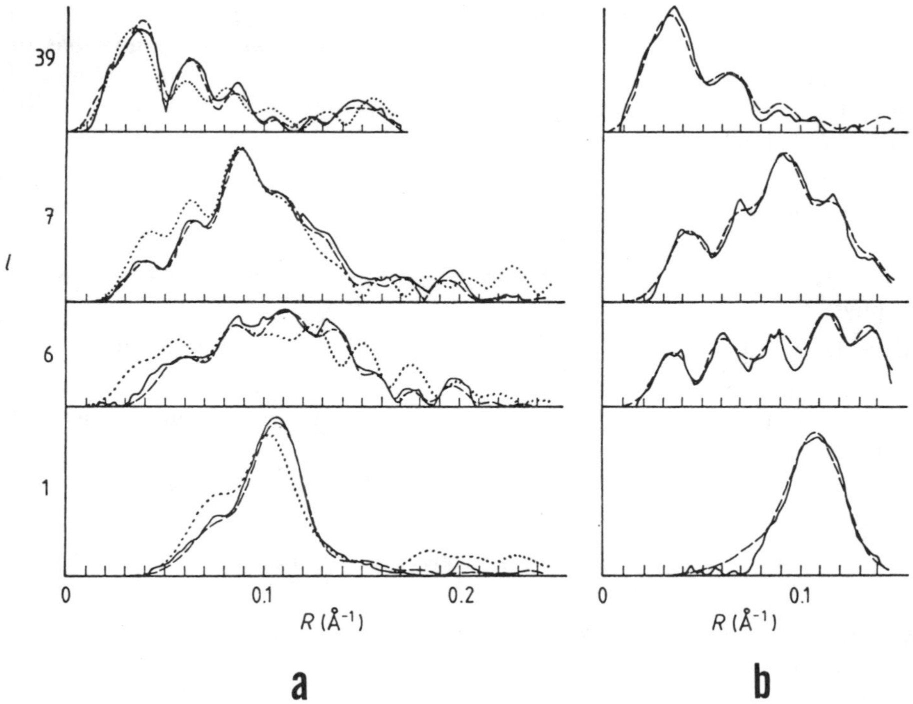

Figure 7. Diffraction amplitudes on layer lines 1, 6, 7, and 39. The solid curves are the observed amplitudes, the dashed are the transform of the maximum entropy map, and the dotted are the transform of the atomic model of Fig. 6. (a) Native amplitudes. (b) Iodine derivative amplitudes. Note the large differences on layer line 6.

To summarize, there are several advantages of using maximum entropy over conventional methods for the fiber diffraction problem. All the layer line data can be used, even if there are too few for conventional phasing methods, and they can be given their correct statistical weights. The maximum radius constraint is easily invoked, and truncation error is reduced since no assumption is made that unmeasured data are zero. The next step will be to apply the method when there are overlapping Bessel function terms on the layer lines. It remains to be seen whether this can be done reliably with considerably fewer isomorphous replacement data than are required by conventional phasing techniques.

6. Acknowledgments

I should like to thank many past and present colleagues at EMBL, in particular Don Marvin and Colin Nave, for their painstaking explanations of the problems involved in the analysis of biological data, to John Skilling and Steve Gull, who first involved me with entropy, and to Steve Provencher for his continued encouragement in this work.

7. References

Ables, J. G. (1974), "Maximum entropy spectral analysis," Astron. Astrophys. Suppl. Ser. 15, pp. 383-393.

Blow, D. M., and F. H. C. Crick (1959), "The treatment of errors in the isomorphous replacement method," Acta Crystallogr. 12, pp. 794-802.

Blundell, T. L., and L. N. Johnson (1976), Protein Crystallography, Academic Press, New York.

Bricogne, G. (1984), "Maximum entropy and the foundations of direct methods," Acta Crystallogr. A40, pp. 410-445.

Bryan, R. K. (1980), "Maximum entropy image processing," Ph.D. thesis, University of Cambridge.

Bryan, R. K., M. Bansal, W. Folkhard, C. Nave, and D. A. Marvin (1983), "Maximum entropy calculation of the electron density at 4 Å resolution of Pf1 filamentous bacteriophage," Proc. Nat. Acad. Sci. USA 80, pp. 4728-4731.

Bryan, R. K., and J. Skilling (1980), "Deconvolution by maximum entropy as illustrated by application to the jet of M87," Mon. Not. R. Astron. Soc. 191, pp. 69-79.

Burch, S. F., S. F. Gull, and J. Skilling (1983), "Image restoration by a powerful maximum entropy method," Computer Vision, Graphics, and Image Processing 23, pp. 113-128.

Collins, D. M. (1982), "Electron density images from imperfect data by iterative entropy maximization," Nature 254, pp. 49-51.

Deas, H. D. (1952), "The diffraction of x-rays by a random assemblage of molecules having partial alignment," Acta Crystallogr. 5, pp. 542-546.

Denhardt, D. T., D. Dressler, and D. S. Ray, eds. (1978), The Single-Stranded DNA Phages, Cold Spring Harbor Laboratory, Cold Spring Harbor, N.Y.

DuBow, M. S., ed. (1981), Bacteriophage Assembly, Liss, New York.

Franklin, R. E., and K. C. Holmes (1958), Tobacco mosaic virus: application of method of isomorphous replacement to the determination of the helical parameters and radial density distribution," Acta Crystallogr. 11, pp. 213-220.

Fraser, R. D. B., T. P. Macrae, A. Miller, and R. J. Rowlands (1976), "Digital processing of fibre diffraction patterns," J. Appl. Crystallogr. 9, pp. 81-94.

Frieden, B. R. (1972), "Restoring with maximum likelihood and maximum entropy," J. Opt. Soc. Am. 62, pp. 511-518.

Frieden, B. R. (1973), "The entropy expression for bounded scenes," IEEE Trans. Inf. Theory IT-19, pp. 118-119.

Gull, S. F., and G. J. Daniell (1978), "Image reconstruction from incomplete and noisy data," Nature 272, pp. 686-690.

Gull, S. F., and J. Skilling (1984), "The maximum entropy method," in Indirect Imaging, J. A. Roberts, ed., Cambridge University Press, pp. 267-279.

Harker, D. (1956), "The determination of the phases of the structure factors of non-centrosymmetric crystals by the method of double isomorphous replacement," Acta Crystallogr. 9, pp. 1-9.

Holmes, K. C. (1982), "The structure and assembly of simple viruses," in Structural Molecular Biology, D. B. Davis, W. Saenger, and S. S. Danyluk, eds., Plenum, New York, pp. 475-505.

Holmes, K. C., and J. Barrington Leigh (1974), "The effect of disorientation on the intensity distribution of non-crystalline fibres. I. Theory," Acta Crystallogr. A30, pp. 635-638.

Holmes, K. C., E. Mandelkow, and J. Barrington Leigh (1972), "The determination of the heavy atom positions in tobacco mosaic virus from double heavy atom derivatives," Naturwissenshaften 6, pp. 247-254.

Johnson, R. W., and J. E. Shore (1983), "Axiomatic derivation of maximum entropy and the principle of minimum cross-entropy—comments and corrections," IEEE Trans. Inf. Theory IT-29, pp. 942-943.

Kemp, M. C. (1980), "Maximum entropy reconstructions in emission tomography," Medical Radionuclide Imaging 1, pp. 313-323.

Klug, A., F. H. C. Crick, and H. W. Wyckoff (1958), "Diffraction by helical structures," Acta Crystallogr. 11, pp. 199-213.

Makowski, L., D. L. D. Casper, and D. A. Marvin (1980), "Filamentous bacteriophage Pf1 structure determined at 7 Å resolution by refinement of models for the α-helical subunit," J. Mol. Biol. 140, pp. 149-181.

Marvin, D. A. (1978), "Structure of the filamentous phage virion," in The Single-Stranded NDA Phages, D. T. Denhardt, D. Dressler, and D. S. Ray, eds., Cold Spring Harbor Laboratory, N.Y., pp. 583-603.

Marvin, D. A., and C. Nave (1982), "X-ray fibre diffraction," in Structural Molecular Biology, D. B. Davis, W. Saenger, and S. S. Danyluk, eds., Plenum, New York, pp. 3-44.

Marvin, D. A., and E. J. Wachtel (1975), "Structure and assembly of filamentous bacterial viruses," Nature 253, pp. 19-23.

Marvin, D. A., R. L. Wiseman, and E. J. Wachtel (1974), "Filamentous bacterial viruses. XI. Molecular architecture of the class II (Pf1, Xf) virion," J. Mol. Biol. 82, pp. 121-138.

Minerbo, G. (1979), "MENT: a maximum entropy algorithm for reconstructing a source from projection data," Computer Graphics and Image Processing 10, pp. 48-68.

Nave, C., R. S. Brown, A. G. Fowler, J. E. Ladner, D. A. Marvin, S. W. Provencher, A. Tsugita, J. Armstrong, and R. N. Perham (1981), "Pf1 filamentous bacterial virus. X-ray fibre diffraction analysis of two heavy atom derivatives," J. Mol. Biol. 149, pp. 675-707.

Nave, C., A. G. Fowler, S. Malsey, D. A. Marvin, and H. Siegrist (1979), "Macromolecular structural transitions in Pf1 filamentous bacterial virus," Nature 281, pp. 232-234.

Provencher, S. W., and J. Gloeckner (1982), "Rapid analytic approximations for disorientation integrals in fiber diffraction," J. Appl. Crystallogr. 15, pp. 132-135.

Shore, J. E., and R. W. Johnson (1980), "Axiomatic derivation of maximum entropy and the principle of minimum cross-entropy," IEEE Trans. Inf. Theory IT-26, pp. 26-37.

Siegman, A. E. (1977), "Quasi fast Hankel transform," Opt. Lett. 1, pp. 13-15.

Skilling, J. (1981), "Algorithms and applications," in Maximum-Entropy and Bayesian Methods in Inverse Problems, C. R. Smith and W. T. Grandy, Jr., eds., Reidel, Dordrecht (1985), pp. 83-132.

Skilling, J. (1983), "Recent developments at Cambridge," these proceedings.

Skilling, J., and R. K. Bryan (1984), "Maximum entropy image reconstruction: general algorithm," Mon. Not. R. Astron. Soc. 211, pp. 111-124.

Skilling, J., and A. K. Livesey (1984), "Maximum entropy as a variational principle and its practical computation," presented at the EMBO Workshop on Maximum Entropy Methods in the X-Ray Phase Problem, Orsay, France.

Skilling, J., A. W. Strong, and K. Bennett (1979), "Maximum entropy image processing in gamma-ray astronomy," Mon. Not. R. Astron. Soc. 187, pp. 145-152.

Stubbs, G. J. (1974), "The effect of disorientation on the intensity distribution of non-crystalline fibres. II. Applications," Acta. Crystallogr. A30, pp. 639-645.

Stubbs, G. J., and R. Diamond (1975), "The phase problem for cylindrically averaged diffraction patterns. Solution by isomorphous replacement and application to tobacco mosaic virus," Acta. Crystallogr. A31, pp. 709-718.

Wilkins, S. W., J. N. Varghese, and M. S. Lehmann (1983), "Statistical geometry. I. A self-consistent approach to the crystallographic inversion problem based on information theory," Acta Crystallogr. A39, pp. 49-60.

Willingale, R. (1981), "Use of the maximum entropy method in x-ray astronomy," Mon. Not. R. Astron. Soc. 194, pp. 359-364.

A METHOD OF COMPUTING MAXIMUM ENTROPY PROBABILITY VALUES FOR EXPERT SYSTEMS

Peter Cheeseman

SRI International, 333 Ravenswood Road, Menlo Park, CA 94025

This paper presents a new method for calculating the conditional probability of any attribute value, given particular information about the individual case. The calculation is based on the principle of maximum entropy and yields the most unbiased probability estimate, given the available evidence. Previous methods for computing maximum entropy values are either very restrictive in the probabilistic information (constraints) they can use or are combinatorially explosive. The computational complexity of the new procedure depends on the interconnectedness of the constraints, but in practical cases it is small.

C. R. Smith and G. J. Erickson (eds.),
Maximum-Entropy and Bayesian Spectral Analysis and Estimation Problems, 229–240.

1. Introduction

Recently, computer-based expert systems have been developed that store probabilistic knowledge obtained from experts and use this knowledge to make probabilistic predictions in specific cases. Similarly, analyses of data, such as questionnaire results, can reveal dependencies between the variables that can also be used to make probabilistic predictions. The essential problem in such systems is how to represent all known dependencies and relationships, and how to use such information to make specific predictions. For example, knowledge of interrelationships among such factors as age, sex, diet, cancer risk, etc., should allow the prediction of, say, an individual's cancer risk, given information about his or her diet, age, etc. However, because of possible interactions among the factors, it is not sufficient merely to combine the effects of the separate factors.

The major problem faced by all such probabilistic inference systems is that the known constraints usually underconstrain the probability space of the domain. For example, if the space consists of 20 predicates, then 2^{20} joint probability constraints are needed to fully specify all the probabilities. When a space is underconstrained, any desired probability usually has a range of possible values consistent with the constraints. The problem is to find a unique probability value within the allowed range that is the best estimate of the true probability, given the available information. Such an estimate is given by the method of maximum entropy (ME) and yields a probability value that is the least commitment value, subject to the constraints. To choose any other value has been shown by Shore and Johnson [1980] to be inconsistent, because any other choice would imply more information than was given in the problem.

This paper focuses on a type of expert system in which all the probability constraints are in the form of conditional probabilities or joint probabilities (sometimes called marginal probabilities because they occur in the margins of contingency tables). Such probability constraints may have come from an expert or from an analysis of data that has shown that particular subsets of factors are significantly correlated. The problem of making probabilistic predictions in underconstrained probability spaces is of sufficient importance that many solutions have been tried. One method is to acknowledge that the desired probabilities are underconstrained and return the range of possible values consistent with the known constraints (rather than a point value). Such an approach is implicit in the method proposed by Shafer [1979].

Another method is to make the strong assumption of conditional independence when combining different evidence. This is the assumption behind PROSPECTOR [Duda et al., 1976] and Dependence Trees [Chow and Liu, 1968] and used most recently by Pearl [1982]. Use of the conditional independence assumption with given conditional probabilities is usually sufficient to constrain the desired probabilities to a unique value. However, this assumption is not always satisfied by actual data and can lead to inconsistent and overconstrained probability values, as pointed out by Konolige [1979].

The main purpose of this paper is to introduce a new method for computing the maximum entropy probability of a predicate or attribute of interest, given specific evidence about related attributes, and subject to any linear probability constraints. This method avoids the combinatorial explosion inherent in previous methods without imposing strong limitations on the constraints that can be used, and it is therefore useful for computer-based expert systems.

2. The Maximum Entropy Method

The method of maximum entropy was first applied by Jaynes to the statistical mechanics problem of predicting the most likely state of a system given the physical constraints (for example, conservation of energy). In Jaynes [1968] the maximum entropy method was used to provide prior probabilities for a Bayesian analysis. Lewis [1959] applied the method of least information (an equivalent method) to the problem of finding the best approximation to a given probability distribution based on knowledge of some of the joint probabilities (that is, constraints on the possible distributions). Ireland and Kullback [1968] applied the minimum discrimination information measure (yet another method equivalent in this case to maximum entropy) to find the closest approximating probability distribution consistent with the known marginals in contingency table analysis. Konolige [1979] applied the least information method to expert systems, and this analysis has been extended by Lemmar and Barth [1982].

The mathematical framework used in this paper is defined below. Although the definitions are for a space of four parameterized attributes, the framework applies to any number of attributes. The attributes are A, B, C, and D, where

A has possible values A_i $\quad i = 1$ to I
B has possible values B_j $\quad j = 1$ to J
C has possible values C_k $\quad k = 1$ to K
D has possible values D_l $\quad l = 1$ to L

and P_{ijkl} is the probability that A has value A_i, B has value B_j, C has value C_k, and D has value D_l. For example, A might be the attribute (function name) "soil-type," where A_1 has the value "clay," A_2 is "silt," and so on.

Each value (category) of an attribute is assumed to be mutually exclusive and exhaustive of the other categories. Any attribute that is not currently exhaustive can be made so by adding an extra category, "other," for anything that does not fit the existing categories. In terms of these attributes, the entropy function, H, is defined as

$$H = -\sum_{ijkl} P_{ijkl} \log P_{ijkl} \,. \tag{1}$$

H is a measure of the uncertainty inherent in the component probabilities. For example, if one of the P_{ijkl} values is 1 (and so all the rest are zero), then H is zero; that is, there is no uncertainty, as expected. Conversely, H can be shown to be a maximum when all the P_{ijkl} values are equal (if there are no constraints); this represents a state of maximum uncertainty.

Because the P_{ijkl} values are probabilities, they must obey the constraint

$$\sum_{ijkl} P_{ijkl} = 1 . \tag{2}$$

In addition, any subset of the following constraints may be asserted:

Joint probabilities, such as

$$\sum_{jkl} P_{ijkl} = P_i^A \tag{3}$$

$$\sum_{ij} P_{ijkl} = P_{kl}^{CD} \tag{4}$$

$$\sum_{i} P_{ijkl} = P_{jkl}^{BCD} \tag{5}$$

$$\vdots$$

$$P_{ijkl} = \text{specific value}$$

Conditional probabilities, such as

$$P(A_2|B_3) = \frac{P_{23}^{AB}}{P_3^B} = \frac{\sum_{kl} P_{23kl}}{\sum_{ikl} P_{i3kl}} \tag{6}$$

And probability assignment to arbitrary logical functions, for example,

$$P(A_2 \Rightarrow B_3) = x \quad \text{(logical implication)} \tag{7}$$

$$\text{implying} \sum_{j} P_{2j}^{AB} = 1 - x \qquad (j \neq 3) \tag{8}$$

$$\text{implying} \sum_{\substack{jkl \\ j \neq 3}} P_{2jkl} = 1 - x \qquad (9)$$

These constraints are given explicitly because their values differ significantly from their default ME value. Such significant constraints could be specified by experts or found by a program that examines data looking for significant combinations. The main reason for calculating ME probability values on the basis of the known constraints is to be able to find any probability value without having to store the entire probability space. Only linear constraints involving equalities have been given above, but the ME method can be extended to include nonlinear constraints as well. Note that Eq. (3), for example, is itself a set of constraints—one for each value of i given. Also, it is assumed here that if either the numerator or the denominator of a conditional probability constraint is given separately (as a joint probability constraint), then the conditional probability constraint is replaced by the equivalent joint probability (marginal) constraints. The last constraint indicates that a probability assignment to any logical formula is equivalent to a probability assignment to a subset of the total probability space, and so forms a simple linear constraint.

The principle of maximum entropy requires that a unique set of values for P_{ijkl} be found that satisfies the given constraints and at the same time maximizes the value of H given by Eq. (1). A method for calculating this ME distribution is discussed in Sec. 3. The reasons for accepting ME probability values as the best estimate of the true probability are discussed in Jaynes [1979] and Lewis [1959] and may be summarized as follows. In expert system applications, when all the significant constraints (for example, marginals and conditionals) have been found, all the information about the domain is contained in these constraints. Any ME probability value calculated with these constraints has distributed the uncertainty (H) as evenly as possible over the underlying probability space in a way consistent with the constraints. Returning any non-ME value implies that extra information is being assumed because H is no longer a maximum.

The shape of a particular H distribution around the ME value indicates how well the particular calculated probability is constrained by the available information. The difference between H for an assumed probability and H maximum (that is, the ME value) gives the amount of information assumed by choosing a non-ME probability value. The variance of the distribution of H around the maximum is determined by the sample size on which the constraints have been induced.

3. A New Method of Calculating Maximum Entropy Distributions

The first use of maximum entropy (least information) for estimating probability distributions in computer science is due to Lewis [1959]. He showed that, if the given probabilities are conditionally independent, then the underlying probability space can be represented by simple product formulas and that this is the maximum entropy distribution. This product form is the basis of Dependence Trees [Chow and Liu, 1968] and the tree-based Bayesian update method of Pearl [1982]. An iterative technique for computing the ME distribution given some of the joint probabilities without requiring conditional independence was developed by Brown [1959]. This method was extended by Ku and Kullback [1969], but, like Brown, they put strong restrictions on the constraints that must be given, and their method combinatorially explodes if the space of attributes is large. The new method of computing ME distributions presented in this section avoids these difficulties.

The problem of optimizing a continuous function subject to constraints is well known in applied mathematics, and a general solution is the method of Lagrange multipliers. The specific problem of maximizing the entropy function (1) subject to constraints was first applied to the domain of statistical mechanics, and specifically to joint marginal constraints, by Gokhale and Kullback [1978]. This section derives the necessary formulas in a form suitable for efficient computation.

The first step is to form a new entropy function, as defined by

$$H' = -\sum_{ijkl} P_{ijkl} \log P_{ijkl} + \lambda\Big(1 - \sum_{ijkl} P_{ijkl}\Big) + \lambda_i \Big(P_i^A - \sum_{jkl} P_{ijkl}\Big)$$

$$+ \lambda_{23}^{AB} \Big(P(A_2|B_3) \sum_{ikl} P_{i3kl} - \sum_{kl} P_{23kl}\Big) + \cdots \tag{10}$$

The next step is to equate the derivative of Eq. (10) (with respect to each variable) to zero, giving:

$$\frac{\partial H'}{\partial P_{ijkl}} = -\log P_{ijkl} - 1 - \lambda - \lambda_i - \cdots - \lambda_{ij} - \cdots - \lambda_{23}^{AB}[1 - P(A_2|B_3)] - \cdots$$

$$= 0 \tag{11}$$

implying

$$\log P_{ijkl} = -\{\lambda_0 + \lambda_i + \cdots + \lambda_{ij} + \cdots + \lambda_{23}^{AB}[1 - P(A_2|B_3)] + \cdots\} \tag{12}$$

where $\lambda_0 = \lambda + 1$, or

$$P_{ijkl} = \exp -\{\lambda_0 + \lambda_i + \cdots + \lambda_{ij} + \cdots$$

$$+ \lambda_{23}^{AB}[1 - P(A_2|B_3)] + \cdots\} \tag{13}$$

and

$$\frac{\partial H'}{\partial \lambda_0} = 0 \Rightarrow \sum_{ijkl} P_{ijkl} = 1 \tag{14}$$

$$\frac{\partial H'}{\partial \lambda_i} = 0 \Rightarrow \sum_{jkl} P_{ijkl} = P_i^A \tag{15}$$

$$\vdots$$

$$\frac{\partial H'}{\partial \lambda_{23}^{AB}} = 0 \Rightarrow \sum_{kl} P_{23kl} = P(A_2|B_3) \sum_{ikl} P_{i3kl} \tag{16}$$

etc.

Equation (13) gives the ME distribution in terms of the λ's, so if the values of all λ's can be found, the ME space is known implicitly. Note that Eq. (12) is the so-called loglinear form, but here this form is a direct consequence of the maximization of H rather than an ad hoc assumption. From Eq. (10) it is clear that there is only one λ per constraint and that these are the only unknowns. If Eq. (13) is substituted into Eqs. (14), (15), etc. (that is, into each of the given constraints), then the resulting set of simultaneous equations can be solved for the λ's. It is more convenient first to apply the following transformations:

$$\alpha_0 = e^{-\lambda_0}, \quad \alpha_i = e^{-\lambda_i}, \quad \alpha_{ij} = e^{-\lambda_{ij}}, \quad \alpha_{23}^{AB} = e^{-\lambda_{23}^{AB}[1-P(A_2|B_3)]}, \quad \text{etc.}$$

$$\Rightarrow P_{ijkl} = \alpha_0\alpha_i \cdots \alpha_{ij} \cdots \alpha_{23}^{AB} \cdots . \tag{17}$$

Thus, the basic distribution P_{ijkl} is given implicitly as a product of α's. Equation (17) is the key to the new ME calculation method, as it implicitly gives the underlying probability space in terms of a product of parameters (the α's), and there are only as many α's as there are constraints. Note that for any particular P_{ijkl}, only those α's with the corresponding indices appear in the product. With these substitutions, Eq. (14) becomes

$$\alpha_0 \sum_{ijkl} \alpha_i\alpha_j \cdots \alpha_{ij}\alpha_{ik} \cdots \alpha_{ijk} \cdots = 1 \tag{18}$$

and Eq. (15) becomes

$$\alpha_2^A \alpha_{23}^{AB} \sum_{kl} \alpha_k\alpha_l\alpha_{kl}\alpha_{2k}^{AC} \cdots = P(A_2|B_3) \sum_{ikl} \alpha_i\alpha_k\alpha_l\alpha_{ik}\alpha_{3k}^{BC} \cdots \tag{19}$$

and so on (one equation for each constraint).

This set of simultaneous equations can be solved by any standard numerical techniques. However, in practice it is more common to need to update an existing solution by adding a new constraint. Such an update introduces a new corresponding (non-unity) α, and causes adjustments to some of the existing α's. Even when a set of constraints is to be added, they can be introduced sequentially; thus an update method is always sufficient to compute the α's. A suitable update method is to assume initially that all the α's have their old value, then calculate a value for the new α from the new constraint equation. This new value is inserted into each of the existing constraint equations in turn, and revised α values are calculated for the α corresponding to each constraint. This process is repeated until all the α values have converged on their new values. Current investigations have shown that the convergence of the α's is reasonably rapid, and the only divergences that have been found are the result of trying to add an inconsistent constraint.

4. Probabilistic Inference

The previous section described a method for representing (implicitly) the underlying ME probability distribution. This section describes how to use such a representation to calculate the conditional probability of any desired attribute, given information about a specific case. Such a computation requires summing over the probability space without creating a combinatorial explosion, as shown below.

Consider a hypothetical example with attributes A_i, B_j, C_k, D_l, and E_m, where each A_i, for example, could represent different age categories. If prior probabilities (constraints) are given for some of the attribute values and prior joint and conditional probabilities are given for combinations of values of different attributes, then the corresponding α values can be computed as explained in the previous section. The resulting probability space for a particular combination of given prior probabilities might be

$$P_{ijklm} = \alpha_0 \alpha_i \alpha_j \alpha_k \alpha_l \alpha_m \alpha_{ij} \alpha_{ik} \alpha_{jl} \alpha_{jm} \alpha_{km} \alpha_{lm} \alpha_{jlm} \,. \tag{20}$$

If the prior probability of, say, a particular A_i is required (that is, it is not one of the given priors), then the value is given by

$$\begin{aligned} P(A_i) &= \alpha_0 \alpha_i \sum_{jklm} \alpha_j \alpha_k \alpha_l \alpha_m \alpha_{ij} \alpha_{ik} \alpha_{jl} \alpha_{jm} \alpha_{km} \alpha_{lm} \alpha_{jlm} \\ &= \alpha_0 \alpha_i \sum_k \alpha_k \alpha_{ik} \Big(\sum_m \alpha_m \alpha_{km} \Big(\sum_j \alpha_j \alpha_{jm} \Big(\sum_l a_l \alpha_{lm} \alpha_{jl} \alpha_{jlm} \Big)\Big)\Big) \end{aligned} \tag{21}$$

Here, the full summation has been recursively decomposed into its component partial sums, allowing each partial sum to be computed as soon as possible; the resulting matrix then becomes a term in the next outermost sum. In the above example, this summation method reduces the cost of evaluating Σ_{jklm} from O(J*K*L*M) (where J,...,M are the ranges of j,...,m respectively) to O(J*L*M)—that is, the cost of evaluating the innermost sum. Note that a different order of decomposition can produce higher costs; that is, the cost of the evaluation of sums is dependent on the evaluation order and is a minimum when the sum of the sizes of the intermediate matrices is a minimum. When there are a large number of attributes, the total computational cost of evaluating a sum is usually dominated by the largest intermediate matrix, whose size is partly dependent on the degree of interconnectedness of the attribute being summed over and the order of evaluation. The above summation procedure is also necessary for updating the previous α's when given new prior probabilities. In the above, α_0 is a normalization constant that can be determined, once all values of $P(A_i)$ have been evaluated, from the requirement that $\Sigma_i P(A_i) = 1$. Such a normalization makes prior evaluation of α_0 unnecessary.

To find the conditional probability of an attribute (or joint conditional probability of a set of attributes), all that is needed is to sum over the values of all attributes that are not given as evidence of the target attribute. For example, to find the conditional probability of A_i given that D_2 and E_3 are true, the correct formula is

$$P(A_i|D_2,E_3) = \beta\alpha_i \sum_{jk} \alpha_j \alpha_k \alpha_{ik} \alpha_{j2}^{BD} \alpha_{j3}^{BE} \alpha_{k3}^{CE} \alpha_{j23}^{BDE}$$

$$= \beta\alpha_i \left(\sum_j \alpha_j \alpha_{j2}^{BD} \alpha_{j3}^{BE} \alpha_{j23}^{BDE}\right)\left(\sum_k \alpha_k \alpha_{ik} \alpha_{k3}^{CE}\right) \quad (22)$$

where β is a normalization constant and the summations are over the remaining attributes (B and C). Note that the more evidence there is concerning a particular case, the smaller the resulting sum. Also, the conditional probability evaluation procedure is nondirectional because, unlike other expert systems, this procedure allows the conditional probability of any attribute to be found for any combination of evidence. That is, it has no specially designated evidence and hypothesis attributes. Also note that in this case the initial sum has been factored into two independent sums.

The above probability evaluation method can be extended to include the case where the evidence in a particular case is in the form of a probability distribution over the values of an attribute that is different from the prior distribution, rather than being informed that a particular value is true. In this case, it is necessary to compute new α's that correspond to the given distribution and use these new α's in place of the corresponding prior α's in probability evaluations such as those above. For instance, if a new distribution is given for $P(A_i)$, then the new α's are given by

$$\alpha_i^A(\text{new}) = \alpha_i^A(\text{old}) \frac{P_i^A(\text{new})}{P_i^A(\text{old})} . \tag{23}$$

The revised α values above are equivalent to the multiplicative factors used by Lemmar and Barth [1982], who showed that the minimum discrimination information update of a particular probability subspace is given by introducing multiplicative factors equivalent to the above. The major difference is that the summation procedure described in this paper will work even when the space cannot be partitioned. The relationship between minimum discrimination information update and conditional probability update where an "external" evidence node is given as true, thus inducing a probability distribution on a known node, is being investigated. In the extreme case where the minimum discrimination update is based on information that a particular value is true, the two approaches give the same results.

The above conditional probability evaluation procedure (a type of expert system inference engine) has been implemented in LISP and has been tested on many well known ME examples. In ME conditional probability calculations when specific evidence is given, it has been found that only short, strong chains of prior joint or conditional probabilities can significantly change the probability of an attribute of interest from its prior value.

When a joint probability value is computed by the proposed method, it is useful to estimate its accuracy as well. There are two sources of uncertainty in a computed ME value. One is the possibility that the known constraints used are not the only ones operating in the domain. This type of uncertainty is hard to quantify and depends on the methods used to find the known constraints. If a constraint search is systematic (over the known data), then we can be confident that we know all the dependencies that can contribute to a specific ME value. If a constraint search is ad hoc, it is always possible that a major contributing factor has been overlooked. If any important factors are missing, the calculated ME probability values will differ significantly from the observed values in many trials. If such systematic deviations are found, it indicates that constraints are missing, and an analysis of the deviations often gives a clue to these missing factors [Jaynes, 1979].

The other source of uncertainty is in the accuracy with which the constraints are known. This accuracy depends on the size of the sample from which the constraints were extracted or the accuracy of the expert's estimates. This uncertainty is also hard to quantify when expert's estimates are used, but when the constraints are induced from data, the accuracy is dependent on the sample size and randomness. In the analysis given here, the constraints were assumed to be known with complete accuracy. When particular conditional probability values are computed, the uncertainty of the conditioning information can introduce additional uncertainty.

5. Summary

This paper presents a new method of computing maximum entropy distributions and shows how to use these distributions. It also presents some specific evidence to calculate the conditional probability of an attribute of interest. Previous methods of computing maximum entropy distributions are either too restrictive in the constraints allowed, or too computationally costly in nontrivial cases. The new method avoids both these difficulties, and provides a framework for estimating the certainty of the computed maximum entropy conditional probability values.

Further research is necessary to improve the efficiency of this method, particularly by automatically finding the optimal summation evaluation order and discovering approximations that would allow the exclusion from the summation of any attributes that could not significantly affect the final result, and by improving the convergence of the α update method. Such improvements should increase the usefulness of this ME computation technique as an expert system inference engine.

6. References

Brown, D. T. (1959), "A note on approximations to discrete probability distributions," Inf. and Control 2, pp. 386-392.

Chow, C. K., and C. N. Liu (1968), "Approximating discrete probability distributions with dependence trees," IEEE Trans. Inf. Theory IT-14, pp. 462-467.

Duda, R. O., P. E. Hart, and N. J. Nilsson (1976), "Subjective Bayesian methods for rule-based inference systems," Proc. AFIPS Nat. Compt. Conf. 47, pp. 1075-1082.

Gokhale, D. V., and S. Kullback (1978), The Information in Contingency Tables, Marcel Dekker, New York.

Ireland, C. T., and S. Kullback (1968), "Contingency tables with given marginals," Biometrika 55, pp. 179-188.

Jaynes, E. T. (1968), "Prior probabilities," IEEE Trans. Syst. Sci. Cybernet. SSC-4, pp. 227-241.

Jaynes, E. T. (1979), "Where do we stand on maximum entropy," in The Maximum Entropy Formalism, R. D. Levine and M. Tribus, eds., MIT Press.

Konolige, K. (1979), "A Computer-Based Consultant for Mineral Exploration," SRI Artificial Intelligence Report, App. D.

Ku, H. H., and S. Kullback (1969), "Approximating discrete probability distributions," IEEE Trans. Inf. Theory, IT-15, No. 4.

Lemmar, J. F., and S. W. Barth (1982), "Efficient Minimum Information Updating for Bayesian Inferencing in Expert Systems," Proc. National Conference on Artificial Intelligence, Pittsburgh, pp. 424-427.

Lewis, P. M. (1959), "Approximating probability distributions to reduce storage requirements," Inf. and Control 2, pp. 214-225.

Pearl, J. (1982), "Distributed Baysian Processing for Belief Maintenance in Hierarchical Inference Systems," Rept. UCLA-ENG-CSL-82-11, University of California, Los Angeles.

Shafer, G. (1979), A Mathematical Theory of Evidence, Princeton University Press.

Shore, J. E., and R. W. Johnson (1980), "Axiomatic derivation of the principle of maximum entropy and the principle of minimum cross-entropy," IEEE Trans. Inf. Theory IT-26, pp. 26-37.

SPECIAL-PURPOSE ALGORITHMS FOR LINEARLY CONSTRAINED ENTROPY MAXIMIZATION

Yair Censor

Department of Mathematics, University of Haifa, Mt. Carmel, Haifa 31999, Israel

Tommy Elfving

National Defence Research Institute, Box 1165, S-581 11 Linköping, Sweden

Gabor T. Herman

Department of Radiology, Hospital of the University of Pennsylvania, Philadelphia, Pennsylvania 19104

We present a family of algorithms of the row-action type that are suitable for solving large and sparse entropy maximization problems with linear constraints. The algorithms are designed to handle equality constraints, inequality constraints, or interval constraints (that is, pairs of inequalities). Six iterative algorithms are discussed, but for one of them convergence of the process is not yet proven. No experimental results are presented.

C. R. Smith and G. J. Erickson (eds.),
Maximum-Entropy and Bayesian Spectral Analysis and Estimation Problems, 241–254.

1. Introduction

The "x log x" entropy functional, ent x, maps the nonnegative orthant $\mathbf{R}^n_+$ of $\mathbf{R}^n$ (the n-dimensional Euclidean space) into $\mathbf{R}$ according to

$$\text{ent } x = -\sum_{j=1}^{n} x_j \log x_j . \tag{1}$$

By definition, 0 log 0 = 0.

The entropy optimization problems that will be addressed here are of the form

$$\left.\begin{array}{l} \text{Maximize ent } x \\ \text{subject to } x \in Q \\ \text{and } \sum_{j=1}^{n} x_j = 1 \text{ and } x \geq 0 . \end{array}\right\} \tag{2}$$

Here $x \geq 0$ is meant componentwise. $Q \subseteq \mathbf{R}^n$ is the constraints set, which takes one of the following forms (where A is a real m × n matrix, and $b = (b_i)$ and $c = (c_i)$ are vectors in $\mathbf{R}^m$):

$$Q_1 = \{x \in \mathbf{R}^n \mid Ax = b\} , \tag{3}$$

$$Q_2 = \{x \in \mathbf{R}^n \mid Ax \leq b\} , \tag{4}$$

$$Q_3 = \{x \in \mathbf{R}^n \mid c \leq Ax \leq b\} . \tag{5}$$

Accordingly, the optimization problem (2) is called entropy maximization with linear equality, inequality, or interval constraints, respectively.

Such linearly constrained entropy optimization problems arise in various fields of applications:

Transportation planning (the gravity model). See for example Lamond and Stewart [1981].

Statistics (adjustment of contingency tables, maximum-likelihood estimation). See for example Darroch and Ratcliff [1972].

Linear numerical analysis (preconditioning of a matrix prior to calculation of eigenvalues and eigenvectors). See for example Elfving [1980].

Chemistry (the chemical equilibrium problem). See for example the remark and references mentioned in Erlander [1981].

Geometric programming (the dual problem). See for example Wong [1981].

Image processing (image reconstruction from projections, image restoration). See for example Herman [1980], Censor [1983], Frieden [1980].

As sources for further general information we mention Jaynes [1982], Kapur [1983], and Frieden [1983].

The use of entropy is rigorously founded in several areas [see, for example, Levine and Tribus, 1978, or Jaynes, 1982] while in other situations entropy optimization is used on a more empirical basis. In image reconstruction from projections, from which our own motivation to study entropy optimization comes, arguments in favor of maximum entropy imaging have been given, but typically they express merely the conviction that the maximum entropy approach yields the most probable solution agreeing with the available data, that is, a solution that is most objective or maximally uncommitted with respect to missing information.

In this report we present a family of algorithms for solving the entropy maximization problems mentioned above. The common denominator of these algorithms is their row-action nature in the sense of Censor [1981a, definition 3.1]. By this we mean that only the original data A, b, c appearing in Eq. (3), (4), or (5), to which no changes are made during iterations, are used. In a single iterative step, access is required to only one row of the system of equations or inequalities. In addition, only x^k, the immediate predecessor of the next iterate, x^{k+1}, is needed at any given iterative step. Algorithms for norm-minimization having such properties were found to perform well on sparse and large systems such as those arising in the field of image reconstruction from projections. Limited experimental experience with some of the algorithms for entropy optimization presented below is also available [for example, Censor et al., 1979; Lent, 1977; Herman, 1982a,b], but it is safe to say that all these represent preliminary tests and that more experimental work is needed to assess their practical value. We do not intend this paper to be an overall review of algorithms for entropy optimization; for example, we do not discuss the MENT algorithm [Minerbo, 1979], and probably leave out several others. Rather, we concentrate on a specific family of algorithms to which our attention and efforts were attracted recently. For mathematical details we refer the reader to the appropriate original publications, and we definitely make no claims here about potential advantages of these algorithms in practical applications. Following the definition of entropy projection in Section 2, we describe row-action algorithms for entropy maximization in Section 3.

2. Entropy Projections onto Hyperplanes

Several of the algorithms presented here use entropy projections onto hyperplanes, which are defined as follows:

Let $H = \{x \in \mathbf{R}^n \mid \langle a,x\rangle = b\}$ be a given hyperplane and let $y \in \mathbf{R}^n_+$ be a given nonnegative vector. Then the system

$$\begin{aligned} y'_j &= y_j \exp(B\, a_j)\,, \quad j = 1,2,\ldots,n \\ \langle a,y'\rangle &= b \end{aligned} \tag{6}$$

determines a point $y' \in \mathbf{R}^n$ and a real number B. It follows from the general theory presented in Censor and Lent [1981], particularly from Lemma 3.1 there, that these y' and B are uniquely determined, and therefore we call them the entropy projection of y onto H and the entropy projection coefficient associated with entropy projecting y onto H, respectively.

The entropy projection onto a hyperplane is a concrete realization of Bregman's notion of D-projection [Bregman, 1967a; Censor and Lent, 1981]. Since D-projections include as a special case also common orthogonal projections, some of the properties of the latter carry over to entropy projections. For example, if

$$H_1 = \{x \mid \langle a,x \rangle = b_1\} \tag{7}$$

and

$$H_2 = \{x \mid \langle a,x \rangle = b_2\} \tag{8}$$

are two parallel hyperplanes in $\mathbf{R}^n$, and $y \in \mathbf{R}^n$ is a given point, then

$$P_{H_2}(P_{H_1}(y)) = P_{H_2}(y) \tag{9}$$

obviously holds, where $P_H(y)$ stands for the orthogonal projection of a point y onto a hyperplane H. In this case

$$d_1 + \overline{d}_2 = d_2 \tag{10}$$

holds, where d_i, i = 1,2, is the distance from y to H_i and $\overline{d}_2$ is the distance from $P_{H_1}(y)$ to H_2.

It turns out that Eq. (9) continues to hold if P is replaced throughout by $\hat{P}$, where $\hat{P}_H(y)$ is the entropy projection of y onto H, and if $y \in \mathbf{R}^n_+$. In that case, Eq. (10) is replaced by

$$B_1 + \overline{B}_2 = B_2, \tag{11}$$

where B_i, i = 1,2, is the entropy projection coefficient associated with entropy projecting y onto H_i, and $\overline{B}_2$ is the coefficient associated with entropy projecting $\hat{P}_{H_1}(y)$ onto H_2. These claims follow from Lemma 3.3 in Censor and Lent [1981]. Further similarities, for example Lemmas 3.2 and 3.4 in that paper and their common source from D-projections, justify the term "entropy projection."

3. Row-Action Algorithms

We organize the presentation according to the type of entropy optimization problem that the algorithms are supposed to solve. All algorithms are row-action methods in the sense described in the introduction, and so their iterations are controlled by a control sequence. This is a sequence of indices $\{i(k)\}_{k=0}^{\infty}$ according to which the rows of the matrix A (equivalently, the equations or inequalities) are taken up; that is, in the iterative step $k \to$

k+1, the i(k)-th row (abbreviated by i wherever possible) is used in the iteration process. Several such control sequences were given by Censor [1981a], but for the sake of simplicity we henceforth assume that all the algorithms used a cyclic control, that is,

$$i(k) = k(\mathrm{mod}\ m) + 1\ , \tag{12}$$

where m is the number of rows in the matrix A.

Some of the algorithms permit the use of relaxation parameters. However, since these parameters are not yet incorporated into all algorithms, we prefer not to include them at all and thereby further emphasize the striking resemblance of the algorithms. We stress, however, that such relaxation parameters [see, for example, Censor, 1981a, p. 449] are important in practice because of the extra degree of freedom they add to the way in which a method might actually be implemented. Further simplification in presentation is achieved by specifying restrictions on initialization of iterations, if any, in the convergence theorems, thus leaving for comparison between the various algorithms only the description of the typical iterative step.

3.1. Algorithms for Entropy Maximization over Linear Equality Constraints

Here we consider two algorithms for solving Eq. (2) with $Q = Q_1$ as in Eq. (3). The first algorithm, called MART (for Multiplicative Algebraic Reconstruction Technique), was first proposed in the area of image reconstruction by Gordon et al. [1970].

Algorithm 1: MART

$$x_j^{k+1} = x_j^k \left(\frac{b_i}{\langle a^i, x^k \rangle}\right)^{a_j^i}, \qquad j = 1,2,\ldots,n\ . \tag{13}$$

Here $a^i = (a_j^i)$ is the ith column of A^T (the transpose of A), and the ith equation of Eq. (3) reads $\langle a^i, x\rangle = b_i$. Also, in Eq. (13), $i = i(k)$, which is a cyclic control. Convergence of MART was proved by Lent [1977] in the following theorem:

Theorem 1

If the following assumptions hold, then Algorithm 1 converges to the unique vector that solves the problem (2), (3):

(i) $Q_1 \cap R_+^n \neq 0$

(ii) All entries of A are $0 \leq a_j^i \leq 1$, and $b_i > 0$, $i = 1,2,\ldots,m$.

(iii) Iteration is initialized at $x_j^0 = \exp(-1)$, $j = 1,2,\ldots,n$.

Assumption (i) is a feasibility requirement on the constraints; assumption (ii) is not too restrictive because frequently the system $Ax = b$ can be preprocessed to conform with it (for example, in image reconstruction from projections or with respect to the gravity model in transportation planning).

Experimental work with this algorithm is reported by Lent [1977], Minerbo and Sanderson [1977], Minerbo [1979], and Herman [1982a].

In our efforts to determine the historical origin of this algorithm, we came across a similar algorithm, applicable to the special case that all entries of the matrix A are 0 or 1 [Bregman, 1967b]. Bregman attributes the algorithm to a Leningrad architect named G. V. Sheleikhovskii, who used the algorithm in the 1930s for calculating passenger flow. Another variant of this algorithm dates back to Kruithof [1937] and is described by Krupp [1979].

Algorithm 2: Bregman's Algorithm

$$x_j^{k+1} = x_j^k \exp(B_k, a_j^i), \quad j = 1,2,\dots,n, \tag{14}$$

where B_k is the entropy projection coefficient associated with entropy projecting x^k onto the ith hyperplane ($i \equiv i(k)$) of the system $Ax = b$, that is, onto

$$H_i = \{x \in \mathbf{R}^n \mid \langle x, a^i \rangle = b_i\} . \tag{15}$$

Geometrically, this means that the algorithm successively performs entropy projections onto the hyperplanes defined by the constraints, which are taken up in a cyclic order. With $\hat{P}$ denoting entropy projection, as defined above, Algorithm 2 actually means that

$$x^{k+1} = \hat{P}_{H_i}(x^k) . \tag{16}$$

It is worthwhile to mention here that if entropy projections are replaced by orthogonal projections onto the hyperplanes in Eq. (16), that is,

$$x^{k+1} = P_{H_i}(x^k) , \tag{17}$$

then we have the well known algorithm for norm-minimization over linear equalities due to Kaczmarz [see for example Censor, 1981a]. The similarity in structure between Eqs. (16) and (17) is more than coincidental. In fact, both algorithms are special cases of a general scheme of Bregman [1967a, Theorem 3]. From Bregman's theorem the convergence of Algorithm 2 follows under the additional assumption that the iterates x^k all stay in $\mathbf{R}^n_+$.

It is well known and easy to check that if matrix A in Eq. (3) has only 0 and 1 entries, then Algorithms 1 and 2 produce precisely the same sequence $\{x^k\}$ of iterates, provided they are initialized at the same point and controlled by the same $\{i(k)\}$ control sequence.

The relationship between these two algorithms for a general A matrix eluded Lamond and Stewart [1981, p. 248] but has recently been investigated by Censor, De Pierro, Elfving, Herman, and Iusem [1986].

In each iterative step, Algorithm 2 requires the calculation of B_k obtained by solving the system of n+1 equations, n of which are nonlinear, given by

$$\left.\begin{aligned} x_j^{k+1} &= x_j^k \exp(B_k a_j^i), \quad j = 1,2,\ldots,n, \\ b_i &= \sum_{j=1}^{n} a_j^i x_j^{k+1} . \end{aligned}\right\} \tag{18}$$

This would be approximated in practical implementation with several iterations of some standard method like Newton-Raphson (thereby causing the practical algorithm to inevitably deviate from the conceptual algorithm). We also remark that if Eqs. (18) are replaced by a single nonlinear equation for B_k [by substituting x_j^{k+1} into the second of Eqs. (18)] and if that equation is solved by Newton's method, the resulting recursive scheme is identical to the SOR-Newton method [see Elfving, 1980].

MART, on the other hand, calls at each iterative step for a closed-form calculation of the number

$$M_k \triangleq \log \frac{b_i}{\langle a^i, x^k \rangle} , \tag{19}$$

which then goes into the iterative step

$$x_j^{k+1} = x_j^k \exp(M_k a_j^i), \quad j = 1,2,\ldots,n, \tag{20}$$

which is identical to Eq. (13) and similar in structure to Eq. (14).

These observations lead us to develop and study some new algorithms for entropy maximization problems over linear inequalities and over linear interval constraints.

3.2. Algorithms for Entropy Maximization over Linear Inequalities

For problem (2) with $Q = Q_2$ of Eq. (4), one may use the following algorithm due to Bregman:

<u>Algorithm 3</u>

$$\left.\begin{aligned} x_j^{k+1} &= x_j^k \exp(c_k a_j^i), \quad j = 1,2,\ldots,n, \\ z^{k+1} &= z^k - c_k e^i , \end{aligned}\right\} \tag{21}$$

where

$$c_k \triangleq \min(z_i^k, B_k) . \tag{22}$$

B_k is the entropy projection coefficient associated with the entropy projection of x^k onto the hyperplane $H_i = \{x \mid \langle a^i,x\rangle = b_i\}$, which is the bounding hyperplane to the half-space $L_i = \{x \mid \langle a^i,x\rangle \le b_i\}$ defined by the ith ($i = i(k)$) inequality of the system

$$\langle a^i,x\rangle \le b_i , \quad i = 1,2,\ldots,m . \tag{23}$$

Once initialized appropriately at both $x^0 \in \mathbf{R}^n$ and $z^0 \in \mathbf{R}^m$, the algorithm iteratively updates not only the sequence $\{x^k\}$ of the so-called primal iterates, but also a sequence $\{z^k\}$ of dual iterates. Only one component of z^k is changed to produce z^{k+1}, since e^i stands in Eq. (21) for the ith unit-basis-vector in $\mathbf{R}^m$ (that is, it has all components 0 except the ith, which is 1).

The convergence of Algorithm 3 to the solution of problem (2) with $Q = Q_2$ as in Eq. (4) follows from the general theory of Bregman [1967a]. The reader might also wish to consult Censor and Lent [1981], specifically Algorithm 4.1 and Theorem 4.1 there, which, by choosing $f(x) = -\text{ent } x$, give rise to Algorithm 3 here. An illustrative example of image reconstruction using Algorithm 3 is given in Censor et al. [1979], but we are unaware of any other experimental work. Regarding the primal iteration [the first of Eqs. (21)], if $c_k = B_k$ in Eq. (22) then Eq. (21) calls for an entropy projection

$$x^{k+1} = \hat{P}_{H_i}(x^k) . \tag{24}$$

If, however, $c_k = z_i^k$, then Eq. (21) means that x^k is entropy-projected onto a hyperplane $\overline{H}_i$ that is parallel to H_i [see, for example, Fig. 6 in Censor, 1981a]. Without writing down $\overline{H}_i$ explicitly, we rewrite Eq. (21) as

$$x^{k+1} = \begin{cases} \hat{P}_{H_i}(x^k) , & \text{if } c_k = B_k , \\ \hat{P}_{\overline{H}_i}(x^k) , & \text{if } c_k = z_i^k , \end{cases} \tag{25}$$

for the primal iteration, where $\hat{P}$ stands, as before, for entropy projection. In this notation, if we replace $\hat{P}$ by orthogonal projection P in Eq. (25) and B_k by the distance d_k of x^k from the hyperplane H_i in Eq. (22), we obtain the Hildreth algorithm. This is a row-action method for norm-minimization over linear inequalities. Introduced by Hildreth [1957] and studied further by Lent and Censor [1980], this method also follows from Algorithm 4.1 and Theorem 4.1 in Censor and Lent [1981], if we choose $f(x) = \frac{1}{2}\|x\|^2$ there.

Algorithm 3 again requires at each iterative step an inner-loop calculation to find the approximate value of B_k. Only with B_k at hand can the choice of c_k in Eq. (22) be made, and the iteration proceeds according to Eq. (21). Motivated by the resemblance of MART [Eqs. (19) and (20)] and Algorithm 2 [Eq. (14)] for the entropy optimization problem over linear equations, we propose a new algorithm for entropy optimization over linear inequalities. This algorithm preserves the structure of Algorithm 3 but calls for the replacement of B_k by M_k.

Algorithm 4

$$\left.\begin{aligned} x_j^{k+1} &= x_j^k \exp(g_k a_j^i), \quad j = 1,2,\ldots,n, \\ z^{k+1} &= z^k - g_k e^i, \end{aligned}\right\} \tag{26}$$

where

$$g_k \triangleq \min(z_i^k, M_k), \tag{27}$$

and M_k is defined as in Eq. (19).

Except for the obvious advantage of relieving the user from the need to iteratively calculate B_k at each iterative step, it is not currently known how Algorithms 1 and 2 or Algorithms 3 and 4 compare in practice when applied to real or test-problem data. A study of the mathematical behavior of the new Algorithm 4 has been published by Censor, De Pierro, Elfving, Herman, and Iusem [1986].

3.3. Algorithms for Entropy Maximization over Linear Interval Constraints

In practical applications it is often the case that consistency of the system of equations $Ax = b$, which represent the measurements model of the problem, cannot be guaranteed. If there is no solution to this system at all, there is no point in seeking a maximum entropy solution of the system. Even if feasibility, that is, Assumption (i) of Theorem 1, holds, we might well suspect that the measurements are inaccurate, the data are noisy, and our model deviates from reality owing to idealizing assumptions such as discretization. Such considerations have led to the replacement of the system of equations $\langle a^i, x\rangle = b_i$, $i = 1,2,\ldots,m$, by a system of interval inequalities

$$b_i - \varepsilon_i \leq \langle a^i, x\rangle \leq b_i + \varepsilon_i, \quad i = 1,2,\ldots,m. \tag{28}$$

Further details about this approach may be found in Herman [1975], Herman and Lent [1978], or Censor [1981a].

The next algorithm is capable of solving the entropy maximization problem (2) with $Q = Q_3$ as in Eq. (5). It follows from the development in Censor and Lent [1981], which was inspired by the algorithm for norm-minimization over interval inequalities proposed by Herman and Lent [1978].

Algorithm 5

$$\left.\begin{aligned} x_j^{k+1} &= x_j^k \exp(h_k a_j^i), \quad j = 1,2,\ldots,n, \\ z^{k+1} &= z^k - h_k e^i, \end{aligned}\right\} \tag{29}$$

where

$$H_k \triangleq \operatorname{mid}(z_i^k, \Delta_k, \Gamma_k) \tag{30}$$

and Δ_k and Γ_k are the entropy projection coefficients associated with the entropy projection of x^k onto the hyperplanes

$$\{x \mid \langle a^i,x\rangle = b_i + \varepsilon_i\} \tag{31}$$

and

$$\{x \mid \langle a^i,x\rangle = b_i - \varepsilon_i\} \tag{32}$$

respectively.

In Eq. (30), "mid" stands for the median of three real numbers, and to determine h_k both Δ_k and Γ_k have to be calculated in every iterative step. The convergence of this algorithm to the desired solution, under some acceptable conditions and appropriate initialization, follows from Algorithm 5.1 and Theorem 5.1 in Censor and Lent [1981]. The practical reasons for preferring Algorithm 5 to the application of Algorithm 3 to the system of one-sided inequalities arising from Eq. (28) by multiplication of one side by (-1) are explained in detail in Herman and Lent [1978] and in Censor and Lent [1981] and are not repeated here.

Reasoning along the same lines that led to Algorithm 4, we propose now a new algorithm for entropy maximization over linear interval constraints.

<u>Algorithm 6</u>

$$\left.\begin{aligned} x_j^{k+1} &= x_j^k \exp(\ell_k a_j^i), \quad j = 1,2,\ldots,n, \\ z^{k+1} &= z^k - \ell_k e^i, \end{aligned}\right\} \tag{33}$$

where

$$\ell_k \stackrel{\Delta}{=} \mathrm{mid}(z_i^k, F_k, G_k), \tag{34}$$

and

$$F_k \stackrel{\Delta}{=} \log \frac{b_i + \varepsilon_i}{\langle a_i,x_k\rangle}, \tag{35}$$

$$G_k \stackrel{\Delta}{=} \log \frac{b_i - \varepsilon_i}{\langle a_i,x_k\rangle}. \tag{36}$$

No practical experience with this algorithm is available. We are presently studying its convergence and will publish the results of our analysis elsewhere. Algorithm 5 follows from the development of Censor and Lent [1981], which was inspired by the norm-minimization over interval inequalities proposed by Herman and Lent [1978].

4. Concluding Remarks

We have described here a family of special-purpose algorithms for linearly constrained entropy maximization problems. Little is known about the practical behavior of these algorithms, and several questions are still open concerning their mathematical analysis.

We have recently established the convergence of Algorithm 4 and have done some further work related to algorithms for entropy maximization [Censor, De Pierro, and Iusem, 1986; Censor and Segman, 1986]. In the process, we have acquired a better understanding of the geometrical nature of entropy projections and their properties. The introduction of relaxation parameters into these algorithms is important for practical implementations and experimental studies, but it is also through them that a precise relationship between Algorithms 1 and 2 (and thereafter between Algorithms 3 and 4 and Algorithms 5 and 6) can be formulated and proven. Recently, some work has been done concerning the behavior of norm-minimization algorithms when they are applied to inconsistent systems [see, for example, Eggermont et al., 1981, and Censor, Eggermont, and Gordon, 1983]. It is worthwhile to study similar questions regarding the entropy maximization algorithms presented here or in Censor, Elfving, and Herman [1983].

5. Acknowledgments

Part of this work was done while the first two authors were visiting the Medical Image Processing Group, Department of Radiology, Hospital of the University of Pennsylvania, Philadelphia, and was supported by NSF Grant ECS-8117908 and NIH Grant HL 28438. Further progress was made during a visit of the first author to the Department of Mathematics at the University of Linköping, Sweden; we are grateful to professors Åke Björck and Kurt Jörnsten and to Dr. Torleiv Orhaug for making this visit possible. We thank Mrs. Anna Cogan for her excellent work in typing the manuscript.

6. References

Bregman, L. M. (1967a), "The relaxation method of finding the common point of convex sets and its application to the solution of problems in convex programming," USSR Computational Mathematics and Mathematical Phys. 7(3), pp. 200-217.

Bregman, L. M. (1967b), "Proof of convergence of Sheleikhovskii's method for a problem with transportation constraints," USSR Computational Math. and Mathemat. Phys. 7, pp. 191-204.

Censor, Y. (1981a), "Row-action methods for huge and sparse systems and their applications," SIAM Review 23, pp. 444-466.

Censor, Y. (1981b), "Intervals in linear and nonlinear problems of image reconstruction," in Mathematical Aspects of Computerized Tomography, G. T. Herman and F. Natterer, eds., Lecture Notes in Medical Informatics, Springer-Verlag, Berlin, Vol. 8, pp. 152-159.

Censor, Y. (1982), "Entropy optimization via entropy projections," in System Modeling and Optimization, R. F. Drenick and F. Kozin, eds., Lecture Notes in Control and Information Science, Springer-Verlag, Berlin, Vol. 38, pp. 450-454.

Censor, Y. (1983), "Finite series expansion reconstruction methods," Proc. IEEE 71, pp. 409-419.

Censor, Y., A. R. De Pierro, T. Elfving, G. T. Herman, and A. N. Iusem (1986), "On iterative methods for linearly constrained entropy maximization," Tech. Rept. MIPG112, Medical Image Processing Group, Department of Radiology, Hospital of the University of Pennsylvania, Philadelphia, Pa.

Censor, Y., A. R. De Pierro, and A. N. Iusem (1986), "On maximization of entropies and a generalization of Bregman's method for convex programming," Tech. Rept. MIPG113, Medical Image Processing Group, Department of Radiology, Hospital of the University of Pennsylvania, Philadelphia, Pa.

Censor, Y., P. P. B. Eggermont, and D. Gordon (1983), "Strong underrelaxation in Kaczmarz's method for inconsistent systems," Numerische Mathematik 41, pp. 83-92.

Censor, Y., T. Elfving, and G. T. Herman (1983), "Methods for entropy maximization with applications in image processing," in Proceedings of the Third Scandinavian Conference on Image Analysis, P. Johansen and P. W. Becker, eds., Chartwell-Bratt, Ltd., Lund, Sweden, pp. 296-300.

Censor, Y., A. V. Lakshminarayanan, and A. Lent (1979), "Relaxational methods for large-scale entropy optimization problems with application in image reconstruction," in Information Linkage Between Applied Mathematics and Industry, P. C. C. Wang et al., eds., Academic Press, New York, pp. 539-546.

Censor, Y., and A. Lent (1981), "An iterative row-action method for interval convex programming," J. Optimization Theory and Applications 34, pp. 321-353.

Censor, Y., and J. Segman (1986), "On block-iterative entropy maximization," J. Inf. Optimization Sci. (in press); also available as Tech. Rept. MIPG111, Medical Image Processing Group, Department of Radiology, Hospital of the University of Pennsylvania, Philadelphia, Pa.

Darroch, J. N., and D. Ratcliff (1972), "Generalized iterative scaling for log-linear models," Ann. Math. Statis. 43, pp. 1470-1480.

Eggermont, P. P. B., G. T. Herman, and A. Lent (1981), "Iterative algorithms for large partitioned linear systems with applications to image reconstruction," Linear Algebra and Its Applications 40, pp. 37-67.

Elfving, T. (1980), "On some methods for entropy maximization and matrix scaling," Linear Algebra and Its Applications 34, pp. 321-339.

Erlander, S. (1981), "Entropy in linear programs," Mathematical Programming 21, pp. 137-151.

Frieden, B. R (1980), "Statistical models for the image restoration problem," Computer Graphics and Image Processing 12, pp. 40-59.

Frieden, B. R. (1983), Probability, Statistical Optics, and Data Testing, Springer-Verlag, New York.

Gordon, R., R. Bender, and G. T. Herman (1970), "Algebraic reconstruction techniques (ART) for three-dimensional electron microscopy and x-ray photography," J. Theoret. Biol. 29, pp. 471-481.

Herman, G. T. (1975), "A relaxation method for reconstructing objects from noisy x-rays," Mathemat. Programming 8, pp. 1-19.

Herman, G. T. (1980), Image Reconstruction from Projections: The Fundamentals of Computerized Tomography, Academic Press, New York.

Herman, G. T. (1982a), "Mathematical optimization versus practical performance: a case study based on the maximum entropy criterion in image reconstruction," Mathemat. Programming Study 20, pp. 96-112.

Herman, G. T. (1982b), "The application of maximum entropy and Bayesian optimization methods to image reconstruction from projections," Tech. Rept. MIPG73, Medical Image Processing Group, Dept. Radiology, Hospital of the University of Pennsylvania, Philadelphia.

Herman, G. T., and A. Lent (1978), "A family of iterative quadratic optimization algorithms for pairs of inequalities, with application in diagnostic radiology," Mathemat. Programming Study 9, pp. 15-29.

Hildreth, C. (1957), "A quadratic programming procedure," Naval Res. Logistic Quart. 4, pp. 79-85; erratum, p. 361.

Jaynes, E. T. (1982), "On the rationale of maximum-entropy methods," Proc. IEEE 70, pp. 939-952.

Kapur, J. N. (1983), "Twenty-five years of maximum-entropy principle," J. Mathemat. Phys. Sci. 17, pp. 103-156.

Kruithof, J. (1937), "Telefoonverkeersrekening" (Calculation of telephone traffic"), De Ingenieur 52, pp. E15-E25.

Krupp, R. S (1979), "Properties of Kruithof's projection method," Bell Syst. Tech. J. 58, pp. 517-538.

Lamond, B., and N. F. Stewart (1981), "Bregman's balancing method," Transport. Res. 15B, pp. 239-248.

Lent, A. (1977), "A convergent algorithm for maximum entropy image restoration with a medical x-ray application," in Image Analysis and Evaluation, R. Shaw, ed., Society of Photographic Scientists and Engineers (SPSE), Washington, D.C., pp. 238-243.

Lent, A., and Y. Censor (1980), "Extensions of Hildredth's row-action method for quadratic programming," SIAM J. on Control and Optimization 18, pp. 444-454.

Levine, R. D., and M. Tribus, eds. (1978), The Maximum Entropy Formalism, MIT Press, Cambridge, Mass.

Minerbo, G. N. (1979), "MENT: a maximum entropy algorithm for reconstructing a source from projection data," Comput. Graphics and Image Process. 10, pp. 48-68.

Minerbo, G. N., and J. G. Sanderson (1977), "Reconstruction of a source from a few (2 or 3) projections," Tech. Rept. LA-6747-MS, Los Alamos Scientific Laboratories, New Mexico.

Wong, D. S. (1981), "Maximum likelihood, entropy maximization, and the geometric programming approaches to the calibration of trip distribution models," Transport. Res. 15B, pp. 329-343.

BAYESIAN APPROACH TO LIMITED-ANGLE RECONSTRUCTION IN COMPUTED TOMOGRAPHY*

Kenneth M. Hanson and George W. Wecksung

Los Alamos National Laboratory, Los Alamos, NM 87545

An arbitrary source function cannot be determined fully from projection data that are limited in number and range of viewing angle. There exists a null subspace in the Hilbert space of possible source functions about which the available projection measurements provide no information. The null-space components of deterministic solutions are usually zero, giving rise to unavoidable artifacts. It is demonstrated that these artifacts may be reduced by a Bayesian maximum a posteriori (MAP) reconstruction method that permits the use of significant a priori information. Since normal distributions are assumed for the a priori and measurement-error probability densities, the MAP reconstruction method presented here is equivalent to the minimum-variance linear estimator with nonstationary mean and covariance ensemble characterizations. A more comprehensive Bayesian approach is suggested in which the ensemble mean and covariance specifications are adjusted on the basis of the measurements.

*Reprinted from Journal of the Optical Society of America 73, pp. 1501-1509 (November 1983).

C. R. Smith and G. J. Erickson (eds.),
Maximum-Entropy and Bayesian Spectral Analysis and Estimation Problems, 255–272.

1. Introduction

The problem of obtaining an artifact-free computed-tomographic (CT) reconstruction from projection data that are limited in number and possibly in angular coverage is, in general, a difficult one to solve. This difficulty arises from a fundamental limitation inherent in incomplete data sets. It is seen that this limitation may be viewed as arising from an essential lack of information about the unknown source function, that is, its null-space components [1]. The Bayesian approach allows one to incorporate a priori information about the source function based on the properties of the ensemble of source functions realizable in the specified imaging situation. If the a priori information is specific enough, reasonable estimates of the null-space components of the source function can be obtained, thereby reducing the artifacts in the reconstruction. The results of this maximum a posteriori (MAP) method [2] are compared with the fit and iterative reconstruction (FAIR) technique [3]. We propose in the discussion that the incorporation of the Bayesian approach in the FAIR technique can provide a more flexible algorithm.

2. Measurement Space-Null Space

The CT problem may be stated as follows: Given a finite set of projections of a function of two dimensions $f(x,y)$ with compact support, obtain the best estimate of that function. The projections may generally be written as a weighted two-dimensional (2-D) integral of $f(x,y)$,

$$g_i = \iint h_i(x,y)\, f(x,y)\, dx\, dy\,, \tag{1}$$

where the h_i are the weighting functions and $i = 1,2,\ldots,N$ for N individual measurements. We refer to the h_i as response functions. In the CT problem the h_i typically have large values within a narrow strip and small or zero values outside the strip. If the h_i are unity within a strip and zero outside, Eq. (1) becomes a strip integral. For zero strip width, it becomes a line integral. These last two cases are recognized as idealizations of the usual physical situation. The generality of Eq. (1) allows it to closely represent actual physical measurements since it can take into account response functions that vary with position. Note that Eq. (1) is applicable to any discretely sampled, linear-imaging system. Thus the concept of null space and the Bayesian methods proposed for overcoming its limitations are relevant to a large variety of image-restoration problems.

The unknown function $f(x,y)$ is usually restricted to a certain class, for example, the class of all integrable functions with compact support. Consider the Hilbert space of all acceptable functions and assume that all the h_i belong to that space. Equation (1) is an inner product of h_i with f. Thus the measurements g_i may be thought of as a projection of the unknown vector f onto the response vector h_i. Only those components of f that lie on the subspace spanned by the set of all h_i contribute to the measurements. We call this subspace the measurement space. The components of f in the remaining orthogonal subspace, the null space, do not contribute to the measurements. Hence the null-space contribution to f cannot be determined from the measurements alone. Since the deterministic (measurement) subspace of f is spanned by the response functions, it is natural to expand the estimate of f in terms of them:

$$\hat{f}(x,y) = \sum_{i=1}^{N} a_i\, h_i(x,y)\,. \tag{2}$$

Thus the null-space components of $\hat{f}$ are zero, which is a necessary condition for the minimum-norm solution. This leads to artifacts in $\hat{f}$ because it does not possess those components of f that lie in the null space. Further reading on the null-space—measurement-space concept may be found in papers by Twomey [4-6].

The response-function expansion [Eq. (2)] is formally identical to the familiar backprojection process in which the value a_i is added to the image along the strip function h_i. Thus the backprojection process affects only the measurement-space components of the reconstruction. Most of the well-known CT reconstruction algorithms incorporate backprojection, including filtered backprojection [7], the algebraic reconstruction technique (ART) [8], the simultaneous iterative construction technique (SIRT) [9], SIRT-like algorithms (least-squares [10] and other variants [11]), and the natural pixel matrix formulation by Buonocore et al. [12, 13]. Such algorithms can alter only the measurement-space part of the initial estimate. When the initial estimate lies solely in the measurement space, as is normally the case, so will the final estimate.

Various augmentations to deterministic algorithms, such as consistency, analytic continuation, and global constraints (including maximum entropy), have been considred by Hanson [1]. These seem to be ineffective in overcoming the measurement-space restrictions presented above for the solution of the general problem. Other authors have mentioned in passing the concept of the measurement-space—null-space dichotomy [14,15] but have not explicitly considered its effect on reconstructions from limited-projection data. As an aside, the range of the transpose of the projection-measurement matrix A, referred to in Ref. 15, is the measurement space in the square pixel representation. Smith et al. [16] considered the null space of a finite number of projections to explore the convergence rate of the ART algorithm. This work was extended by Hamaker and Solmon [17]. Katz made extensive use of the null-space concept to determine the conditions for uniqueness of a reconstruction [18]. Louis [19] developed an explicit expression for functions belonging to the null space corresponding to a finite set of projection data and showed that ghosts from the null space can appear as lesions. Medoff et al. [20] recognized the consequences of the null space associated with limited data and introduced a method to eliminate null-space ghosts through the application of known constraints on the reconstructed image. Further references on the limited-angle CT problem may be found in Ref. 1.

The restriction of deterministic solutions to the measurement space should not be viewed as a negative conclusion. Rather, it is simply a statement of what is possible for a given set of measurements in the absence of further information. It allows one to state formally the goal in an improved limited-angle CT reconstruction as that of estimating the null-space contribution through the use of further information about the function to be reconstructed.

3. Bayesian Solution

The Bayesian approach to CT reconstruction [21] is based on the assumption that the image to be reconstructed belongs to an identifiable ensemble of similar images. In the following discussion, the image f and the set of all projections g are considered to be vectors. The best estimate for the reconstruction is taken to be that particular image f that maximizes the a posteriori conditional probability density of f given the measurements g. This probability is given by Bayes' law,

$$P(f\,|\,g) \;=\; \frac{P(g\,|\,f)\,P(f)}{P(g)}\,, \tag{3}$$

in terms of the conditional probability of g given f and the a priori probability distributions of f and g separately. We assume that the measurement noise is additive with a probability distribution that has zero mean and is Gaussian distributed. P(f) is assumed to be a Gaussian distribution with a mean value $\overline{f}$. The covariance matrices of the noise and the ensemble image vectors are R_n and R_f, respectively. Under these assumptions, the MAP solution is easily shown to satisfy [22]

$$R_f^{-1}(\overline{f}-f) + H^T R_n^{-1}(g-Hf) \;=\; 0\,, \tag{4}$$

where H is the linear operator (matrix) corresponding to the projection process described by the integral in Eq. (1). The transpose of H is the familiar backprojection operation. It can be seen from Eq. (4) that the desired solution strikes a balance between its difference with the ensemble mean $\overline{f}$ and the solution to the measurement equation (g = Hf). This balance is determined by the covariance matrices R_f and R_n that specify the confidence with which each difference is weighted as well as possible correlations between the differences.

We have adopted an iterative approach to the solution of Eq. (4) based on the scheme proposed by Herman and Lent [21]. The nth estimate of f^n is given by the iteration scheme:

$$f^0 = \overline{f} \tag{5a}$$

$$f^{n+1} = f^n + c^n r^n \tag{5b}$$

$$r^n = \overline{f} - f^n + R_f H^T R_n^{-1}(g - Hf^n) \tag{5c}$$

$$c^n = \frac{r^{nT} s^n}{s^{nT} s^n} \tag{5d}$$

$$s^n = (I + R_f H^T R_n^{-1} H)\, r^n \tag{5e}$$

where vector r^n is the residual of Eq. (4) (multiplied by R_f) and the scalar c^n is chosen to minimize the norm of r^{n+1}. This iterative scheme is similar to the one proposed by Hunt [22] for nonlinear MAP-image restoration. We have found that this technique works well, although convergence typically requires 10 to 20 iterations.

It is easy to see from the form of this iterative procedure that significant null-space contributions to f can arise from the a priori information. First, the zero-order estimate is $\overline{f}$, which can contain null-space contributions. Second, in Eq. (5c), R_f can generate null-space contributions when it operates on the results of the backprojection (H^T) process, which lies wholly in the measurement space. R_f, in effect, weights the backprojection of the measurement residuals. If R_f is chosen as zero in certain regions of the reconstruction, these regions will not be changed throughout the iteration scheme. It must be emphasized that the choices for $\overline{f}$ and R_f are exceedingly important since it is only through them that a nonzero null-space contribution to the reconstruction arises. As was stated earlier, this is the major advantage of the Bayesian approach over deterministic algorithms. Trivial choices, such as using for $\overline{f}$ a constant or a filtered backprojection reconstruction based on the same projections or assuming R_f to be proportional to the identity matrix [22,23], are not helpful for reducing artifacts.

The iteration scheme given by Eqs. (5) is SIRT-like [11] in that the reconstruction update [Eq. (5b)] is accomplished only after all the ray sums (Hf^n) have been calculated. It is known that ART-like algorithms converge much faster than SIRT-like ones [11]. Herman et al. [23] have proposed an ART-like reconstruction algorithm that converges to the solution of the MAP equation [Eq. (4)] under the assumption that R_f and R_n are proportional to the identity matrix. This algorithm is worth exploring, as it is likely to converge much more rapidly than the one used here. However, as was stated above, their algorithm should be extended to include nontrivial choices for R_f. The iterative scheme used here, although it may be slower than necessary, does provide a solution to the MAP equation, which is the important thing. We have found its convergence to be such that the norm of the residual [Eq. (5c)] behaves as the iteration number raised to the -1.5 to -2.0 power.

It is well known that the assumption of normal probability density distributions leads to a MAP solution [Eq. (4)] that is equivalent to the minimum-variance linear estimator [24]. Application of this estimator in a matrix formalism to tomographic reconstruction has been pursued by Wood et al. [25,26] and Buonocore [27]. These authors stressed the importance of a priori information in limited-angle reconstruction. However, the main thrust of their work was toward improving the computational efficiency of the required matrix operations, to which end they were successful.

Tam et al. [28] introduced a method to use the a priori known region of support of the source distribution. This method is the 2-D counterpart of the celebrated Gerchberg-Papoulis algorithm for obtaining superresolution. It is an iterative technique in which the known properties of the image in the spatial and the Fourier domains are alternately invoked. The objective is to use the known spatial extent of source image to extend its 2-D Fourier transform from the known sector into the missing sector. Although this method has been studied extensively [29-32] and has been shown to have some merit when either the region of support is restrictive or the angular region of the missing projections is fairly narrow, the region-of-support constraint may be incorporated more directly into many reconstruction algorithms. In virtually every iterative algorithm it is possible to invoke constraints on the reconstructed function. Thus it is possible to require the function to be zero outside the region of support at each update step, for example, in computing Eq. (5b) in the present MAP algorithm. Such an iterative algorithm will yield a solution that satisfies the region-of-support constraint, and Tam's procedure is not required. One may enforce this constraint through a redefinition of the response functions h_i in Eq. (1) to make them zero outside the region of support. Then the backprojection operation [Eq. (2)] affects only the reconstruction within the region of support. With this redefinition of h_i, the measurement space includes only functions that fulfill the region-of-support constraint. From this standpoint, Tam's iterative procedure does not affect the null space associated with the available measurements. The natural pixel formulation of Buonocore et al. [12,13] may be revised in a similar manner, but that would probably ruin the properties of the measurement matrix that they exploited to provide efficient matrix calculations.

4. Fit and Iterative Reconstruction

The preceding MAP solution is compared in Section 5 with an alternative method to incorporate a priori information in the reconstruction process, namely, the FAIR technique introduced by Hanson [3]. In this algorithm the a priori information about the source function is used to construct a parameterized model of the unknown function. As the first step in this algorithm, the parameters in the model are determined from the available projection data by a least-squares (or minimum chi-squared) fitting procedure. In the second step of FAIR an iterative reconstruction procedure is performed using the fitted parametric model as the initial estimate. Although, in the past, the ART algorithm [8] was used in the second iterative step of FAIR, other iterative algorithms, such as the MAP algorithm above, can be used advantageously (see Section 6). The iterative reconstruction procedure forces the result to agree with the measurements through alteration of its measurement-space contribution. The null-space contribution to the FAIR reconstruction arises solely from the parametric model fitted in the first step and hence from the a priori information used in specifying the model.

5. Results

The results of applying various reconstruction methods to a specific 2-D, limited-angle reconstruction problem are compared. Algorithms that are useful for handling incomplete data through the use of a priori information must possess the following important characteristics: (1) they must significantly reduce artifacts that arise from inappropriate null-space contributions, (2) they must gracefully respond to inconsistencies between the actual source function and the assumptions about it, and (3) they must tolerate noise in the projection data. We demonstrate that the MAP and the FAIR algorithms conform to these requirements.

The relevant reconstruction techniques have been applied to an example source function consisting of a fuzzy annulus with variable amplitude, Fig. 1(a), that roughly emulates the nuclear-isotope distribution in the cross section of a heart. The peak value of the distribution is 1.24. The available projection data consist of 11 views covering 90° in projection angle. At first no noise was added to the projections. Each projection contained 128 samples. All reconstructions contain 128 × 128 pixels. The measurement-space reconstruction obtained using ART [8], Fig. 1(b), shows severe artifacts that tend to obscure much of the source distribution.

Figure 1(c) shows the reconstruction obtained by using the maximum entropy algorithm (MENT) provided to us by Minerbo [33]. This algorithm provides a modest improvement over ART, particularly in regard to the detection of the dip in the annulus at 50°. However, MENT does not have much effect on the splaying of the reconstruction along the central axis of the available views. In our experience the principal advantage of the maximum-entropy constraint is its implicit constraint of nonnegativity. ART reconstructions that are constrained to be nonnegative are similar to the MENT results. The nonnegativity constraint amounts to the incorporation of a priori knowledge about the source function. This constraint is generally applicable and is effective in the reconstruction of certain types of source distributions, such as pointlike objects on a zero background. However, there are many source distributions and data-collection geometries for which nonnegativity provides little benefit, such as the present test case. We do not apply the nonnegativity constraint to any other reconstructions here, simply to avoid confusion about its role in providing improvement in the reconstructions as opposed to the role of the use of other a priori knowledge.

It was assumed that the a priori information consisted of the knowledge that the source function had an annular structure with known position, radius, and width. Thus, in the MAP approach (Fig. 2), $\overline{f}$ was chosen to be an annulus with constant amplitude and with Gaussian cross section. The mean radius and width of the annulus were chosen to be the same as the unknown source function. The covariance matrix R_f was assumed to be diagonal and was hence an image proportional to the ensemble variance about the mean $\overline{f}$. The covariance image R_f was large (1.0) at the peak of

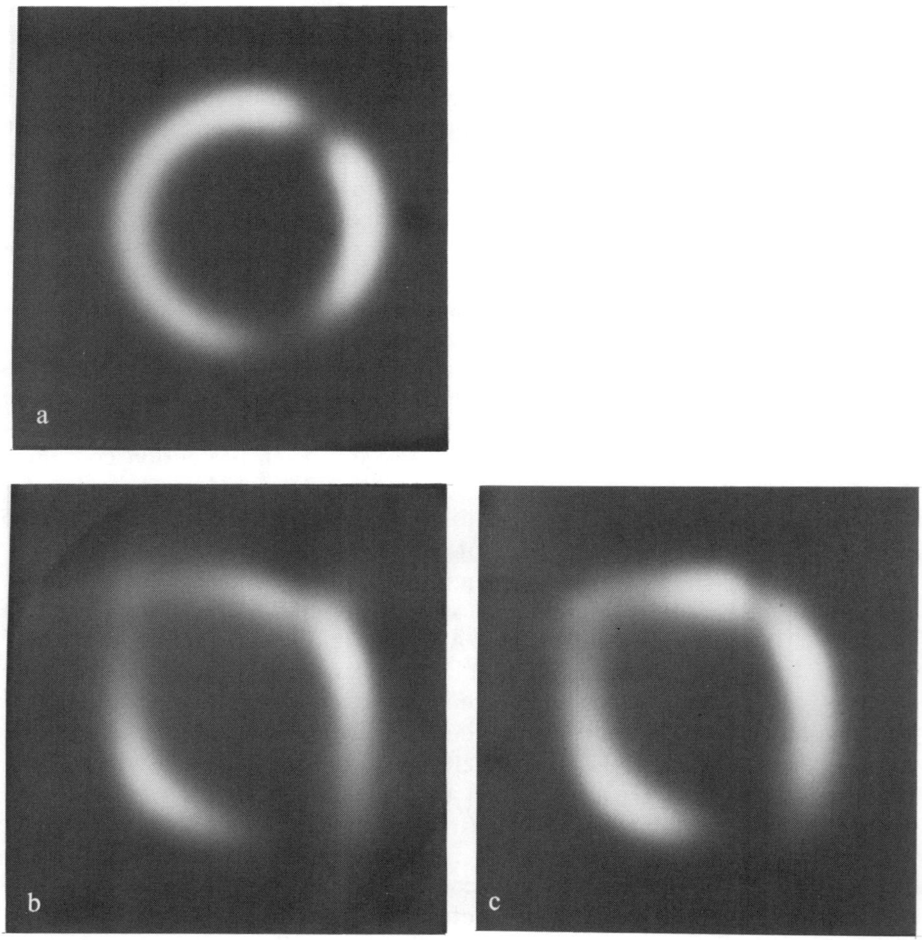

Figure 1. (a) Source distribution used for the first example. (b) ART and (c) MENT reconstructions obtained using 11 views covering 90° in projection angle. Unconstrained ART was used, whereas MENT has an implicit nonnegativity constraint.

the annulus and small (0.2) inside and outside [Fig. 2(a)]. Since noiseless projections were used, the measurement noise was assumed to be uncorrelated, constant, and low in value. The resulting MAP reconstruction [Fig. 2(b)] is vastly superior to the ART and MENT results, eliminating

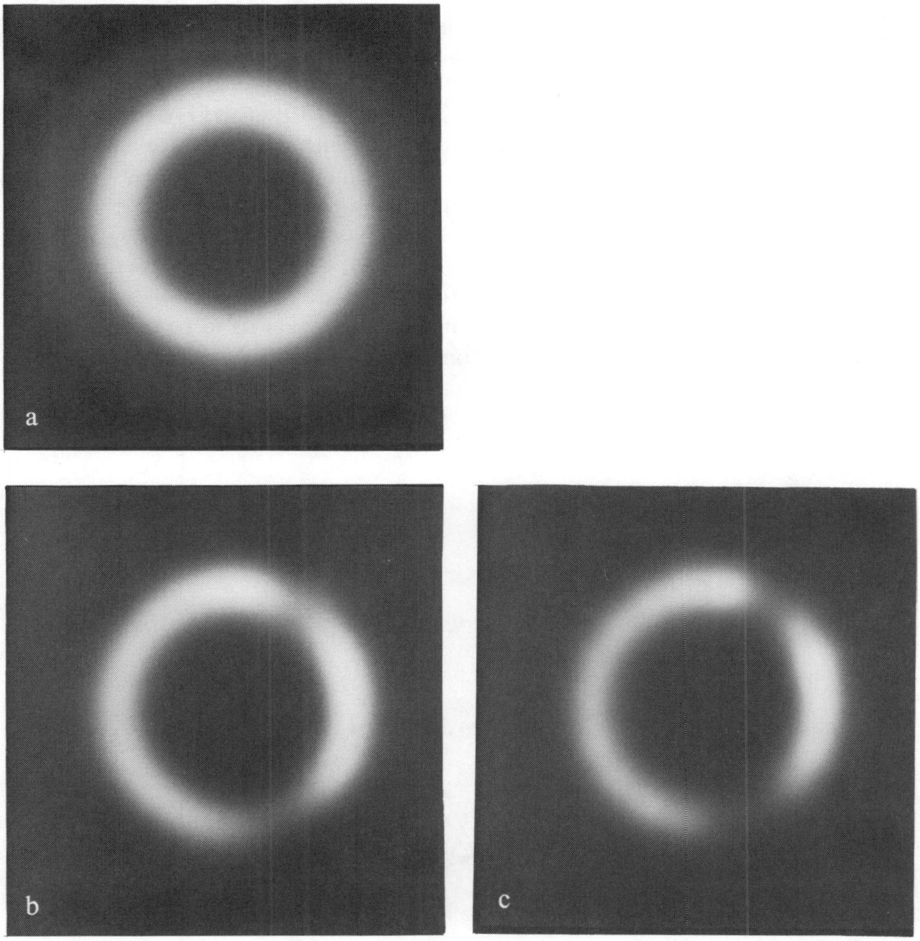

Figure 2. Reconstructions using the a priori information that the unknown source function is a fuzzy annulus with known radius and width. (b) MAP reconstruction was obtained using a flat annulus for $\overline{f}$ and the variance image (a) as the diagonal entries of R_f (nondiagonal entries assumed to be zero). (c) The FAIR result was based on a model of the image consisting of 18 Gaussian functions distributed around the circle. The use of a priori knowledge significantly reduces the artifacts present in the deterministic reconstructioning of Fig. 1.

essentially all the artifacts present in these deterministic solutions. The parametric model chosen for the FAIR method consisted of 18 2-D Gaussian functions evenly distributed on a circle. The radius of the circle and the width of the Gaussians were chosen to be the same as those of the source function. The fitting procedure determined the amplitudes of each of the Gaussian functions. The resulting fitted function was used as the initial estimate in ART to obtain the final result [Fig. 2(c)]. This FAIR reconstruction is comparable to the MAP result.

For a quantitative comparison, Fig. 3 shows the maximum reconstruction value obtained along radii as a function of angle for the various reconstruction methods presented in Figs. 1 and 2. The FAIR method is seen to follow the original source dependence most closely, with the MAP result a close second. The ART reconstruction has many quantitatively serious defects. The computation times on a CDC 7600 computer for the algorithms presented here are ART (10 iterations), 17 sec; MAP (10 iterations), 73 sec; FAIR (3 iterations), 25 sec; MENT (6 iterations), 105 sec. The corresponding execution time for filtered backprojection is 5 sec.

A slightly different source function, Fig. 4(a), was used to test the ability of the algorithms to deal with inconsistencies between assumptions about

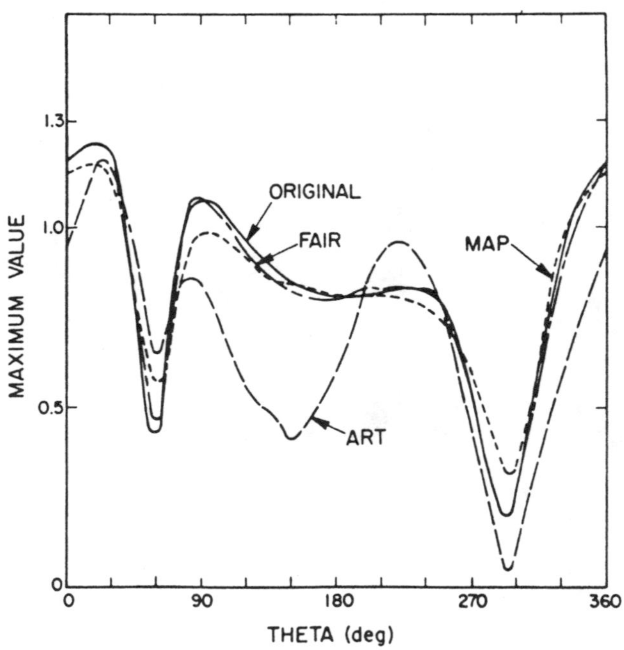

Figure 3. Angular dependence of the maximum values along various radii for the ART, MAP, and FAIR reconstructions in Figs. 1 and 2 compared with that for the original function [Fig. 1(a)]. This graph quantitatively demonstrates the improvement afforded by MAP and FAIR.

the source function and its actual distribution. This source function is the same as Fig. 1(a) with a narrow, 0.6-amplitude, 2-D Gaussian added outside the annulus at 330° and a broad, 0.1-amplitude Gaussian added underneath the annulus at 162°. The reconstructions obtained using the same assumptions as above are shown in Figs. 4(b) through 4(d). Both MAP and FAIR handle the inconsistencies similarly. The angular dependence of the maxi-

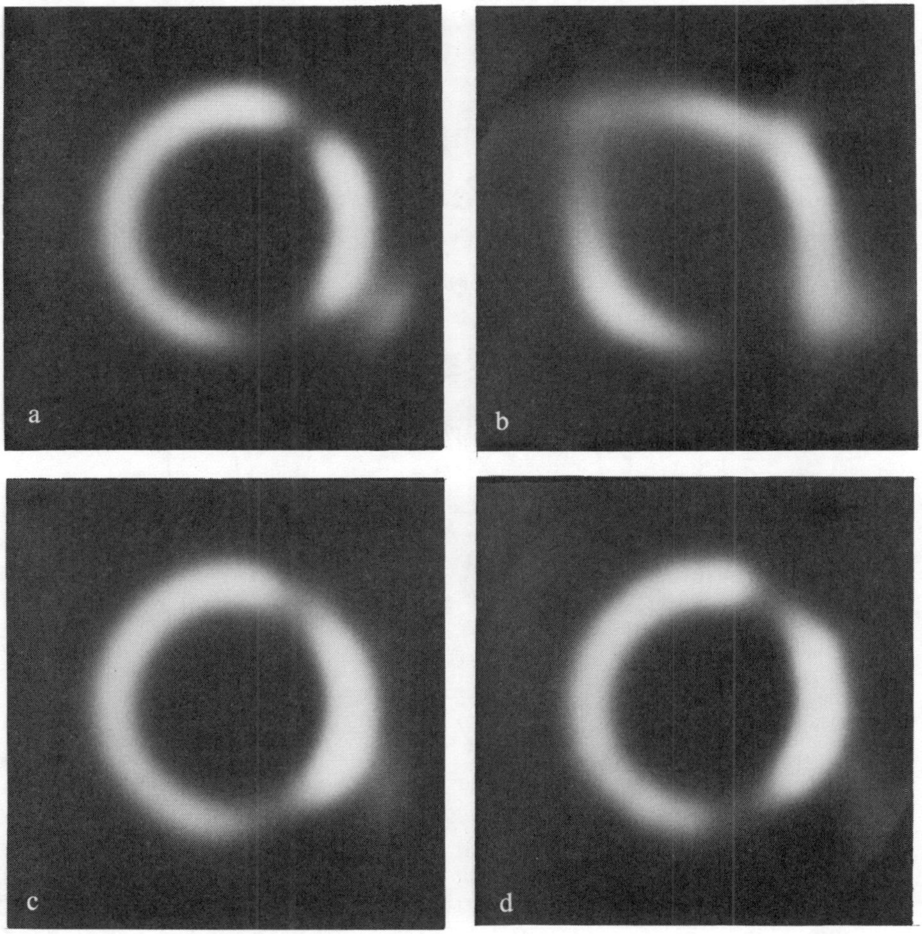

Figure 4. (a) Source distribution that does not conform to the annular assumption. (b) ART, (c) MAP, and (d) FAIR reconstructions obtained from 11 views subtending 90°. Both MAP and FAIR tend to move the added source outside the annulus onto the annulus. However, they provide indications in the reconstructions that there is some exterior activity.

mum reconstruction value, Fig. 5, shows that both algorithms produce an excess near 330° since they have tried to shift the discrepant exterior source to the annulus, which is consistent with the a priori assumptions. However, both methods have a significant response in the region of the exterior source and therefore provide some information about the discrepancy. This would not be the case for the MAP algorithm were R_f chosen to be zero outside the annulus. This points out the need to be conservative in placing restrictions on the reconstruction that may be violated by the actual source distribution. The second iterative reconstruction step in the FAIR method is needed for this same reason, as it allows correction to be made to the fitted model, if indicated by the available projections.

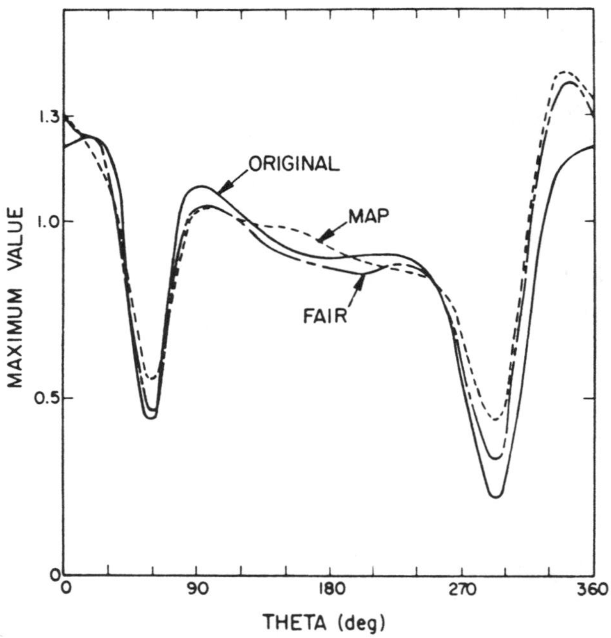

Figure 5. Angular dependence of the maximum values in the MAP and the FAIR reconstructions of Fig. 4.

The final example is the reconstruction of Fig. 1(a) from noisy data. The same 11 projections were used as before but with random noise added with an rms deviation of 10% relative to the maximum projection value. The reconstructions in Fig. 6 demonstrate that both MAP and FAIR simply yield noisy versions of those obtained from noiseless projections. There is no diastrous degradation, as would be expected for algorithms based on analytic continuation [34,35]. Although the FAIR result appears to be much noisier than the MAP reconstruction, they both possess nearly identical noise in the annular region. The rms difference between the projection measurements and the ray sums of the MAP and FAIR reconstructions, respectively, are roughly 0.8 and 0.5 times the actual rms deviation of the noise in the projections. This indicates that both algorithms have attempted to solve the measurement equations beyond what is reasonable. The MAP algorithm does balance the rms error in the projections against the deviation from $\overline{f}$. However, ART simply attempts to reduce the rms projection error to zero.

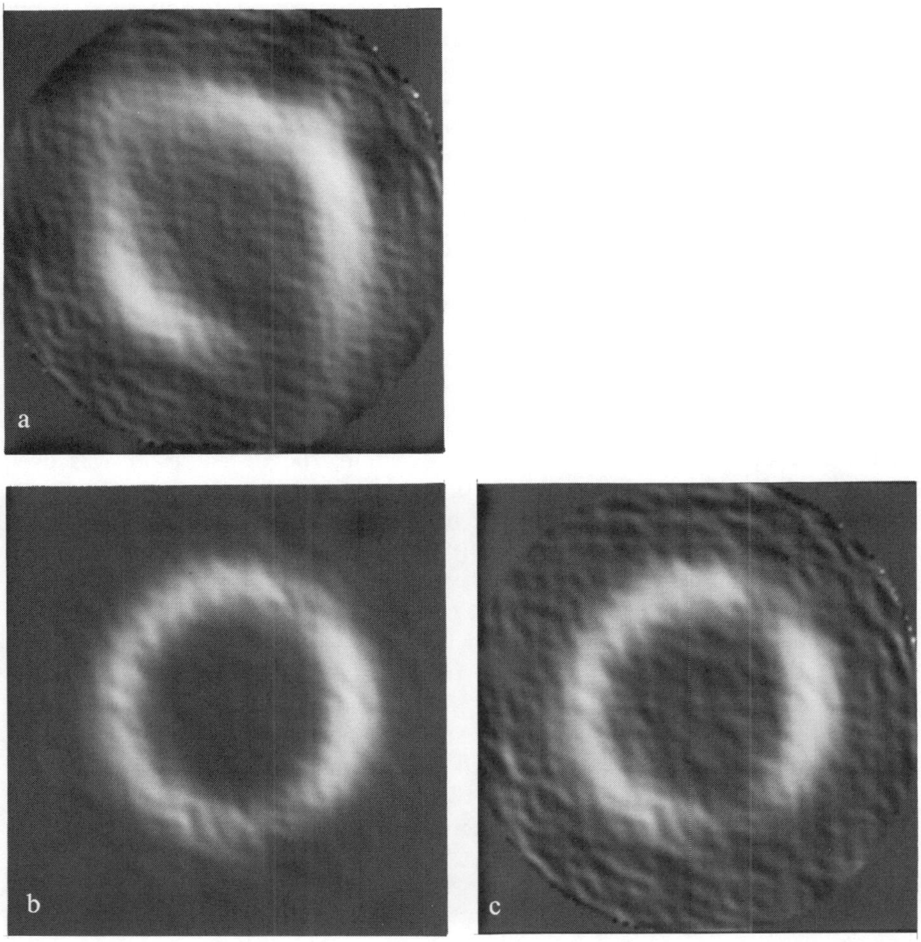

Figure 6. Reconstructions of the source in Fig. 1(a) from 11 noisy projections using (a) ART, (b) MAP, and (c) FAIR algorithms show that the latter two algorithms are tolerant of noise.

6. Discussion

In past comparisons of MAP results to more standard techniques in the areas of CT [21,23] and image restoration [36-38], the MAP approaches yielded few or no benefits. The reasons for the success of the MAP approach in the above limited-angle CT problem are: (1) The solution is severely underdetermined because of the limited data set. (2) The a priori assumptions about $\overline{f}$ and R_f can be made quite restrictive. It is expected that the MAP analysis will be most useful in situations in which these two conditions hold.

The incorporation of a priori knowledge in the MAP algorithm presented above is quite restricted. It does not readily accommodate source distributions that vary in size, shape, or location. However, the fitting procedure used in the first step of FAIR can easily handle such variations by including them as variables to be determined from the data. In the spirit of the Bayesian approach, constraints on these variables may be introduced to guide the fitting procedure toward a reasonable result. The use of ART in the second iterative portion of FAIR has the disadvantage that ART tries to reduce the discrepancy in the measurement equations to zero without regard for the estimated uncertainties in the data. Thus the FAIR result shown in Fig. 6(c) is quite noisy and is substantially further from the actual source distribution (rms deviation = 0.154) than the intermediate fitted result (rms deviation = 0.031).

In a more global Bayesian approach to the problem, the fitting procedure in FAIR may be used to estimate $\overline{f}$ and R_f for input to a MAP algorithm. The fitting procedure may be viewed as defining a subensemble appropriate to the available data. For the present example, the fitted result was used for $\overline{f}$, and R_f was assumed to be 0.1 times the identity matrix. R_f is assumed to be much smaller than that used in the preceding MAP calculation to reflect the supposition that the fit is much closer to the desired result than is the annulus of constant amplitude used for $\overline{f}$ previously. The resulting MAP reconstruction using the 10%-rms noise data, shown in Fig. 7, is better than any of the results shown in Fig. 6. The rms deviation of this reconstruction relative to the source function is 0.035, whereas that for the earlier MAP result, Fig. 6, is 0.060. When this FAIR-MAP method is applied to the projections of the inconsistent source function shown in Fig. 4(a), the result is similar to that obtained with FAIR using ART [Fig. 4(c)]. These examples demonstrate the power of this global Bayesian approach in which the MAP algorithm is used for the second, iterative step of FAIR.

7. Acknowledgments

This work was supported by the U.S. Department of Energy under contract W-7405-ENG-36. The authors wish to acknowledge helpful and interesting discussions with James J. Walker, William Rowan, Michael Kemp, Gerald Minerbo, Michael Buonocore, Barry Medoff, and Jorge Llacer.

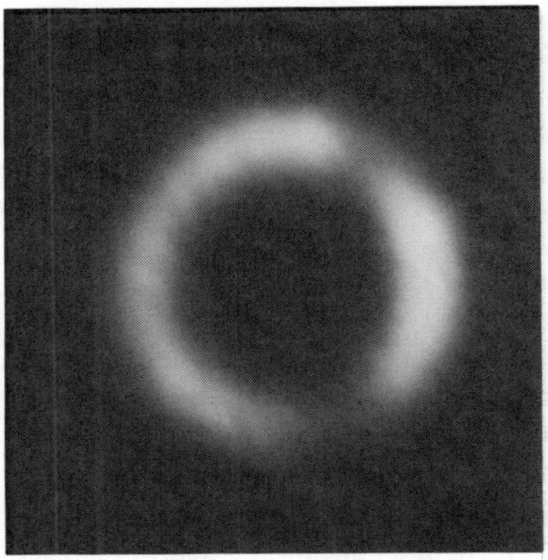

Figure 7. Reconstruction from the same data as used in Fig. 6, obtained by employing MAP as the second step in the FAIR procedure. This global Bayesian approach yields the best estimate of the original function and provides flexibility in the use of a priori information.

8. References

1. Hanson, K. M. (1982), "CT reconstruction from limited projection angles," Proc. Soc. Photo.-Opt. Instrum. Eng. 374, pp. 166-173.

2. Hanson, K. M., and G. W. Wecksung (1983), "Bayesian approach to limited-angle CT reconstruction," in Digest of the Topical Meeting on Signal Recovery and Synthesis with Incomplete Information and Partial Constraints, Optical Society of America, Washington, D.C., pp. FA6-FA14.

3. Hanson, K. M. (1982), "Limited-angle CT reconstruction using a priori information," in Proceedings of the First IEEE Computer Society International Symposium on Medical Imaging and Image Interpretation, Institute of Electronics and Electrical Engineers, New York, pp. 527-533.

4. Twomey, S. (1974), "Information content in remote sensing," Appl. Opt. 13, pp. 942-945.

5. Twomey, S., and H. B. Howell (1967), "Some aspects of the optical estimation of microstructure in fog and cloud," Appl. Opt. 6, pp. 2125-2131.

6. Twomey, S. (1965), "The application of numerical filtering to the solution of integral equations encountered in indirect sensing measurement," J. Franklin Inst. 279, pp. 95-109.

7. Shepp, L. A., and B. F. Logan (1974), "The Fourier reconstruction of a head section," IEEE Trans. Nucl. Sci. NS-21, pp. 21-43.

8. Gordon, R., R. Bender, and G. T. Herman (1970), "Algebraic reconstruction techniques for three-dimensional electron microscopy and x-ray photography," J. Theor. Biol. 29, pp. 471-481.

9. Gilbert, P. (1972), "Iterative methods for the three-dimensional reconstruction of an object from projections," J. Theor. Biol. 36, pp. 105-117.

10. Goitein, M. (1972), "Three-dimensional density reconstruction from a series of two-dimensional projections," Nucl. Instrum. Methods 101, p. 509.

11. Herman, G. T., and A. Lent (1976), "Iterative reconstruction algorithms," Comput. Biol. Med. 6, pp. 273-294.

12. Buonocore, H. B., W. R. Brody, and A. Macovski (1981), "Natural pixel decomposition for two-dimensional image reconstruction," IEEE Trans. Biomed. Eng. BME-28, pp. 69-78.

13. Buonocore, H. B., W. R. Brody, and A. Macovski (1981), "Fast minimum variance estimator for limited-angle CT image reconstruction," Med. Phys. 8, pp. 695-702.

14. Logan, B. F., and L. A. Shepp (1975), "Optical reconstruction of a function from its projections," Duke Math. J. 42, pp. 645-659.

15. Lakshminarayan, A. V., and A. Lent (1976), "The simultaneous iterative reconstruction technique as a least-squares method," Proc. Soc. Photo.-Opt. Instrum. Eng. 96, pp. 108-116.

16. Smith, K. T., P. L. Solomon, and S. L. Wagner (1977), "Practical and mathematical aspects of the problem of reconstructing objects from radiographs," Bull. Am. Math. Soc. 83, pp. 1227-1270.

17. Hamaker, C., and D. C. Solmon (1978), "The angles between the null spaces of x rays," J. Math. Anal. Appl. 62, pp. 1-23.

18. Katz, M. B. (1979), "Questions of uniqueness and resolution in reconstruction from projections," in S. Levin, ed., Lecture Notes in Biomathematics, Springer-Verlag, Berlin.

19. Louis, A. K. (1981), "Ghosts in tomography—the null space of the Radon transform," Math. Meth. Appl. Sci. 3, pp. 1-10.

20. Medoff, B. P., W. R. Brody, and A. Macovski (1982), "Image reconstruction from limited data," in Digest of the International Workshop on Physics and Engineering in Medical Imaging, Optical Society of America, Washington, D.C., pp. 188-192.

21. Herman, G. T., and A. Lent (1976), "A computer implementation of a Bayesian analysis of image reconstruction," Inf. Control 31, pp. 364-384.

22. Hunt, B. R. (1977), "Bayesian methods in nonlinear digital image restoration," IEEE Trans. Comput. C-26, pp. 219-229.

23. Herman, G. T., H. Hurwitz, A. Lent, and H. Lung (1979), "On the Bayesian approach to image reconstruction," Inf. Control 42, pp. 60-71.

24. Sage, A. P., and J. L. Melsa (1979), Estimation Theory with Applications to Communications and Control, Krieger, Melbourne, Fla., p. 175.

25. Wood, S. L., A. Macovski, and M. Morf (1978), "Reconstructions with limited data using estimation theory," in J. Raviv, J. F. Greenleaf, and G. T. Herman, eds., Computer Aided Tomography and Ultrasonics in Medicine, Proc. IFIP, TC-4 Working Conf., Haifa, Israel, August 1978, North-Holland, Amsterdam, 1979, pp. 219-233.

26. Wood, S. L., and M. Morf (1981), "A fast implementation of a minimum variance estimator for computerized tomography image reconstruction," IEEE Trans. Biomed. Eng. BME-28, pp. 56-68.

27. Buonocore, M. H. (1981), "Fast minimum variance estimators for limited-angle CT image reconstruction," Tech. Rept. 81-3, Advanced Imaging Techniques Laboratory, Department of Radiology, Stanford University, Stanford, Calif.

28. Tam, K. C., B. Macdonald, and V. Perez-Mendez (1979), "3-D object reconstruction in emission and transmission tomography with limited angular input," IEEE Trans. Nucl. Sci. NS-26, pp. 2797-2805.

29. Tam, K. C., and V. Perez-Mendez (1981), "Tomographic imaging with limited-angle input," J. Opt. Soc. Am. 71, pp. 582-592.

30. Tam, K. C., and V. Perez-Mendez (1981), "Limits to image reconstruction from restricted angular input," IEEE Trans. Nucl. Sci. NS-28, pp. 179-183.

31. Grunbaum, F. A. (1980), "A study of Fourier space methods for limited-angle image reconstruction," Numerical Funct. Anal. Optim. 2, pp. 31-42.

32. Sato, T., S. J. Norton, M. Linzer, O. Ikeda, and M. Hirama (1981), "Tomographic image reconstruction from limited projections using iterative revisions in image and transform spaces," Appl. Opt. 20, pp. 395-399.

33. Minerbo, G. (1979), "MENT: a maximum entropy algorithm for reconstructing a source from projection data," Comput. Graphics Image Process. 10, pp. 48-68.

34. Inouye, T. (1979), "Image reconstruction with limited-angle projection data," IEEE Trans. Nucl. Sci. NS-26, pp. 2666-2669.

35. Inouye, T. (1982), "Image reconstruction with limited-view angle projections," in Digest of the International Workshop on Physics and Engineering in Medical Imaging, Optical Society of America, Washington, D.C., pp. 165-168.

36. Trussell, H. J., and B. R. Hunt (1979), "Improved methods of maximum a posteriori restoration," IEEE Trans. Comput. C-28, pp. 57-62.

37. Trussell, H. J. (1976), "Notes on linear image restoration by maximizing the a posteriori probability," IEEE Trans. Acoust. Speech Signal Process. ASSP-26, pp. 174-176.

38. Cannon, T. M., H. J. Trussell, and B. R. Hunt (1978), "Comparison of image restoration methods," Appl. Opt. 17, pp. 3384-3390.

APPLICATION OF THE MAXIMUM ENTROPY PRINCIPLE TO RETRIEVAL FROM LARGE DATA BASES

Paul B. Kantor

Tantalus, Inc., Cleveland, OH 44118

C. R. Smith and G. J. Erickson (eds.),
Maximum-Entropy and Bayesian Spectral Analysis and Estimation Problems, 273–281.

This problem was suggested by Cooper [1978] and Cooper and Huizinga [1982]. Their work is not very technical and, in elaborating the mathematics [Kantor, 1983a, b, c], I thought, "Boy, I am really discovering the wheel!" Sitting at this conference for three and a half days, I am now sure that everything I have to say will be well known to many of you, or wrong, or both.

The problem: There is a collection of retrievable items—meaning "all the books in the library," "all the articles in some literature," or "all the facts in some intelligence data base." The number of items is large. A real human being, which in the library business we call a "patron (P)," submits a query (Q) and we assume that this pair, (P,Q), defines a point function on all these retrievable items x belonging to a set X, which might be called the value of the item x, as an answer to that query, for that person.

We assume that the pair (P,Q) defines a point function v (value; utility; chance of relevance) with values in the closed interval [0,1]:

$$(P,Q):x \rightarrow v(x;P,Q) \ . \tag{1}$$

The items have been examined, and are equipped with "descriptor vectors" that indicate whether they have or lack various properties a,b,... . The subset of elements having properties a and b but not c is denoted by abc'. The entire "reservoir" of elements that we denoted by X is the union (we use $\cup$ to represent logical union); for example, for three properties,

$$X = a'b'c' \cup a'b'c \cup \ldots \cup abc \ . \tag{2}$$

You can of course argue that this isn't true. For example, maybe the value does not reside in individual items, but in suitable subsets of items. But perhaps it's true enough to take as a starting point. The value can also be called a "utility," "probability or chance of relevance," and so forth. I am going to work in a framework where we assume these have been normalized so that they lie between zero and one (the usual arguments for scaling utility into a compact interval apply here). From now on, I'll just talk about v(x), the value of an item, and suppress P and Q.

A retrieval system, if I may speak anthropomorphically, "wants" to provide most valuable items first because there are so many. In other words, it wants to find, among all the 2^N components of the reservoir, the ones for which the values are highest.

The items have previously been examined as they were installed in the system. In principle they might be undergoing continual re-examination by some computer program. In any case, they are equipped with descriptor vectors. They might have a property a or b or c, etc. In conventional bibliographic data bases, like the on-line retrieval systems, properties are the presence or absence of a particular keyword. You can do more, though. You can talk about the number of times a keyword occurs, or you can define measures of relatedness between documents and use them as descriptors. In any case, a document will or won't have a property a, b, or c, and so forth.

Fuzzy properties can, in principle at least, be reduced by some kind of binning to this case. So the whole reservoir is the union of all the different things that could happen to all those properties. However complex the situation really may be, I am going to pretend that the properties can be shown in some kind of order as the rows of a table (Fig. 1).

Component	Fraction	v = 0	v = 1
a'b'c'	n(a'b'c')		
a'b'c	n(a'b'c)		
a'bc	n(a'bc)		
abc	n(abc)		
α	n(α)		
ab'c	n(ab'c)		
⋮	⋮		

Figure 1. Properties of documents in a reservoir, represented as a table.

The general row is called alpha, and by introspection the system knows exactly how many items there are in each row. What it doesn't know is how many of them have a particular value. Whatever the complexity, we conceive of a list as shown in the figure. Somewhat dangerously, we use lower-case n to represent the fraction of the reservoir having each descriptor; n is not an integer:

$$\sum_{\alpha} n(\alpha) = 1 . \tag{3}$$

In old-fashioned discussions of retrieval, it was customary to say an item had value either 0 or 1. I don't know how this developed, but one just said an item is "either relevant or not relevant." I don't think this is reasonable, and I will drop it shortly. In the simplest case the question is: "Which of these rows have more of their items in the right-hand column (v = 1) than the left-hand column (v = 0)?" The number of items in a particular cell (c) can be described as the total size of the reservoir times the relevant fraction q(c)—or, more conveniently for our purposes, the total number in the reservoir N, times the fraction that sit in that particular row or component n(α), times the fraction that sit in that cell p(α,c). For example, the number of items in the cell labeled a'b'c;1 (that is, the number of items having descriptor a'b'c and value 1) is the product Nq(a'b'c;1) or Nn(a'b'c)p(a'b'c;1).

Constraints are assumed to be either given or generated during the retrieval process. Each constraint, K_1, K_2, and so forth, will apply to some set of rows R_1, R_2. (Note that K_n is the name of a constraint, while R_n is the name of a set of rows.) And it will in general take this form:

Constraints $K_1, K_2, \ldots$
Apply to rows $R_1, R_2, \ldots$

$$\sum_{\alpha \in R_j} n(\alpha)\, v(\alpha) = V_j\,, \qquad j = 1,\ldots,M\,. \tag{4}$$

The sum of the number of items in the row times the average value of the items in the row summed over all the rows that the constraint applies to is the total value that we expect to find associated with that condition. A typical example would be to say that "all the items that have descriptor a are expected together to have a certain value."

These can be given (terribly) as a priori estimates (that is, the patron P says, "I think that items reported since 1982 have a 90% chance of being relevant to my search"). More persuasively (I think this point of view has been pioneered by Salton), this kind of information should be derived by a feedback process: As you pull things out, you show them to the ultimate evaluator and take note of whether he or she says they are good or not. That's conceptually compelling. I don't know whether it's really practical because it takes most of us a little while to decide on relevance. In any case, let $K(\alpha)$ be the set of constraints that apply to the row α:

$$K(\alpha) = \{j\colon \alpha \text{ is in } R_j\}\,. \tag{5}$$

The nice suggestion that Cooper and Huizinga made was: Given these kinds of clues and the need to have a definite answer, postulate that the distribution of v, the value (I talk of it as a distribution of v although it's really a distribution of items into cells with corresponding values), is precisely the one that maximizes the entropy.

I'll talk a little bit at the end of the talk about why this might plausibly be true, but I'm not a "convert," and I really feel, in this particular situation, that maximum entropy is an assumption to be tested empirically. We'll figure out how to do it: build the systems, work with real patrons, and eventually either succeed or fail. There are some built-in difficulties because almost anything you do using a computer pleases the patrons. They are getting more than they can possibly cope with, and it may be quite a while before we know if maximum entropy improves the situation.

At the end of the first day of this workshop, I felt kind of bad that I was doing plain old entropy maximization instead of minimization of cross-entropy. But the more talks I have heard, the more I have come to feel that the choice of a prior for cross-entropy is really a two-edged sword that can easily be grabbed by the blade. It is a way of putting some fact in and pretending you really firmly believe in it, therefore muddling it up with constraints. Certainly at this point there isn't anything that I believe that firmly.

The steps of the mathematics will be very clear. The entropy is sensibly broken down into a part over which we have no control, that is, the distribution of items among the components, and the part that is the expected value over all the components of the entropy of the individual components:

$$S = - \sum_{\text{cells } c} q(c) \ln q(c)$$

$$= - \sum_{\text{rows}} n(\alpha) \ln n(\alpha) + \sum_{\alpha} n(\alpha) \overbrace{\sum_{c \in \alpha} [-p(c) \ln p(c)]}^{S(\alpha)} . \tag{6}$$

Whether or not I am a true believer, I can use an entropy-like formulation for the case where the values are either 0 or 1. Write the distribution using exponents so the occupancy of the cells is proportional to either 1 or exp(r), where r depends on the row:

$$v \in \{0,1\} \qquad p(c) = \left\{ \begin{matrix} 1 \\ e^{r(\alpha)} \end{matrix} \right\} \frac{1}{1 + e^{r(\alpha)}} \tag{7}$$

$$v(r(\alpha)) = v(\alpha) = \frac{e^{r(\alpha)}}{1 + e^{r(\alpha)}} \tag{8}$$

where $v(\alpha)$ is the average value (relevance) of an item in row α. It took me a long time to recognize that the parameter r can very sensibly be thought of as the "richness" of the row, which makes sense to people I usually talk to. Saying that the best components of the table are "the ones with negative temperature" doesn't. The mean value v(r) is some function that falls to 0 as r goes to $-\infty$ and goes to +1 as r goes to $+\infty$.

More realistically, I am working with a model in which the value can lie anywhere in the interval [0,1] with uniform prior, so we get the standard exponential divided by normalization, and $p(\alpha;c)$ becomes

$$p(\alpha;c) = \frac{e^{r(\alpha)\, v(c)}}{\int dv\, e^{r(\alpha)\, v(c)}} . \tag{9}$$

The 'value function' for each component is a function that occurred very long ago in statistical mechanics of a dipole in a magnetic field:

$$v(\alpha) = \int vp(v)\,dv = \langle v \rangle$$

$$= \frac{1}{2}\left[1 + \coth\frac{r}{2} - \frac{2}{r}\right]. \tag{10}$$

Lee Schick discusses a more general case in his paper in this volume. But whatever the form of the function v, the well known result is that maximizing the entropy subject to the constraints leads to the result that the 'richness' of the row α is just the sum of the Lagrange multipliers (with their signs changed) corresponding to all the constraints that apply to the row

$$r_\alpha = \sum_{k \in K(\alpha)} \lambda_k. \tag{11}$$

The resulting equations can be directly applied to study some questions of interest in information retrieval—particularly questions like: 'Is the overlap of two desirable properties enhanced in richness as compared to either of those properties?' We can also ask whether there exists something called 'term weight.' 'Term weight' is a positive number that can be attached to a term, such that the retrieval value of a document can be estimated just by adding up all the weights of the terms that appear in it. We can certainly show that term weights don't exist in general, and I think the situation is really kind of bleak for term weighting—which is all right because it looks quite good for maximum entropy. The equations to be solved are those for each constraint K_n; the sum over all components that are constrained by K_n of the richness of that component (which is determined by all the constraints that apply to that component) multiplied by the size of that component is equal to some given number:

$$\sum_{\alpha \in R(k)} v\left(\sum_{j \in K(\alpha)} \lambda_j\right) n(\alpha) = V_k. \tag{12}$$

I think that any of the methods we've heard described could be applied to it. I differ from everyone else in not having done a single computer run. I currently believe that something called the homotopy method will be very effective [Zangwill and Garcia, 1981]. In this technique we take a problem to be solved (here the problem is completely controlled by the vector of constraints, V) and try to continuously deform it into a problem that we can solve on the back of an envelope. When we have that one solved, we just re-deform it back in the other direction, keeping careful track of the solution. The idea is that, step by step, we will have a very good starting point for whatever technique we apply to solve these rather horribly nonlinear equations.

In this particular case it is easy to find a starting point. If all the λ's are 0, then it is easy to calculate what all the V_k are. And so we know the solution for those particular constraints: V_k are

$$V_k^0 = \sum_{\alpha \in R(k)} n(\alpha)\frac{1}{2} \tag{13}$$

Solution: All $\lambda_k^0 = 0$.

We then deform all the $V_k(t)$, varying t from zero to one.

Homotopy method:

The deformation of the left-hand side of the equations is

$$(1-t)\begin{bmatrix} V_1^0 \\ \cdot \\ \cdot \\ \cdot \\ V_n^0 \end{bmatrix} + t\begin{bmatrix} V_1 \\ \cdot \\ \cdot \\ \cdot \\ V_n \end{bmatrix} = \begin{bmatrix} V_1^t \\ \cdot \\ \cdot \\ \cdot \\ V_n^t \end{bmatrix}. \tag{14}$$

I am hopeful that this will provide really quick solutions. There is a real-time constraint on this because, if it is to work, it has to work while the patron is at the keyboard expecting the computer to respond instantly. We can put a little dancing demon on the screen to hold his attention for a while, but even early results show that although people are tickled pink with using computers for information retrieval, they are very impatient. To really be successful it has to work in the order of two or three minutes. [Note added March 1985: We have a Pascal program running on an 8088 PC with an 8087 math chip that traces solutions for five constraints in a few seconds.]

Question: "Couldn't you divert them with information—the Encyclopaedia Brittanica or something?"

Question: "Or can you ask them if they want to learn about cars? Or can you change the subject?"

PBK: There are some interesting results about what people will tolerate from a computer system. It depends on how much they already know about the computer approach. If they don't know anything, they will put up with all kinds of nonsense. They'll read through menu after menu. If they know the answers to all those menus, they are less patient.

Paul B. Kantor

There are some real objections to applying the maximum entropy principle to an information reservoir, which are immediately raised by any librarians who understand what the maximum entropy principle means. They say: "We're not dumb, and we've worked really hard to put all these things in there right. How can you possibly argue that you are going to do better by using some technique that is used with molecules scattered around the room?" This is disturbing. In the case of molecules, dynamical theory has advanced to a point where even those of us who are determinists can be shown that maximum entropy will hold. Furthermore, I have never interviewed a molecule, and it doesn't exhibit any volition.

But how can we feel this way about activities of human beings, and the classification of items put into the reservoir? There are three different arguments that encourage me.

The first goes back to Emile Durkheim who, I think, is regarded as the founder of sociology. He pointed out that human beings, with all of their volition and intelligence, behave in horrifyingly predictable ways. The example he gave was the number of suicides in France. He argued very convincingly that these are not the same people prone, year after year, to plunge into the Seine. Instead, there are underlying regularities that govern the behavior of the French. I believe it is the most elegant argument that has ever been given in the social sciences.

The second argument is rather technical. The terminology, descriptors, and what constitutes a sensible query change fairly rapidly. Think of the descriptor "solid-state," which might now be used to retrieve the catalog of an expensive audio house but, 15 years ago, would only have been used to retrieve things from technical journals. Because descriptors and queries are "moving" all the time, it may well be that the kind of shaking up, the mixing that goes on in a dynamical system, is actually going on in the very well intentioned efforts to assign exactly the right descriptors to the retrievable items.

The third argument is the result proved by Van Campenhout and Cover [1981], which shows that without "believing" anything about minimum cross-entropy one can show that if the average of independent identically distributed (i.i.d.) variables is constrained, then the marginal distribution of any one of them turns out to be the same as that given by the entropy principle. Since I am willing to believe that the values in the cells are i.i.d. and they seem to have a constraint on their sum, this argument makes it easier for me to behave like a believer. [Note added in proof: The result can be derived, at least heuristically, from the following line of argument: If there are N i.i.d. variables with a constrained sum, then the distribution of any one of them, say the first, is the same as the distribution of the constraint minus the sum of all the others. But the sum of all the others is approximately governed by the normal distribution (by the central limit theorem), and when you do the arithmetic you find the maximum entropy result.]

I always talk fast and apparently discourage questions, so I have 3 minutes to speculate. I don't really have 3 minutes worth of speculation, but I have a lot of hopes. In the course of doing this, and doing it with my background as a physicist, I couldn't help noticing that after pages of arithmetic I was rediscovering relations of the form

$$dV = TdS - dW \tag{15}$$

as I calculated these Lagrange multipliers—which meant that the value function was playing, in the mathematical picture, the role of an internal energy in thermodynamics, when there is no external work. So I am encouraged to speculate that it may be possible to link into this formalism something having to do with the actual work or effort required to maintain, retrieve from, or reorganize the information system. I have no results to report. I have a couple of gedanken experiments that need to be performed, filtering a row to try to make it richer and see what kind of effort is involved in that.

Acknowledgment

This work was supported in part by the National Science Foundation under Grant IST-8110510.

References

Cooper, William S. (1978), "Indexing documents by gedanken experimentation," J. Am Soc. Inf. Sci. 29(3), pp. 107-119.

Cooper, William S., and P. Huizinga (1982), "The maximum entropy principle and its application to the design of probabilistic retrieval systems," Inf. Technol. Res. Devel. 1, pp. 99-112.

Kantor, Paul B. (1983a), "Maximum entropy and the optimal design of automated information retrieval systems," Inf. Technol. 3(2), pp. 88-94.

Kantor, Paul B. (1983b), "Minimal Constraint Implementation of the Maximum Entropy Principle in the Design of Term-Weighting Systems," Proceedings of the 46th Annual Meeting of the ASIS, Oct. 1983, pp. 28-31.

Kantor, Paul B. (1983c), "Maximum entropy, continuous value, effective temperature, and homotopy theory in information retrieval," Tantalus Rept. 84/1/3.

Van Campenhout, Jan M., and Thomas M. Cover (1981), "Maximum entropy and conditional probability," IEEE Trans. Inf. Theory IT-27(4), pp. 483-489.

Zangwill, W. I., and C. B. Garcia (1981), Pathways to Solutions and Equilibria, Prentice-Hall, New York.

TWO RECENT APPLICATIONS OF MAXIMUM ENTROPY

Lee H. Schick

Department of Physics and Astronomy, University of Wyoming, Laramie, WY 82070

C. R. Smith and G. J. Erickson (eds.),
Maximum-Entropy and Bayesian Spectral Analysis and Estimation Problems, 283–293.

1. Introduction

Presented here are two applications of the maximum entropy method (MEM) that two of my students have recently worked on. The first of these applications, the quantum mechanical inverse scattering problem, was the subject of a Master's thesis by John P. Halloran, who is now at the Lockheed California Company, in Burbank, California. Some of the results of that work were reported at the Third Rocky Mountain Southwest Theoretical Physics Conference held in Boulder, Colorado, earlier this year. The second application, the processing of seismic time series, is part of a Ph.D. thesis by Kent E. Noffsinger that is in progress.

The basic theoretical ideas behind these applications are due to people such as Ed Jaynes, John Skilling, Steve Gull, Tom Grandy, and maybe half of the rest of this audience. My apologies to those of you who are not properly acknowledged.

2. The Quantum Mechanical Inverse Scattering Problem

We have heard in the past few days the details of many elegant theories. In this section, I shall not bother with such intricacies, but will content myself with a rather broad outline. In the words of that great philosopher John Madden, "Don't worry about the horse being blind. Just load up the wagon." Well, just substitute "theory" for "horse" and "computer" for "wagon" and you have the idea.

One may divide all problems into two types, direct and inverse. A direct problem is "easy," is a problem in deduction, and is thus solved first. To each such problem there corresponds an inverse problem that is "hard," is a problem of inference, and is thus solved later, if at all. Two examples of these types of problems are as follows:

Direct Problem #1

Q: What are the first four integers in the sequence $2+3n$, $n = 0,1,2,\ldots$?

A: 2, 5, 8, 11

Inverse Problem #1

Q: If "2, 5, 8, 11" is the answer, what is the question?

A: What are the first four stops on the Market Street subway?

Direct Problem #2

Q: What did Mr. Stanley say to Dr. Livingston?

A: "Dr. Livingston, I presume."

Inverse Problem #2

Q: If "Dr. Livingston, I presume" is the answer, what is the question?

A: What is your full name, Dr. Presume?

As amusing as these examples may, or may not, be, they do illustrate three important properties of inverse problems: (1) In both inverse problems, the answer was certainly not unique. This is the case in many such problems. (2) The first inverse problem might have been easily solved by a Philadelphian used to riding that city's subway system. In other words, prior information is important in solving the inverse problem. (3) In the second inverse problem, the correct answer is made even more difficult to infer than it might otherwise have been because there is misinformation in the information given to us; that is, the answer for which we were asked to infer the question should have been written: "Dr. Livingston I. Presume." In other words, the presence of noise in the data can influence the answer to the inverse problem.

I shall now consider a simplified quantum mechanical inverse scattering problem that embodies considerations (2) and (3).

We are given the prior information—that is, information made available before the experiment in question is performed--that we are dealing with the quantum problem of a particle of mass m being scattered by a central potential $V(r)$ that satisfies $V(r) = 0$ for all $r > 8$ fm and $V(r) > 0$ for all $r < 8$ fm.

In addition, an experiment has been performed, the results of which are a set of noisy Born scattering amplitudes for a set of values of momentum transfer q. Thus the data $D(q)$ are given by

$$D(q) = -\frac{2m}{\hbar^2 q}\int_0^\infty r\,V(r)\sin(qr)\,dr + e(q) \tag{1}$$

where the noise distribution, $e(q)$, is known to be Gaussian. In fact, we have data only at a discrete number of values of momentum transfers, so we can write Eq. (1) in discretized form

$$D_i = \sum_{j=1}^{M} F_{ij} V_j + e_i \,. \tag{2}$$

Finally, since the only values of V_j that contribute to the sum are (from our prior information) all positive, we may consider the distribution of V_j's to be a distribution of probabilities p_j. This reduces the problem to the type of image restoration problem analyzed by Gull and Daniell [1978] using a MEM applied to the p_j. The prior distribution of the p_j's is assumed to be uniform for $r < 8$ fm, zero for $r > 8$ fm, and of course normalized to 1, consistent with the prior information on the $V(r)$. The entropy of the distribution of the p_j's is maximized subject to the constraint that the χ^2 per degree of freedom, given by χ^2/M, with

$$\chi^2 = \sum_{i=1}^{M} \frac{e_i^2}{\sigma_i^2}, \tag{3}$$

is equal to 1, where σ_i is the experimental random error of the ith data point, and e_i is given by Eq. (2).

To test the efficacy of the MEM in this case, synthetic data D_i were constructed by first calculating the Born amplitudes B_i for a simple potential shape at a number of different values of q_i. For each calculated amplitude, a standard deviation σ_i was chosen from a Gaussian distribution of standard deviations, and the experimental value of the amplitude, D_i, was chosen by picking a random value from a Gaussian distribution that was centered on the calculated value B_i and whose standard deviation was the chosen value σ_i. The signal-to-noise ratio (SNR) for the data set, defined by

$$\mathrm{SNR} = \sum_k |D_k| \Big/ \sum_\ell |D_\ell - B_\ell|, \tag{4}$$

could be varied by varying the center and width of the distribution of standard deviations from which the σ_i were chosen.

Figures 1 through 4 are the results of sample calculations. These calculations were carried out for synthetic data derived from an exponential potential, $\exp(-r/r_0)$, with $r_0 = 1$ fm. The momentum of the particle was taken to be 200 MeV/c. Ninety values of q_i were chosen by varying the scattering angle from 20° to 160°. Ninety values of the r variable from 0 to 8 fm were used to evaluate the radial integrals.

Even for the rather poor data of Fig. 3(a), the reconstruction of the exponential potential is quite good. There is little improvement in the reconstruction in going from Fig. 3 to Fig. 4, even though the data, in the sense of SNR, are much better. The reason is that the maximum value of momentum transfer used puts a natural limit on the accuracy of the reconstruction for $r < 1$ fm.

3. Processing Time-Domain Seismic Data

Motivated by the success of the MEM in astrophysics [Gull and Daniell, 1978; Bryan and Skilling, 1980; Willingale, 1981], Ramarao Inguva and I [Inguva and Schick, 1981] utilized the MEM image processing formalism to analyze seismic data directly in the time domain. I shall discuss here some modifications of the formalism designed to take account of the fact that photographic intensities are always positive whereas seismic trace values are both positive and negative. These modifications were developed by Mr. Noffsinger as a way of adapting a formalism used in statistical mechanics [Grandy, 1981] to the processing of seismic time series.

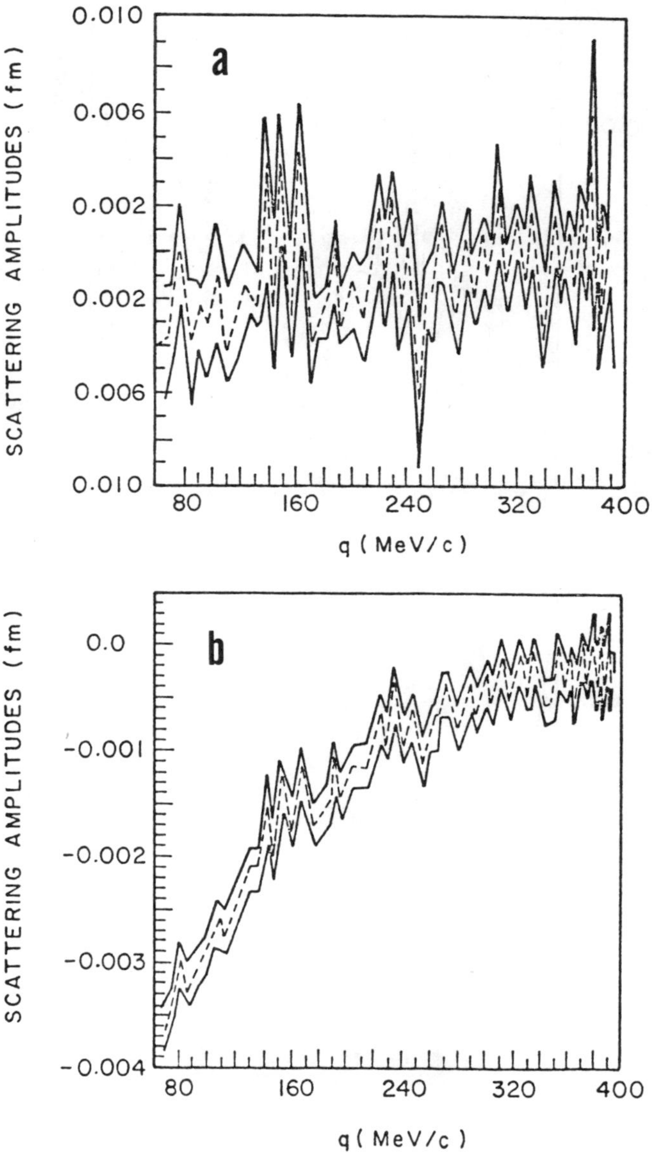

Figure 1. The noisy Born amplitudes, D_i, as a function of momentum transfer (dashed line), with the $\pm\sigma_i$ limits (solid line). (a) SNR = 1.3. (b) SNR = 6.0.

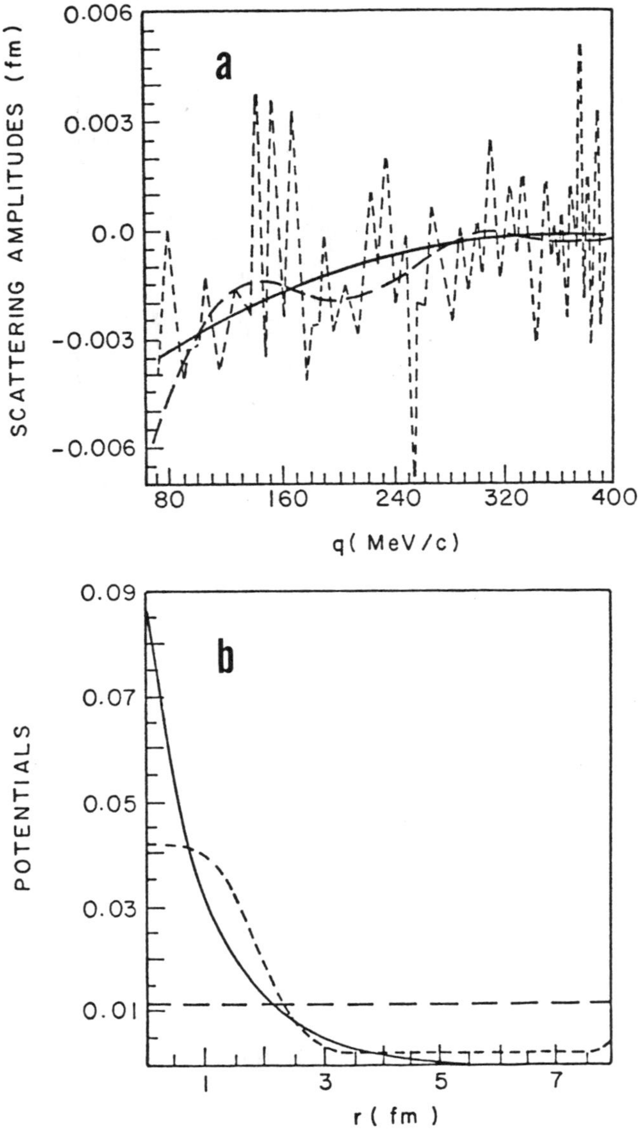

Figure 2. (a) The noisy Born amplitudes, with SNR = 1.3 (lighter dashed line), exact Born amplitudes (solid line), and Born amplitudes reconstructed using the MEM (heavier dashed line). (b) The exact exponential potential (solid line), prior distribution for the potential (heavy dashed line), and potential distribution reconstructed using the MEM and the noisy amplitudes of (a) (light dashed line).

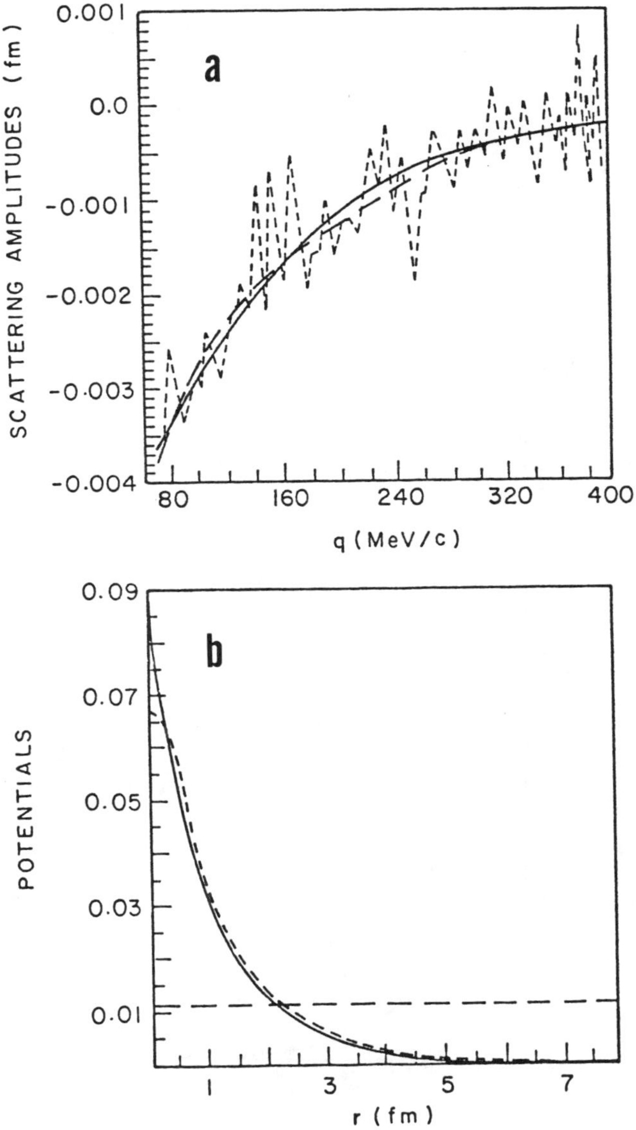

Figure 3. Same as Fig. 2, but with SNR of 3.6.

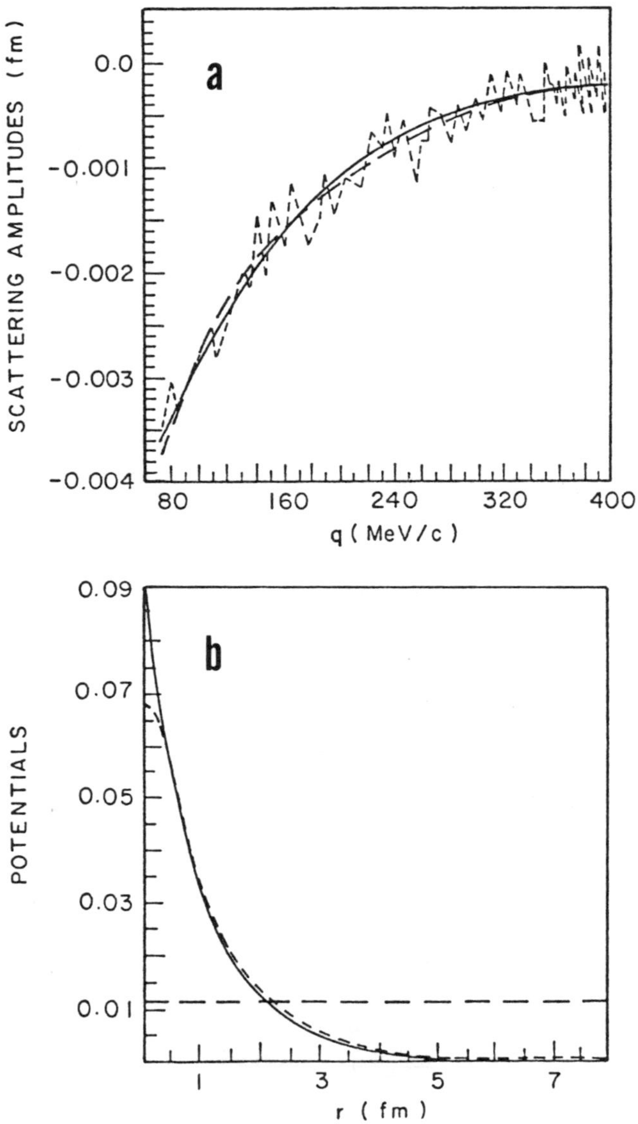

Figure 4. Same as Fig. 2, but with SNR of 6.0.

A seismic section can be treated to some extent as a photographic image and can, with the purpose of enhancing the SNR, be processed accordingly. Here, however, we limit the discussion to the processing of individual traces that are assumed to be already deconvolved. The extension of the method to more complicated cases is straightforward.

In this application, q_i, the MEM estimate of the true amplitude of the seismic signal at the ith time-point on a given trace, is represented not by a probability, as would be the case with a photograph, but by an expectation value over a set of possible values that q_i might have, each such value being weighted by its probability of occurrence. It is the entropy of the distribution of these probabilities over the entire set of time-points comprising the trace that is to be maximized.

For example, if q_0 ($q_0 > 0$) is taken to be the basic unit of amplitude, and if q_m, the maximum value of q_i over all $i = 1,2,3,\ldots,N$ sample points, is such that for every positive integer R, $q_m < Rq_0$, then every q_i may be written

$$q_i = q_0 \sum_{r=-R}^{R} r\, p_r(i) \,. \tag{5}$$

Here, $p_r(i)$ is the probability that at the ith time-point the trace amplitude has the value $q_0 r$. It is important to realize that $p_r(i)$ is not a joint probability, but rather for each time-point i there is a separate normalization condition

$$\sum_{r=-R}^{R} p_r(i) = 1 \,. \tag{6}$$

With no further information available, the constrained entropy for the trace takes the form

$$S = -\sum_{i,r} p_r(i) \log p_r(i) - \sum_i \mu_i \left[\sum_r p_r(i) - 1 \right], \tag{7}$$

where each μ_i is a Lagrange multiplier. S may be maximized with respect to each $p_r(i)$ and with respect to each μ_i. The result is that $p_r(i) = 1/(2R+1)$, for all i, so that from Eq. (5) $q_i = 0$, for all i. In other words, without any data having been taken, the MEM estimate for the trace is a "dead" trace. This is certainly the "flattest," most noncommittal, and safest estimate to be made under the circumstances.

Now, for a noisy trace, in which the noise is additive, the MEM may be used to obtain an estimate q_i of the signal, as follows. The measured trace amplitude at the ith time-point, t_i, may be written

$$t_i = q_i + n_i \,, \tag{8}$$

where n_i is the (unknown) noise amplitude at the ith time-point. If the n_i are all independent and each is a sample from a Gaussian distribution with known variance σ_i^2, then the appropriate constraint is a χ^2 constraint [Jaynes, 1981]; that is, we set

$$\chi^2 = \sum_{i=1}^{N} \frac{(t_i - q_i)^2}{\sigma_i^2} = \chi_c^2 , \tag{9}$$

where χ_c^2 is the value of χ^2 that is assumed known from whatever our confidence level in the data is. More generally, for some other type of noise distribution, Eq. (9) will be replaced by

$$F(q_1,q_2,\ldots,q_N) - F_0 = 0 , \tag{10}$$

where F_0 is known. In addition, we may rewrite the sum in Eq. (5) in a more convenient form, namely

$$q_i = q_0 \sum_{s=1}^{R} (s - d_i)\, p_{si} , \tag{11}$$

where s is an integer, p_{si} is the probability that q_i has the value $q_0(s-d_i)$, and the d_i are a known set of constants. (For example, $M = 2R+1$, $d_i = M-1$, for all i, yields the exact form of Eq. (5) with the $r = 0$ term included in the right-hand side.) The result of the MEM with Eq. (10) as an additional constraint is

$$q_i - q_0 d_i = \frac{q_0}{2} \coth\left(\alpha_i \frac{q_0}{2}\right) - M \frac{q_0}{2} \coth\left(\alpha_i M \frac{q_0}{2}\right) , \tag{12}$$

where

$$\alpha_i = \lambda\, q_0 \frac{\partial F}{\partial q_i} \tag{13}$$

and λ is the Lagrange multiplier for the constraint of Eq. (10).

For a given value of M, Eqs. (12) and (13) may be solved, once the constraint Eq. (10) has been chosen. I will not present results of this type of numerical work, but I will note the following. If one looks at the right-hand side of Eq. (12) as a function of $\alpha_i q_0$ for $M = 2,3,\ldots$ with $\alpha_i q_0/M$ kept constant as $M \to \infty$, one finds that: (a) for every $0.001 < \alpha_i < 1.00$ the result for $M = 2$ is within 30% of the $M = \infty$ result, and (b) for every $0.001 < \alpha_i < 1.00$ the result for $M = 4$ is within 6% of the $M = \infty$ result.

Finally, I point out an interesting result for the case of $M = 2$. In this case, we have

$$q_i = q_0(2p_i - 1) \tag{14}$$

with the entropy given by

$$S = -\sum_i p_i \log p_i - \sum_i (1 - p_i) \log (1 - p_i) . \qquad (15)$$

Next, let's change from the variables p_i to δ_i defined by $p_i = (1+\delta_i)/2$, so $q_i = q_0\delta_i$. Now if we choose q_0 large enough, no matter what other constraints we may have, we can accommodate any q_i with a set of δ_i's such that for all i, $\delta_i \ll 1$. In this case, Eq. (15) becomes

$$S = \sum_i [\log 2 - \delta_i^2] . \qquad (16)$$

Except for the constant term, this is just the form of regulator that has been used many times in the literature [Lindley and Smith, 1972; Jaynes, 1974]. But it must be noted, in contradistinction to this earlier work, that the δ_i's in Eq. (16) are not probabilities; that is, they do not have to be positive and they do not satisfy a normalization condition.

4. References

Bryan, R. K., and J. Skilling (1980), "Deconvolution by maximum entropy as illustrated by application to the jet of M87," Mon. Not. R. Astron. Soc. 191, pp. 69-79.

Grandy, W. T. (1981), "Indistinguishability, symmetrization, and maximum entropy," Europ. J. Phys. 2, pp. 86-90.

Gull, S. F., and G. J. Daniell (1978), "Image reconstruction from incomplete and noisy data," Nature 272, pp. 686-690.

Inguva, R., and L. H. Schick (1981), "Information theoretic processing of seismic data," Geophys. Res. Lett. 8, pp. 1199-1202.

Jaynes, E. T. (1974), Probability Theory with Applications in Science and Engineering, unpublished.

Jaynes, E. T. (1981), "What is the problem?" Proceedings of the Second ASSP Workshop on Spectrum Analysis, S. Haykin, ed., McMaster University.

Lindley, D. K., and A. F. M. Smith (1972), "Bayes estimates for the linear model," J. R. Statist. Soc. 34B, pp. 1-40.

Willingale, R. (1981), "Use of the maximum entropy method in x-ray astronomy," Mon. Not. R. Astron. Soc. 194, pp. 359-364.

A VARIATIONAL METHOD FOR CLASSICAL FLUIDS

Ramarao Inguva, C. Ray Smith,* and T. M. Huber

Department of Physics and Astronomy, University of Wyoming, Laramie, WY 82071

Gary Erickson

Department of Electrical Engineering, Seattle University, Seattle, WA 98122

A new variational method is proposed in which we compute the thermodynamic properties of classical fluids using the correlation functions as constraints. As an explicit example, we present the results of carrying out the variational procedure using the hypernetted chain integral equations for the correlation functions as constraints. Such an analysis provides insight into the density-temperature regime over which the hypernetted chain approximation is valid.

*Present address: Advanced Sensors Directorate, Research, Development and Engineering Center, U.S. Army Missile Command, Redstone Arsenal, AL 35898.

C. R. Smith and G. J. Erickson (eds.),
Maximum-Entropy and Bayesian Spectral Analysis and Estimation Problems, 295–304.

1. Introduction

The most successful methods for computing thermodynamic properties of classical liquids are based on various approximate nonlinear integral equations for the radial distribution function. Examples of such equations include the Born-Green-Yvon, the Percus-Yevick, and the hypernetted-chains equations [Balescu, 1975; Hill, 1956; March, 1968; Temperly et al., 1968]. All of these equations correspond to a truncation of the exact equilibrium hierarchy for the reduced distribution functions. Such truncation procedures make no attempt to satisfy the principle of maximum entropy from which the original phase-space distribution is derived. In this paper, we outline a variational method in which any one of the integral equations for the radial distribution function can be used as a constraint in the entropy maximization; this guarantees optimal correlation functions and consistent thermodynamic quantities. We discuss only the hypernetted-chains equation (HNC). The generalization to other cases is straightforward.

2. Formulation

We consider a monatomic system of N particles, the dynamics of which is described by the Hamiltonian

$$H = \frac{1}{2m} \sum_{i=1}^{N} \mathbf{p}_i^2 + \sum_{i<j=1}^{N} V(r_{ij})$$

$$= K + V . \tag{1}$$

The equilibrium phase-space probability distribution $\rho(\mathbf{r}_1,\ldots,\mathbf{r}_N;\ \mathbf{p}_1,\ldots,\mathbf{p}_N)$ describes the state of the system and is the main quantity of interest. With the normalization

$$\int d^{3N}r\, d^{3N}p\ \rho(\mathbf{r}_1,\ldots,\mathbf{r}_N;\ \mathbf{p}_1,\ldots,\mathbf{p}_N) = 1 , \tag{2}$$

the entropy and average energy are given by

$$S = -\kappa \int d^{3N}r \int d^{3N}p\ \rho \ln\rho \tag{3}$$

$$\langle E\rangle = \int d^{3N}r \int d^{3N}p\ (\rho H) \tag{4}$$

where κ is the Boltzmann constant. If Eqs. (2) and (4) are the only constraints on ρ, then maximization of the entropy in Eq. (3) subject to these constraints leads to the familiar Gibbs canonical distribution for ρ. This is not the problem of interest in this paper.

The form of the Hamiltonian in Eq. (1) suggests the following decomposition of the phase-space distribution function [March, 1968]:

$$\rho = \rho_K \rho_V \tag{5}$$

where

$$\rho_K = \prod_{i=1}^{N} a(\mathbf{p}_i) \tag{6}$$

and

$$\rho_V = \prod_{i<j}^{N} \frac{f(r_{ij})}{Z} . \tag{7}$$

The normalizations of ρ_K and ρ_V are chosen to be

$$\int d^3p \, a(\mathbf{p}) = 1 \tag{8}$$

$$\int d^{3N}r \, \rho_V = 1 . \tag{9}$$

The constant Z in Eq. (7) will be determined later; in turn, the $f_{ij} = f(r_{ij})$ must satisfy Eq. (9).

We wish to express S and $\langle E \rangle$ in Eqs. (3) and (4) in terms of $a(\mathbf{p})$ and $f(r)$; it will turn out that we shall also need the radial distribution function. We start with the k-particle reduced distribution function, defined [Balescu, 1975] by

$$P_k(\mathbf{r}_1,\ldots,\mathbf{r}_k) = \frac{N!}{(N-k)!} \frac{\int d^3r_{k+1} \ldots d^3r_N \prod_{i<j} f_{ij}}{\int d^{3N}r \prod_{i<j} f_{ij}} . \tag{10}$$

We shall work in the thermodynamic limit, in which the particle number N and the volume Ω increase without limit while satisfying

$$\lim_{N\to\infty} \lim_{\Omega\to\infty} (N/\Omega) = n , \tag{11}$$

where n is the particle density. Next, we introduce the quantities

$$g_k(\mathbf{r}_1,\ldots,\mathbf{r}_k) = n^{-k} P_k(\mathbf{r}_1,\ldots,\mathbf{r}_k) . \tag{12}$$

Note that $g_2(\mathbf{r}_1,\mathbf{r}_2) = g_2(r)$ is the radial distribution function. If the normalization in Eq. (9) is applied to $g_2(r)$, there results

$$n \int d^3r \; g_2(r) = N - 1 . \tag{13}$$

Finally, using the above results and definitions, we obtain for the entropy and average energy

$$S = -N\kappa \int d^3p \; a(\mathbf{p}) \ln a(\mathbf{p}) - \frac{Nn\kappa}{2} \int d^3r \; g_2(r) \ln[f(r)/Z] \tag{14}$$

$$\langle E \rangle = \frac{N}{2m} \int d^3p \; \mathbf{p}^2 \, a(\mathbf{p}) + \frac{Nn}{2} \int d^3r \; g_2(r) V(r) . \tag{15}$$

Note that these are functionals of $a(\mathbf{p})$, $f(r)$, and $g_2(r)$.

It is straightforward to derive from Eqs. (10) and (12) the following equation for the radial distribution function:

$$\nabla_1 \ln \left[\frac{g_2(r_{12})}{f(r_{12})}\right] = n \int d^3r_3 \; \frac{g_3(r_{12},r_{13},r_{23})}{g_2(r_{12})} \nabla_1 \ln f(r_{13}) , \tag{16}$$

where $\nabla_1 = \partial/\partial r_1$. This is the first of a hierarchy of equations known as the Born-Green-Yvon integral equations. Various truncations of this hierarchy can be used to arrive at approximate nonlinear integral equations for $g_2(r)$. As mentioned above, we consider only the HNC equation for $g_2(r)$.

To derive the HNC integral equation from Eq. (16), we first use the Kirkwood superposition approximation [Balescu, 1975]

$$g_3(r_{12},r_{13},r_{23}) = g_2(r_{12})\, g_2(r_{13})\, g_2(r_{23}) . \tag{17}$$

Substitution of Eq. (17) into Eq. (16) breaks the hierarchy and leads to the following Born-Green equations as a formal solution for $g_2(r)$ [March, 1968]:

$$\ln\left[\frac{g_2(r)}{f(r)}\right] = n \int d^3r'[g_2(r') - 1] \, E(|\mathbf{r} - \mathbf{r}'|) \, , \tag{18}$$

where

$$E(r) = -\int_r^\infty ds \, g_2(s) \, \frac{d}{ds} \ln f(s) \, . \tag{19}$$

The next step in the HNC approximation consists of approximating E(r). To do this, we partially integrate Eq. (19) to obtain

$$E(r) = g_2(r) - 1 - \int_r^\infty ds \, g_2(s) \, \frac{d}{ds} \ln \left[\frac{g_2(s)}{f(s)}\right] , \tag{20}$$

where we used

$$g_2(r) \xrightarrow[r\to\infty]{} 1 \, . \tag{21}$$

Finally, we replace $g_2(s)$ in the integral of Eq. (20) by its asymptotic value in Eq. (21) to obtain

$$E(r) \simeq g_2(r) - 1 - \gamma(r) \tag{22}$$

where

$$\gamma(r) = \ln \left[\frac{g_2(r)}{f(r)}\right] . \tag{23}$$

Using these results in Eq. (18), we obtain the HNC approximation for g(r):

$$\gamma(r) = n \int d^3r'[g_2(r') - 1] \, [g_2(|\mathbf{r} - \mathbf{r}'|) - 1 - \gamma(|\mathbf{r} - \mathbf{r}'|)] \, . \tag{24}$$

There are three quantities, $a(\mathbf{p})$, $f(r)$, and $g_2(r)$, that we wish to determine in some optimal fashion. As mentioned earlier, we wish to make use of the principle of maximum entropy. If we maximize the entropy in Eq. (14) subject to the constraints in Eqs. (8), (9), and (15), we obtain the standard canonical distribution for ρ in Eqs. (5) to (7) with

$$f(r) = \exp[-\beta V(r)] \, , \tag{25}$$

where $\beta = (\kappa T)^{-1}$ and

$$a(\mathbf{p}) = (\beta/2\pi m)^{3/2} \exp(-\beta \mathbf{p}^2/2m) . \tag{26}$$

The radial distribution function $g_2(r)$ is still given by Eqs. (10) and (12) with $k = 2$ where $f(r)$ is obtained from Eq. (25). In the following we wish to consider the entropy maximization considered above with the additional constraint provided by the integral equation in Eq. (24). Using Eq. (24) as a constraint allows us to maximize the entropy by varying independently not only $a(\mathbf{p})$ but also $f(r)$ and $g_2(r)$. Maximization of the entropy in Eq. (14) subject to the constraints in Eqs. (8), (9), (15), and (24) can be achieved by considering the unrestricted variation of the functional defined by

$$\begin{aligned} \Phi[a(\mathbf{p}), U(r), g_2(r)] &= -S + \alpha_K \int d^3p\, a(\mathbf{p}) \\ &+ \alpha_V n \int d^3r\, g_2(r) + \beta\langle E\rangle + \int d^3r\, \lambda(r) \Big\{ \ln g_2(r) - U(r) \\ &- n \int d^3r' [g_2(r') - 1][g_2(|\mathbf{r} - \mathbf{r}'|) - 1 - \gamma(|\mathbf{r} - \mathbf{r}'|)] \Big\}, \end{aligned} \tag{27}$$

where α_K, α_V, β [cf. Eq. (25)], and $\lambda(r)$ are Lagrange multipliers, and

$$U(r) = \ln f(r) . \tag{28}$$

By setting the variations of Φ with respect to $a(\mathbf{p})$, $U(r)$, and $g_2(r)$, respectively, equal to zero, we obtain Eq. (26) for $a(\mathbf{p})$ and

$$-\frac{n}{2} g_2(r) + \lambda(r) + n \int d^3r'\, \lambda(r')[g_2(|\mathbf{r} - \mathbf{r}'|) - 1] = 0 \tag{29}$$

$$\begin{aligned} &\frac{n}{2} U(r) + \alpha_V - \frac{n}{2} \ln Z + \frac{\beta\kappa}{2} V(r) + \frac{\lambda(r)}{g_2(r)} \\ &- \left[1 - \frac{1}{g_2(r)}\right] n \int d^3r'\, \lambda(r')\, [g_2(|\mathbf{r} - \mathbf{r}'|) - 1] \\ &- n \int d^3r'\, \lambda(r')\, [g_2(|\mathbf{r} - \mathbf{r}'|) - 1 - \gamma(|\mathbf{r} - \mathbf{r}'|)] = 0 . \end{aligned} \tag{30}$$

We can solve for the Lagrange multiplier $\lambda(r)$ by introducing the following Fourier transforms:

$$S(k) - 1 = n \int d^3r\, [g_2(r) - 1] \exp(i\mathbf{k}\cdot\mathbf{r}) , \tag{31}$$

$$\overline{\lambda}(k) = \int d^3r\, [\lambda(r) - n/2] \exp(i\mathbf{k}\cdot\mathbf{r}) , \tag{32}$$

$$\gamma(k) = \int d^3r\, \gamma(r) \exp(i\mathbf{k}\cdot\mathbf{r}) = \frac{[S(k) - 1]^2}{nS(k)} . \tag{33}$$

Using Eqs. (31) to (33), we obtain from Eq. (29)

$$\overline{\lambda}(k) = \frac{S(k) - S(0)}{2S(k)} . \tag{34}$$

Using the normalization in Eq. (7) to determine α_V and with these results, we can solve Eq. (30) for f(r). The result is

$$f(r) = \exp[-\beta U(r)] \tag{35}$$

where

$$U(r) = V(r) - \frac{\kappa T}{(2\pi)^3 n} \int d^3k\, \exp(-i\mathbf{k}\cdot\mathbf{r})\, W(k) \tag{36}$$

and

$$W(k) = \frac{[S(k) - 1]\,[S(k) + 1]\,[S(0) - S(k)]}{[S(k)]^2} . \tag{37}$$

From the well known properties of the structure factor S(k) [Balescu, 1975], we observe that Eqs. (35) through (37) imply

$$f(r) \xrightarrow[r\to\infty]{} 1 , \tag{38}$$

as they should.

Next, we eliminate f(r) from Eqs. (34) through (36) to obtain the following integral equation for $g_2(r)$:

$$\ln g_2(r) = \gamma(r) - \beta V(r) + \frac{1}{(2\pi)^3 n} \int d^3k\, \exp(-i\mathbf{k}\cdot\mathbf{r})\, W(k) , \tag{39}$$

where

$$\gamma(r) = (2\pi)^{-3} \int d^3k\ \gamma(k) \exp(-i\mathbf{k}\cdot\mathbf{r}) \tag{40}$$

and $\gamma(k)$ is given by Eq. (33). We can cast Eq. (39) in the alternative form

$$\ln g_2(r) = -\beta V(r) + \frac{1}{(2\pi)^3 n} \int d^3k \exp(-i\mathbf{k}\cdot\mathbf{r}) [\gamma(k) + W(k)] , \tag{41}$$

where $\gamma(k)$ is given by Eq. (33) and $W(k)$ is given by Eq. (37).

3. Results and Discussion

We can solve Eq. (39), or equivalently Eq. (41), by iteration to obtain the optimal correlation function $g_2(r)$ and the associated structure factor $S(k)$. So far we have considered only the Lennard-Jones potential, as we discuss next. For the Lennard-Jones potential,

$$V(r) = 4\varepsilon[(\sigma/r)^{12} - (\sigma/r)^6] , \tag{42}$$

we define the following dimensionless parameters:

$$\begin{aligned} T^* &= \kappa T/\varepsilon \\ n^* &= n\sigma^3 \\ r^* &= r/\sigma . \end{aligned} \tag{43}$$

Figures 1 and 2 show typical solutions of Eq. (39) for $g_2(r)$ and the corresponding structure factor [see Eq. (31)] for $n^* = 0.4$, $T^* = 1.25$. These solutions were obtained by integration using a method by Krylov and Skoblya [Davis and Rabinowitz, 1975]. These curves are similar in form to curves obtained earlier [see Temperly et al., 1968]. For densities $n^* > 0.4$ and temperatures $T^* < 1.1$, we are unable to obtain convergent solutions to Eq. (39), even though the HNC equations possess solutions outside these ranges. In our view, this indicates that the HNC approximation should not be used for $n^* > 0.4$ and $T^* < 1.1$.

To broaden our understanding of this method and this problem, we have undertaken similar studies for the Percus-Yevick equation. Detailed calculations for thermodynamic quantities arising from the optimal correlation functions of both the HNC equation and the Percus-Yevick equation [see also Levesque, 1966] are also in progress.

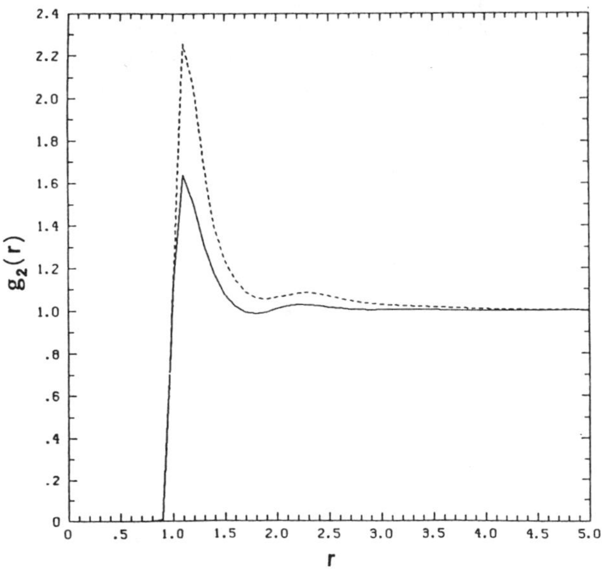

Figure 1. The radial distribution function as a function of r determined by solving Eq. (39) (solid line) for $n^* = 0.1$ and $T^* = 1.25$. The HNC solution (dashed line) is shown for comparison.

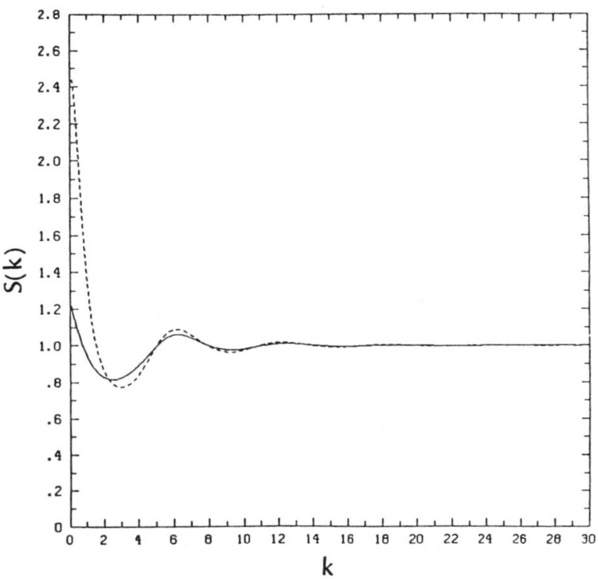

Figure 2. The structure factor as a function of wave number (in A^{-1}) determined by solving Eqs. (39) and (31) for $n^* = 0.1$ and $T^* = 1.25$ (solid line). The HNC solution (dashed line) is shown for comparison.

In Table 1 we compare some features of the maximum entropy solution with the direct HNC solution. The values of $g_2(r)$ at the first peak and also S(0) become considerably different for higher densities at a given temperature or for low temperatures at a given density. So far, we have been unable to locate the precise density-temperature ranges at which strong departures from the HNC solution occur.

Table 1. Comparison of g_{max} [Value of $g_2(r)$ at the First Peak] and S(0) for Maximum Entropy and HNC Solutions at Various Temperatures and Densities

Temp T^*	Density n^*	Max entropy g_{max}	Max entropy S(0)	HNC g_{max}	HNC S(0)
2.0	0.1	1.59	1.06	1.67	1.25
2.0	0.2	1.78	0.98	1.72	1.37
1.5	0.1	1.62	1.18	1.95	1.73
1.5	0.2	1.77	1.10	1.99	3.16
1.25	0.1	1.64	1.28	2.25	2.53

4. References

Balescu, R. (1975), Equilibrium and Nonequilibrium Statistical Mechanics, Wiley, New York, Chaps. 7 and 8.

Davis, P. J., and P. Rabinowitz (1975), Methods of Numerical Integration, Academic Press, New York, pp. 180-185.

Hill, T. L. (1956), Statistical Mechanics, McGraw-Hill, New York, Chap. 6.

Levesque, D. (1966), "Etudes des equations de Percus et Yevick d'hyperchaine et de Born et Green dans le cas de fluides classiques," Physica 32, p. 1985.

March, N. H. (1968), "The Liquid State," in Theory of Condensed Matter, International Atomic Energy Agency, Vienna, pp. 93-174.

Temperly, H. N. V., J. S. Rowlinson, and G. S. Rushbrooke, eds. (1968), Physics of Simple Liquids, North-Holland, Amsterdam.

UPDATING INDUCTIVE INFERENCE

N. C. Dalkey

Cognitive Systems Laboratory, University of California at Los Angeles, Los Angeles, CA 90024

Updating inductive inference cannot be performed in the classic fashion of using the conclusion of one stage as a prior for the next stage. Minimum cross-entropy inference, which uses the prior-posterior formalism, can lead to inconsistencies and violations of the positive value of information principle. A more acceptable updating procedure is based on the notion of proper scoring rule as a figure of merit.

C. R. Smith and G. J. Erickson (eds.),
Maximum-Entropy and Bayesian Spectral Analysis and Estimation Problems, 305–316.

1. Classical vs. Inductive Updating

Most inquiries do not proceed by a single, definitive inference that leads from data to an established conclusion. Rather, the inquiry proceeds by a sequence of steps in which intermediate conclusions are revised on the receipt of additional information. This iterative process—often called updating—is an integral feature of the theory of inference.

In classical Bayesian inference, the updating process is quite straightforward; the conclusion of the initial inference is taken as a prior for the subsequent revision given new information. The procedure is justified on the grounds that the initial conclusion is the correct probability given the available information.

In inductive inference, the situation is quite different. The conclusion is not the "correct probability given the available information." Rather, it is a best guess, or most reasonable estimate, based on the available information. Receipt of new information can affect the "reasonableness" of a conclusion in a radical way. As a result, if the conclusion of an initial inference is used as a prior in a subsequent inference, serious difficulties, including outright inconsistencies, can arise.

2. Inductive Inference

As a framework for discussing the problem of updating inductive inference, the theory of induction presented in Dalkey [1985] will be assumed. The basic idea is to use a proper scoring rule as the figure of merit for an inductive inference. Thus a proper score plays the role for induction that the truth value plays for deduction. In deduction, the truth of the premises guarantees the truth of the conclusion. In induction, the truth of the premises guarantees an appropriate expected score for the conclusion.

An elementary inductive inference involves—

(1) A partition $E = \{e,f,...\}$ of events of interest (event space).
(2) The actual probability distribution $\mathbf{P} = \{\mathbf{P}(e),\mathbf{P}(f),...\}$ on E, which is unknown, but which the analyst would like to estimate.
(3) A set of constraints (available information) that restricts $\mathbf{P}$ to a set K (knowledge set) of potential distributions.
(4) A score rule (payoff function), S(P,e), that determines the reward for asserting distribution P given that event e occurs.
(5) An inference rule that selects a distribution R out of K as the best guess or posit.

Figures of merit are restricted to the family of proper scoring rules. These are rewarded functions S(P,e) that exhibit the reproducing property. The expected score, given that $\mathbf{P}$ is the correct probability distribution on E and that distribution R is asserted, is $\Sigma_E \mathbf{P}(e)S(R,e)$. The reproducing property is expressed by the condition

$$\sum_{E} \mathbf{P}(e)S(R,e) \leq \sum_{E} \mathbf{P}(e)S(\mathbf{P},e) \,. \tag{1}$$

That is, the expectation of the score is a maximum when the correct probability distribution is asserted. In the context of personalistic, or subjective, theories of probability, where the notion of correct probability distribution is not always clear, Eq. (1) is frequently introduced in terms of an honesty promoting score. That is, if an estimator believes **P**, then his subjective expected score is a maximum if he asserts **P** [Savage, 1971].

Condition (1) has been proposed on a variety of grounds. However, it has a particularly clear significance for inductive logic. Suppose K is a unit class (specifying **P** uniquely). Then clearly the conclusion should be **P**. However, if S is not reproducing, a higher expected score would be obtained by asserting some distribution other than **P**.

Proper scoring rules, in effect, establish verifiability conditions for probability statements. In particular, they permit verification of assertions of the probability of single events. The assigned score S(P,e) is a function both of the event e that occurs—giving the required tie to reality—and of the asserted distribution P—giving the required dependence on the content of the statement.

It is convenient to define some auxiliary notions:

$$H(P) = \sum_{E} P(e)\ S(P,e) \tag{2a}$$

$$G(P,R) = \sum_{E} P(e)\ S(R,e) \tag{2b}$$

$$N(P,R) = H(P) - G(P,R) \,. \tag{2c}$$

H(P) is the expected score given that P is asserted and that P is also the correct distribution. G(P,R) is the expected score given that P is the correct distribution but R is asserted. N(P,R) is the net score, and measures the loss if P is correct and R is asserted. (Conversely, N(P,R) can be viewed as the gain resulting from changing an estimate from the distribution R to the correct distribution P.)

In the new notation, Eq. (1) can be written

$$H(P) \geq G(P,R) \,, \tag{1'}$$

and thus we have

$$N(P,R) \geq 0 \,. \tag{3}$$

The net score is always positive.

There is an infinite family of score rules that fulfill condition (1). Only two will be explicitly referred to here: (1) the logarithmic score, $S(P,e) = \log P(e)$, and (2) the quadratic score, $S(P,e) = 2P(e) - \Sigma_E P(e)^2$. $H(P)$ for the logarithmic score is $\Sigma_E P(e)\log P(e)$, that is, the negentropy. $N(P,R)$ for the log score is $\Sigma_E P(e)\log P(e)/R(e)$, the cross-entropy. $H(P)$ for the quadratic score is $\Sigma_E P(e)^2$. $N(P,R)$ for the quadratic score is $\Sigma_E [P(e) - R(e)]^2$, that is, the squared distance between P and R.

Given a set K of potential distributions and a proper score rule S, the inductive rule is: Select the distribution P^0 such that $H(P^0) = \min_K H(P)$. In the general case, the relevant operator may be inf rather than min, and to achieve the guarantee described below, the analyst may have to adopt a mixed posit, that is, select a posit according to a probability distribution on K. These extensions are important for a general theory but are not directly relevant to the logical structure of the inductive procedure.

The inductive rule can be called the min-score rule. The rule is a direct generalization of the maximum entropy rule proposed by Jaynes [1968]. The entropy of a distribution P is the negative of the expected logarithmic score. Thus, maximizing the entropy is equivalent to minimizing the expected logarithmic score. The min-score rule extends the prescription to any proper score rule.

Although the formalism of the min-score rule is analogous to the maximum entropy principle, the justification for the rule is much stronger. The justification stems from two basic properties of the min-score distribution: (1) It guarantees $H(P^0)$; that is, for every $P \in K$, $G(P,P^0) \geq H(P^0)$, or in particular, $G(\mathbf{P},P^0) \geq H(P^0)$. (2) It is the only rule that fulfills the positive value of information (PVI) principle; that is, if $K \subset K'$ (K contains more information than K') and P^0 is the min-score posit for K and Q^0 the min-score posit for K', then $H(P^0) \geq H(Q^0)$. These two properties are derived by Dalkey [1985]. The two properties give a clear content to the notion of validity for induction.

3. Min-Net-Score Updating

The min-score rule as formulated above is a one-stage procedure. In practice, as mentioned in the first section, an inquiry is likely to progress in multiple steps. Before discussing the appropriate adaptation of the min-score rule to iterative inference, it is instructive to examine a cognate procedure that is somewhat closer in spirit to classical Bayesian updating. I call it min-net-score (MNS) inference, for reasons that will become clear shortly. In the literature the topic has been treated in the narrower context of minimum-cross-entropy inference (also referred to as minimum directed divergence, minimum discrimination information, minimum relative entropy, and other terms). The procedure was first suggested by Kullback [1959] and elaborated by Good [1963], Pfaffelhuber [1975], Shore and Johnson [1980], and others.

Given a knowledge set K (current knowledge) and a distribution P (prior information), the MNS rule is to select the member Q of K that minimizes the net score with respect to P; that is, $N(Q,P) = \min_{R \in K} N(R,P)$. As mentioned above, the rule has been investigated only for the logarithmic score. However, it clearly is extendable to any proper score.

Figure 1 illustrates the rule for the quadratic score for an E consisting of three events, where the triangle is the simplex Z of all probability distributions on three events. For the quadratic score, if K is convex and closed, then Q is the point such that a line joining it and P is perpendicular to the hyperplane supporting K at Q. The actual distribution **P** is assumed to be in K, but the fact that it is unknown is indicated by a fuzzy point.

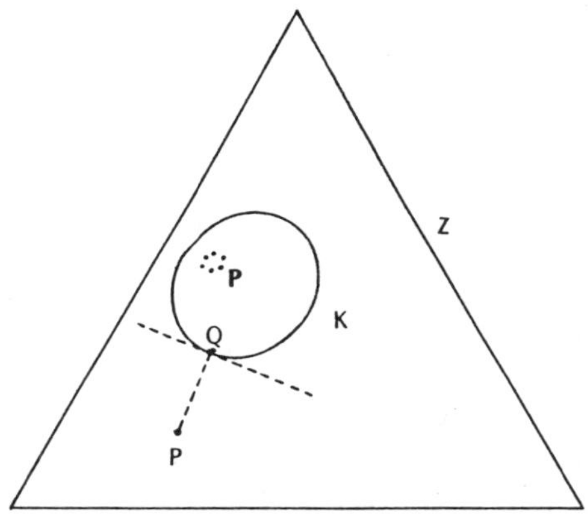

Figure 1. Minimum-net-score inference.

The most persuasive property of the MNS rule has been demonstrated by Shore and Johnson [1980]. By assumption, the actual distribution **P** is in K. Shore and Johnson show, for the logarithmic score, that $N(\mathbf{P},Q) \leq N(\mathbf{P},P)$. In words, the actual distribution is closer to Q than it is to P (in the sense of cross-entropy). Thus, if it is felt that P embodies relevant information about **P** (for example, that P is not a "bad guess"), then Q is a better guess than P; the analyst suffers a smaller loss believing Q than he would staying with P.

Call the fact that $N(\mathbf{P},Q) \leq N(\mathbf{P},P)$ the "better guess" property. It can be shown that the better guess property holds for the MNS rule and any proper score.

Theorem: If K is convex and closed, and $G(P,R)$ is bounded on K, then, for any P in Z, there exists a distribution Q^0 in K such that $N(\mathbf{P},Q^0) \leq N(\mathbf{P},P)$.

Proof: Let $J(R,Q;P) = N(R,P) - N(R,Q) = G(R,Q) - G(R,P)$. J can be interpreted as a game in which "nature," as the minimizing player, selects R, and the analyst, as the maximizing player, selects Q. From Eqs. (2), J is linear in R, and therefore convex in R. Hence, there exists a value for the game, and a pure strategy Q^0 for nature [Blackwell and Girshick, 1954, theorem 2.5.1]. By definition of an optimal strategy, $J(Q^0,Q;P) \lesssim v$ for every Q, and thus $J(Q^0,Q^0;P) = N(Q^0,P) \lesssim v$. On the other hand, $v \lesssim \min_R \max_Q J(R,Q;P) = \min_R J(R,R;P) = \min_R N(R,P) \lesssim N(Q^0,P)$. Thus, $v = N(Q^0,P) \gtrsim 0$, by Eq. (3).

The theorem assures that there is a Q^0 such that $N(R,Q^0) \lesssim N(R,P)$ for every R in K, and thus in particular for **P**. In the general case, a mixed posit may be needed to guarantee the better guess property, and v may be attainable only to within any ε. For "well-behaved" score rules such as the log or quadratic score, there is a pure strategy R^0 for the analyst that guarantees v.

Corollary. With the hypothesis as in the theorem, $G(\mathbf{P},Q^0) \gtrsim G(\mathbf{P},P)$.

Proof: Immediate from the definition of net score.

The corollary states that the actual expected score of the posit Q^0 is at least as great as the actual expected score of the prior P. The corollary is particularly striking if S is an economic score, expressed, for example, in terms of money. The corollary states that the actual monetary expectation for a suitably chosen Q is greater than the monetary expectation of P.

It might be illuminating to point out that, for the quadratic score, the "better guess" property is a simple consequence of an extension of the Pythagorean theorem, namely, that if the angle opposite a given side of a triangle is obtuse, then the square on the given side is greater than the sum of the squares on the opposite sides. If K is convex, as noted above, Q is the point in K such that the line from P to Q is perpendicular to the hyperplane supporting K at Q (see Fig. 2). But then, for the triangle consisting of Q, P, and any other **P** in K, the angle between **P**Q and QP is obtuse. Thus the squared distance (net score) between **P** and P is at least as great as the squared distance between **P** and Q.

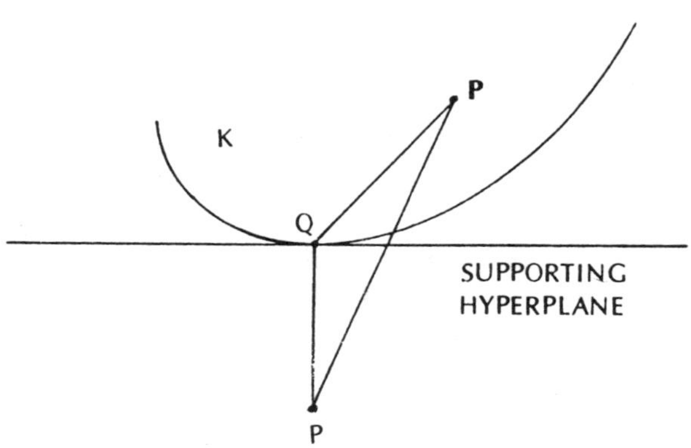

Figure 2. Generalized Pythagoras.

A highly favorable feature of the approach is the fact that the MNS rule is definable for any score rule and that the better guess property holds (at least to within ε) for every score rule. However, there are serious difficulties with the MNS rule. In particular, the rule can mask inconsistencies between the prior distribution P and the knowledge set K. In addition, the rule violates the PVI principle.

To illustrate the first difficulty, consider a case in which the prior distribution P was obtained by a previous MNS inference, but, for whatever reason, the knowlege set on which it was based is not available. To simplify the example, suppose the previous inference was based on a knowledge set K' and a uniform prior P'. The MNS posit for the earlier inference is P, the min-score distribution in K' (at least, for a symmetrical score rule) as illustrated in Fig. 3. Now, suppose we iterate the inference with a new knowledge set K. By assumption, the actual distribution **P** is in K', and also in K. Thus **P** is in the intersection K•K' (shaded in the figure). However, as the figure illustrates, the MNS posit Q, based on the "prior" P, is, in general, not in K•K'. So far, there is no inconsistency; the example merely shows that the MNS procedure does not represent a reasonable use of the knowledge K' that generated a prior P.

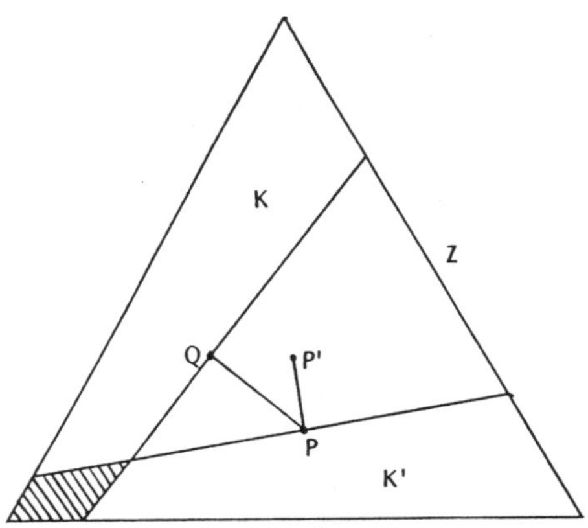

Figure 3. MNS inference (without memory).

Now, consider the situation in Fig. 4. Here the sets K and K' do not intersect, and thus are incompatible: one or the other must be incorrect. Yet, if only the "bare" prior P is known, the MNS procedure allows the inference to proceed, and winds up with Q as the conclusion. In short, the MNS procedure has no provision for identifying incompatibilities between a prior P and a current knowledge set K.

With regard to the PVI principle, consider a case like that in Fig. 5. The circle is one of the family of iso-expected score surfaces (in this case for the quadratic score). The expected score increases outward from the centroid of the triangle. For the set K (the upper triangle) and the prior P, the MNS rule selects Q as a posit. Now suppose the knowledge set is reduced to K' (the shaded quadrangle). Since K' is totally included in K, we can assert that K' is more informative than K. The MNS rule selects Q' as the posit given K' and P. From the illustration, it is clear that H(Q) > H(Q'); restricting the knowledge set to K' leads to a smaller expected score. At the same time, suppose that **P** is at the indicated location. Again, it is clear from the construction that N(**P**,Q) < N(**P**,Q'). Thus, increasing information does not guarantee that the resulting posit is closer to the actual distribution.

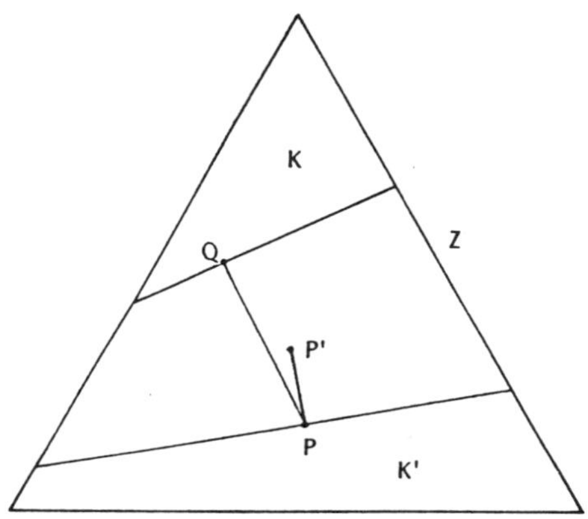

Figure 4. Inconsistent MNS update.

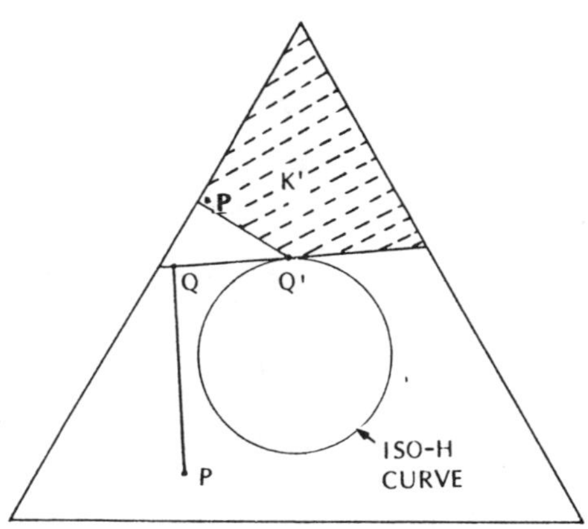

Figure 5. Failure of MNS inference to support positive value of information.

To summarize the illustrated case, for a fixed prior, increasing information (smaller K) neither increases the expectation of the conclusion nor guarantees moving closer to the actual distribution. Thus, the MNS procedure cannot guarantee that increased information will be constructive.

The two difficulties exemplified in the illustrations—failure to guard against inconsistencies, and failure to guarantee the positive value of information—appear sufficient to reject the MNS rule as a method of iterating inductive inferences.

4. Min-Score Updating

Most of the considerations relevant to the application of the min-score rule to iterating inductive inference have shown up in the illustrations of problems with the MNS rule. In a sense, the basic consideration is trivial. If new knowledge K is acquired, but the knowledge set K' from the previous stage is retained, there is no problem. One simply takes the intersection K•K' of the old and new knowledge sets, and selects the min-score distribution in K•K' as the updated posit.

More generally, let K_j, j = 1,2,•••, be a sequence of knowledge sets relevant to a given set E of events, where each K_j is the new knowledge available at stage j of a sequence of inferences. Let $K^i = \prod_{j=1}^{i} K_j$, where $\prod$ is the logical product—that is, K^i is the intersection of all preceding and current knowledge sets at stage i. The basic consistency condition for the inference is that $\mathbf{P} \in K_j$ for every j, and therefore $\mathbf{P} \in K^i$ for every i. However, since **P** is unknown, the operative condition is $K^i \neq 0$.

Call an inference an <u>iteration with memory</u> (at stage i) if, at stage i, K^i is known. It is straightforward to show that, for an iteration with memory, the basic properties of guaranteed expectation and the PVI principle hold. It is also clear that iteration with memory has an automatic consistency check. If at any stage i, $K^i = 0$, one or more of the K_j is incorrect.

In a sense, iteration with memory is the only "reasonable" way to accumulate knowledge, and in this sense the theory is "complete." However, in practice, the requisite memory is not always available. A common situation is that in which a distribution is furnished by expert judgment. Suppose, for example, that you have some (incomplete) information K concerning a given set E of events, and in addition a recognized expert in the field expresses the opinion (not knowing K) that a reasonable guess for the distribution on E is P. You might give a fair amount of credence to the estimate P, but if P is not in K, then some procedure for "adding" P to K might appear reasonable.

An estimate P may arise from information of a sort that is not directly conformable with a knowledge set K. For example, P might be an expert's estimate of the conditional probability of the events E given some observations O. The aggregation of such an estimate with a knowledge set K requires techniques beyond the elementary min-score rule. Thus, I will restrict the discussion to the case that the prior P may be assumed to stem from an inductive inference based on a knowledge set K' that, for whatever reason, is no longer available. In this case, if it can be assumed that the unknown K' is convex, it is feasible to determine the weakest knowledge set that would support the prior estimate P.

Let $K^+(P)$ denote the half space that supports the iso-H surface at P, or in other words, $K^+(P)$ is the half space that is bounded by the hyperplane supporting the convex set $H(P)^- = \{Q \mid H(Q) \lesssim H(p)\}$ at P, and on the opposite side from $H(P)^-$. Assuming that the unknown K' is convex, it must be contained in $K^+(P)$, since H(P) is a minimum in K', and the hyperplane supporting $H(P)^-$ at P is thus a separating hyperplane for K' and $H(P)^-$. $K^+(P)$ is thus the desired "weakest K' that could generate P as a posit." The construction of K^+ is illustrated in Fig. 6.

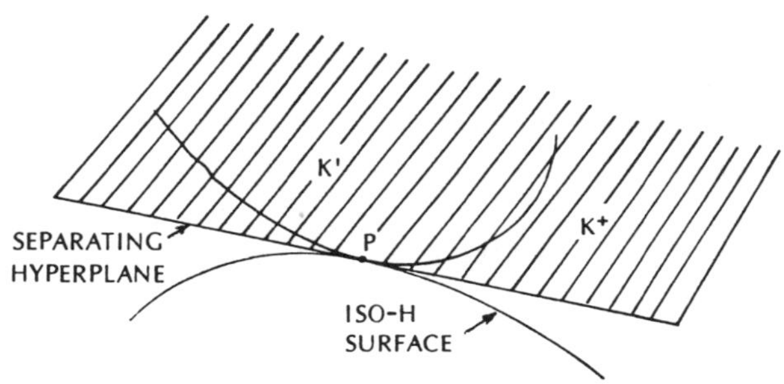

Figure 6. Construction of half-space K^+.

Since any convex K' from which P could be obtained as a min-score posit must be contained in $K^+(P)$, we can use K^+ as a weak surrogate for the missing K'. Since **P**, by assumption, is in K', it is also in K^+. Thus if the current knowledge set K and K^+ do not intersect, that is, $K \cdot K^+ = 0$, we can say P and K are incompatible. The consistency check thus defined is not a sufficient condition, but only necessary; the test could be met and the unknown K•K' be empty. However, it is a partial guard against flagrant inconsistencies.

The updating rule, given $K^+(P)$, is to select the min-score distribution out of $K \cdot K^+$. Again, it is straightforward to show that this rule, assuming consistency, also retains the two basic inductive properties of guaranteed expectation and PVI.

If the assumption that P was derived from a convex knowledge set is implausible, then the K^+ surrogate loses its cogency. It can be shown that for an elementary induction—one not involving iteration—if the knowledge set is not convex, extending it to the convex closure maintains guaranteed expectation and positive value of information. But this expedient is not appropriate for the case of an existent prior estimate with an unknown basis.

5. Discussion

Iteration for inductive inference is more demanding than iteration in deductive inference; in essence, more information must be "remembered" with induction. This requirement becomes especially severe if information of different types is to be aggregated, as in the case of a well defined knowledge set and a "prior" estimate by an expert. As shown above, if the expert estimate can be assumed to stem from an inductive inference with a convex knowledge set, a surrogate "least informative" knowledge set can be introduced. But if the assumption of convexity is not persuasive, it is not clear how the current knowledge and the prior estimate can be combined.

It should be emphasized that any updating procedure has the implicit assumption that the frame of reference or universe of discourse remains constant throughout the sequence of iterations. One condition for a constant frame of reference is that the event set E remain fixed. In many practical problems involving iteration, the relevant event set does not remain constant. Thus, in the classic problem of random sampling with unknown prior, the relevant E is the joint distribution on the events of interest and the sample events. Changing the sample size changes the event space, and thus there may be no direct transition from an inductive inference with sample size n and one with sample size n+1.

6. Acknowledgment

This work was supported in part by National Science Foundation Grant IST 8201556.

7. References

Blackwell, D., and M. A. Girshick (1954), Theory of Games and Statistical Decisions, Wiley, New York.

Dalkey, N. C. (1985), "Inductive inference and the maximum entropy principle," in Maximum-Entropy and Bayesian Methods in Inverse Problems, C. Ray Smith and W. T. Grandy, Jr., eds., D. Reidel, Dordrecht, pp. 351-364.

Good, I. J. (1963), "Maximum entropy for hypothesis formulation, especially for multidimensional contingency tables," Ann. Math. Stat. 34, pp. 911-934.

Jaynes, E. T. (1968), "Prior probabilities," IEEE Trans. Syst. Sci. Cybernet. SSC-4, pp. 227-241.

Kullback, S. (1959), Information Theory and Statistics, Wiley, New York.

Pfaffelhuber, E. (1975), "Minimax information gain and minimum discrimination principle," in Colloquia Mathematica Societatis János Bolyai, vol. 16, Topics in Information Theory, Keszthely (Hungary).

Savage, L. J. (1971), "Elicitation of personal probabilities and expectations," J. Am. Stat. Assoc. 66, pp. 783-801.

Shore, J. E., and R. W. Johnson (1980), "Axiomatic derivation of the principle of maximum entropy and the principle of minimum cross-entropy," IEEE Trans. Inf. Theory IT-26, pp. 26-37.

PARALLEL ALGORITHMS FOR MAXIMUM ENTROPY CALCULATION

Stuart Geman

Division of Applied Mathematics, Brown University, Providence, RI 02912

Maximum entropy extensions of partial statistical information lead to Gibbs distributions. In applications, the utility of these extensions depends upon our ability to perform various operations on the Gibbs distributions, such as random sampling, identification of the mode, and calculation of expectations. In many applications these operations are computationally intractable by conventional techniques, but it is possible to perform these in parallel. The architecture for a completely parallel machine dedicated to computing functionals of Gibbs distributions is suggested by the connection between Gibbs distributions and statistical mechanics. The Gibbs distribution defines a collection of local physical rules that dictate the programming of the machine's processors. The machine's dynamics can be described by a Markov process that is demonstrably ergodic with marginal distribution equal to the specified Gibbs distribution. These two properties are easily exploited to perform the desired operations.

Applications include relaxation techniques for image segmentation and analysis, and Bayesian solutions, without independence assumptions, for medical diagnosis and other "expert system" problems. These applications are discussed in detail in the following references:

Geman, S., and D. Geman (1985), "Stochastic relaxation, Gibbs distributions, and the Bayesian restoration of images," IEEE-PAMI.

Geman, S. (1985), "Stochastic relaxation methods for image restoration and expert systems," in Automated Image Analysis: Theory and Experiments (Proceedings of the ARO Workshop on Unsupervised Image Classification, 1983), D. Cooper, R. Launer, and D. McClure, eds., Academic Press, New York.

C. R. Smith and G. J. Erickson (eds.),
Maximum-Entropy and Bayesian Spectral Analysis and Estimation Problems, 317.

SUBJECT INDEX